D0174884

Signaling and Communication in Plants

Series Editors

František Baluška
Department of Plant Cell Biology, IZMB, University of Bonn, Kirschallee 1, D-53115 Bonn, Germany

Jorge Vivanco
Center for Rhizosphere Biology, Colorado State University, 217 Shepardson Building, Fort Collins, CO 80523-1173, USA

For further volumes:
http://www.springer.com/series/8094

Shaul Yalovsky • František Baluška • Alan Jones
Editors

Integrated G Proteins Signaling in Plants

Editors
Dr. Shaul Yalovsky
Tel Aviv University
Dept. Plant Sciences
69978 Ramat Aviv, Tel Aviv
603 Britannia Bldg.
Israel
shauly@tauex.tau.ac.il

Dr. František Baluška
Department of Plant Cell Biology
IZMB
University of Bonn
Kirschallee 1
D-53115 Bonn
Germany
baluska@uni-bonn.de

Dr. Alan Jones
University of North Carolina
Dept. Biology & Pharmacology
317 Coker Hall
Chapel Hill NC 27599
CB 3280
USA
alan_jones@unc.edu

ISSN: 1867–9048 e-ISSN: 1867–9056
ISBN: 978-3-642-03523-4 e-ISBN: 978-3-642-03524-1
DOI 10.1007/978-3-642-03524-1
Springer Heidelberg Dordrecht London New York

Library of Congress Control Number: 2009934482

Cover design: WMXDesign GmbH, Heidelberg, Germany

Printed on acid-free paper

Springer is part of Springer Science+Business Media (www.springer.com)

Preface

When guanine nucleotide-binding proteins (G proteins) bind GTP, they adopt an activated conformation that leads to activation of downstream signaling elements. In this capacity, G proteins couple, amplify, and integrate upstream signals to downstream cellular changes. One of the most fascinating aspects of G proteins is that they operate like a molecular timer because the GTP-bound, activated state converts to the GDP-bound, resting state after an inherently determined amount of time due to hydrolysis of GTP to GDP by the intrinsic GTPase. The cycling of activated to resting states, known as the G cycle, enables cellular signaling to occur within kinetics of seconds to hours. In eukaryotes, G proteins are divided into two major subgroups: the Ras superfamily of small G proteins and the heterotrimeric G proteins. The Ras superfamily is further divided into the Ras, Rho, Rab, Arf, and Ran subfamilies. The Gα subunits of the heterotrimeric G-protein complex divide the complexes into four subclasses, G_i, G_s, G_q, and $G_{12/13}$. The Ras, Rho, and the hetrotrimeric G proteins are implicated in regulation of signaling, while Rab, Arf, and Ran carry out other cellular functions and are not be covered in this book. Ras proteins have not been identified in plants. This leaves the small G protein from the Rho family, called ROPs or RACs (here after ROPs/RACs), and the heterotrimeric G proteins to comprise the two major groups of signaling G proteins in plants. This book summarizes a decade of research on ROPs/RACs and heterotrimeric G proteins in plants.

In the active state, the small GTPases interact with target proteins commonly referred to as effectors to initiate a signaling process. In most small G proteins, the GDP/GTP exchange is not spontaneous and requires accessory proteins known as *G*unanine nucleotide *E*xchange *F*actors (GEFs). The inefficient GTPase activity is enhance by a second group of proteins known as *G*TPase *A*ctivating *P*roteins (GAPs). GEFs and GAPs provide a means to regulate the activity of the small GTPases in time and space. Subcellular distribution of proteins from Rho and Rab families is regulated by a third group of proteins designated *G*uanine nucleotide *D*issociation *I*nhibitors (GDIs).

Heterotrimeric G protein are composed of three subunits designated α, β, and γ. The α subunit is a GTP-binding protein that contains one domain that resembles small GTPases. The β subunit has a seven-bladed propeller structure and forms a tight dimeric complex with the γ subunit. In metazoans, heterotrimeric G proteins are associated with membrane proteins known as G-Protein-Coupled Receptors (GPCRs) that are the ligand-regulated GEFs. Activation of GPCRs upon ligand binding leads to GDP/GTP exchange and activation of the Gα. In turn, Gβ- and Gγ dissociate from the subunit as a complex and signaling is induced by both the dissociated Gα and Gβ γ complex. Signaling terminates by GTP hydrolysis that leads to reassociation of Gβγ with the Gα. As discussed in the Chapter "Bioinformatics of Seven-Transmembrane Receptors in Plant Genomes," the existence of GPCRs in plants is questioned.

Furthermore, as discussed in the Chapter "Plant Gα Structure and Properties," while GDP release from the Gα subunit is the rate-limiting step in vertebrate G protein complexes, that does not seem to be the case for *Arabidopsis*.The implications of this strange property are described.

The book begins with a chapter from Janice Jones describing these and other G protein signaling principles and then describes the unique properties of plant heterotrimeric G proteins. The chapter "Structure and function of ROPs and their GEFs," by ChristophThomas and Antje Berken takes a similar approach with the small G proteins and thus, these two chapters provide an interesting comparison of the Gα subunit of the heterotrimeric G protein complex and the small G proteins in plants. Physiological aspects are taken up in later chapters. For example, in the Chapter "Heterotrimeric G Proteins and Plant Hormone Signaling in Rice," by Yukimoto Iwasaki and coworkers, evidence is presented that the G protein in rice is mediating fundamantally different signaling than in *Arabidopsis*. Jin-Gui Chen in the Chapter "Heterotrimeric G-Proteins and Cell Division in Plants," builds the case that the heterotrimeric G protein complex controls the rate of the plant cell cycle and consequently cell proliferation. Two chapters ("The Role of Seven Transmembrane Domain MLO Proteins, Heterotrimeric G-Proteins and Monomeric RAC/ROPs in Plant Defense" and "G Proteins and Plant Innate Immunity") by Justine Lorek et al. and Yuri Trusov et al., respectively, deal with the role of small and heterotrimeric G proteins in plant defense against different pathogens. Whether or not G proteins couple multiple plant hormones and environmental signals remains an open question but is a theme throughout the book. As mentioned earlier, the receptors that activate the heterotrimeric G protein complex are poorly conserved at the primary sequence level. Therefore, in the Chapter "Bioinformatics of Seven-Transmembrane Receptors in Plant Genomes," Etsuyko Moriyma and Stephen Opiyo provide strategies to identify 7-transmembrane proteins from divergent genomes. They then apply these tools to 18 genomes of the bikonts, the group that includes higher plants and the algae. The chapter "Evolution of the ROP GTPase Signaling Module" is about the bizaar; Lei Ding and coworkers discuss proteins that share limited sequence similarity to canonical Gα subunits of the heterotrimeric G protein complex.

ROPs/RACs are master regulators of cell polarity, similar to their homologs in yeast and animal cells. Remarkably, these studies showed that regardless of the evolutionary-conserved functions, many of the ROPs/RACs effectors are unique to plants. The chapter "ROP GTPases and the cytoskeleton" by Ying Fu focuses on the function of ROPs/RACs in cytoskeleton organization, highlighting the role of a plant-unique group of proteins designated RICs (*R*OP *I*nteracting *C*RIB containing) as well as by the evolutionary-conserved WAVE and Arp2/3 complexes. In the absence of other signaling small GTPases, ROPs/RACs were suggested to function in diverse signaling cascades. The chapter "RAC/ROP GTPases in the Regulation of Polarity and Polar Cell Growth" by He-Ming Wu and Alice Cheung describes the role of ROPs/RACs in cell polarity, hormonal, and reactive oxygen species signaling. The chapter highlights how conserved mechanisms involving proteins such as ADF/cofilins and formins together with plant-unique proteins such as the RICs and ICR1 (Interactor of Constitutive active ROP1) orchestrate polar cell growth. The chapter "The Role of Seven Transmembrane Domain MLO Proteins, Heterotrimeric G-Proteins and Monomeric RAC/ROPs in Plant Defense" by Justine Lorek et al. discusses the role of ROPs/RACs MLO proteins and hetrotrimeric G proteins in plant defense responses and how they interface with cell polarity, complementing the discussion in the chapter "G proteins and plant innate immunity." Two types of ROP/RAC GEFs are currently known in plants: an evolutionary-conserved Dock180 protein called SPIKE1, which may be associated with the WAVE complex and a family of proteins designated PRONE GEFs that can activate ROPs/RACs but not non-plant Rho proteins. In the chapter "Structure and Function of ROPs and Their GEFs," Christoph Thomas and Antje Berken discuss the structure and function of ROPs/RACs and the PRONE ROPGEFs, highlighting the common and plant-unique features. GAPs and RhoGDIs play pivotal roles in regulation of signaling by Rho GTPases. The chapter "Regulatory and Cellular Functions of Plant RhoGAPs and RhoGDIs" by Benedikt Kost highlights studies showing how spatial distribution of certain GAPs and function of RhoGDIs regulate polar cell growth. The chapter "Evolution of the ROP GTPase Signaling Module" by John Fowler discusses the origin and evolution of ROPs/RACs. ROP/RACs and heterotrimeric G proteins function at the plasma membrane to which they attach by virtue of postranslational lipid modifications and polybasic region comprised of lysine and arginine residues. In the chapter "Protein–lipid Modifications and Targeting of ROP/RAC and Heterotrimeric G Proteins," Nadav Sorek and Shaul Yalovsky describe the lipid modifications and their regulatory roles in function of ROP/RACs and heterotrimeric G proteins.Each chapter of this book offers a different perspective of the state-of-the-art in the field, presenting a well-balanced and an up-to-date description of the current knowledge on G protein signaling in plants. The breadth of the book offers a thourough introduction, and at the same time, a detailed in-depth discussion to those who are new to the field. Thus, we hope to draw the interest of both new and advanced students to this relatively young but fast-progressing and fascinating field of plant cell biology

Tel Aviv, June 2009 Shaul Yalovsky
Bonn, June 2009 František Baluška
Chapel Hill, June 2009 Alan Jones

Contents

Plant Gα Structure and Properties

Janice C. Jones

Abstract G-protein-coupled receptors (GPCRs) are a large family of eukaryotic membrane-spanning proteins that convert signals from the outside of a cell to an appropriate response inside the cell. GPCRs typically associate with a heterotrimeric Gαβγ protein. Activated receptors propagate signals by causing the Gα subunit of the heterotrimer to release GDP and to bind GTP. GTP binding causes a conformational change in the Gα protein that triggers heterotrimer dissociation and downstream signaling. In this way, the guanine nucleotide occupancy of the Gα subunit determines the protein structure and activity. Although most of the G protein paradigm has been established by studying animal G proteins, recent research has revealed diverse roles for G protein signaling in plants. The first part of the chapter (Section "Introduction: Structure-Function Relationships in G Protein Signaling" in Chapter "Plant Gα Structure and Properties") reviews G protein signaling principles, with an emphasis on the information that has been gleaned from atomic structures. The second part (Section "Comparison of Plant Gα Proteins to Mammalian Gα Proteins" in Chapter, "Plant Gα Structure and Properties") compares plant Gα proteins to animal Gα proteins with an emphasis on how structure confers function for these proteins. Although plant and animal Gα proteins share less than 40% identity, the key residues that confer G protein function are nearly invariant across all G protein families. The third part (Section "Properties of Plant Gα Proteins" in Chapter "Plant Gα Structure and Properties") describes the physical properties of plant Gα proteins, including kinetic properties, localization, receptor coupling, and effector activation.

J.C. Jones
Division of Biochemistry and Biophysics, Department of Biology, University of North Carolina, PO Box 7260, Chapel Hill, NC 27599, USA
e-mail: biochemnerd2000@yahoo.com

S. Yalovsky et al. (eds.), *Integrated G Proteins Signaling in Plants*,
Signaling and Communication in Plants,
DOI 10.1007/978-3-642-03524-1_1,

1 Introduction: Structure–function Relationships in G Protein Signaling

Organisms rely on cell surface receptors to connect cues from their extracellular environment to appropriate intracellular responses. G-protein-coupled receptors (GPCRs) serve this purpose by relaying signals through their associated heterotrimeric Gαβγ proteins. Ligand binding to a GPCR causes the Gα subunit to release GDP, bind GTP, and dissociate from the Gβγ dimer. The GTP-activated Gα protein and the liberated Gβγ proteins in turn activate or inhibit downstream effectors such as enzymes (see Section "Candidate Plant Gα Effectors"), ultimately leading to a cellular response. The response is terminated after the intrinsic GTPase activity of the Gα protein hydrolyzes GTP to GDP, and the Gαβγ heterotrimer reassociates. Signal termination can be accelerated by proteins that contain GTPase accelerating activity in their RGS domains (see Section "RGS Interacting Interface").

A number of snapshots along this signaling pathway (Fig. 1) have been captured in crystal structures. First, the Gα protein was crystallized in its GTP-activated form (Noel et al. 1993). This structure reveals that the Gα protein contains a Ras GTPase-like domain and a helical domain, and the guanine nucleotide-binding pocket is situated between these two lobes. Subsequently, the structure of a different Gα protein showed that these basic features are conserved across G protein classes (Coleman et al. 1994). Later, two different Gα proteins were crystallized in their inactive GDP-bound forms (Lambright et al. 1994; Mixon et al. 1995). Comparison of the inactive and active Gα conformations shows that guanine nucleotide exchange and hydrolysis are accompanied by movements in three small highly conserved regions, which have been designated as "switch" domains (see Section "Switch Regions and the Gβγ Interacting Interface").

Gα proteins have also been crystallized in complex with binding partners. The structures of Gα proteins in heterotrimeric complexes (Lambright et al. 1996; Wall

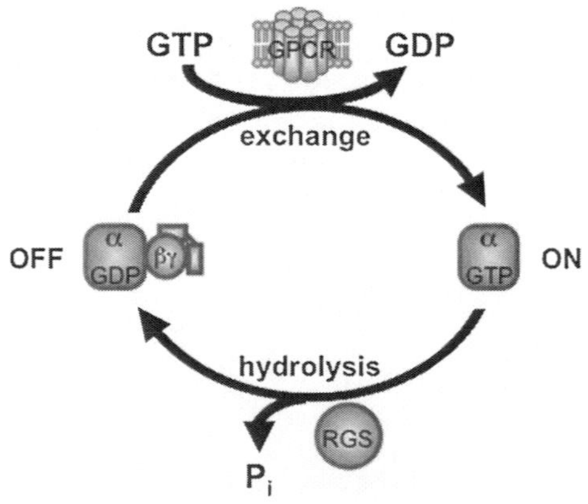

Fig. 1 *Schematic of GTP binding and hydrolysis by a typical G protein*. Gα proteins become activated by releasing GDP and binding GTP. GPCRs promote activation by stabilizing the nucleotide-free Gα protein. Gα proteins become inactivated by hydrolyzing GTP to GDP. RGS proteins promote inactivation by stabilizing the transition state for GTP hydrolysis. Gα inactivation leads to heterotrimer association

et al. 1995; Wall et al. 1998) complete the picture of how activation by GTP leads to Gβγ dissociation. GTP-induced rearrangement of the switch regions alters the Gβγ-interacting interface. The structures of Gα proteins with effectors define effector–Gα interfaces and show a possible mechanism for effector activation by Gα proteins (Lutz et al. 2007; Slep et al. 2001; Tesmer et al. 1997b). Finally, a crystal structure helped define the basis for signal termination by the RGS proteins. The structure of a Gα protein bound to an RGS protein (Tesmer et al. 1997a) shows how they accelerate the GTPase activity of Gα proteins by binding to and stabilizing the transition state for GTP hydrolysis (Berman et al. 1996a). Together these atomic structures reveal how G protein structure confers activity.

2 Comparison of Plant Gα Proteins to Mammalian Gα Proteins

While mammals have 23 Gα proteins, 5 Gβ proteins, 12 Gγ proteins, and nearly 1,000 GPCRs, plants have a much smaller repertoire of G protein signaling components. Animal Gα proteins are subdivided into four classes ($G\alpha_s$, $G\alpha_{i/o}$, $G\alpha_q$, $G\alpha_{12/13}$) based on sequence similarities and effector activation (see Section "Candidate Plant Gα Effectors"). In contrast, plants typically only have one or two Gα proteins, a single Gβ protein, and at least two Gγ proteins (Temple and Jones 2007). The number of plant GPCRs is debatable since GPCR sequences can be quite divergent and difficult to identify with traditional sequence alignment tools (see Section "Candidate Plant GPCRs").

Although there is no published crystal structure for a plant Gα protein, sequence comparison with mammalian Gα proteins with solved crystal structures affords predictions of structural elements for plant Gα proteins. This text will mainly focus on comparing plant Gα proteins to animal $G\alpha_i$ because much of the published structural information has come from this protein. Plant Gα proteins share about 38% identity (56% similarity) with mammalian $G\alpha_i$, but a more detailed comparison of these proteins reveals that the regions that confer function are more highly conserved across these families. The sequence alignment (Fig. 2) compares two well-characterized representative plant Gα proteins (AtGPA1 and OsRGA1) with representative Gα proteins from each of the animal classes. Note that the residues discussed in the following sections are marked in Fig. 2. Secondary structural elements are also marked, and these α-helices and β-sheets are labeled in the structural model of the *Arabidopsis* Gα protein (Fig. 3).

2.1 Switch Regions and the Gβγ Interacting Interface

Comparison of the crystal structures of GDP-bound and GTPγS-bound Gα proteins revealed three "switch" regions, which rearrange depending on whether the Gα protein is inactive or active. Switch regions contain residues that interact with Gβ proteins, RGS proteins (see Section "RGS interacting Interface") and downstream effectors (see Section "Candidate Plant Gα effectors). Moreover, residues that are necessary for GTP binding and hydrolysis activities are found in switch regions.

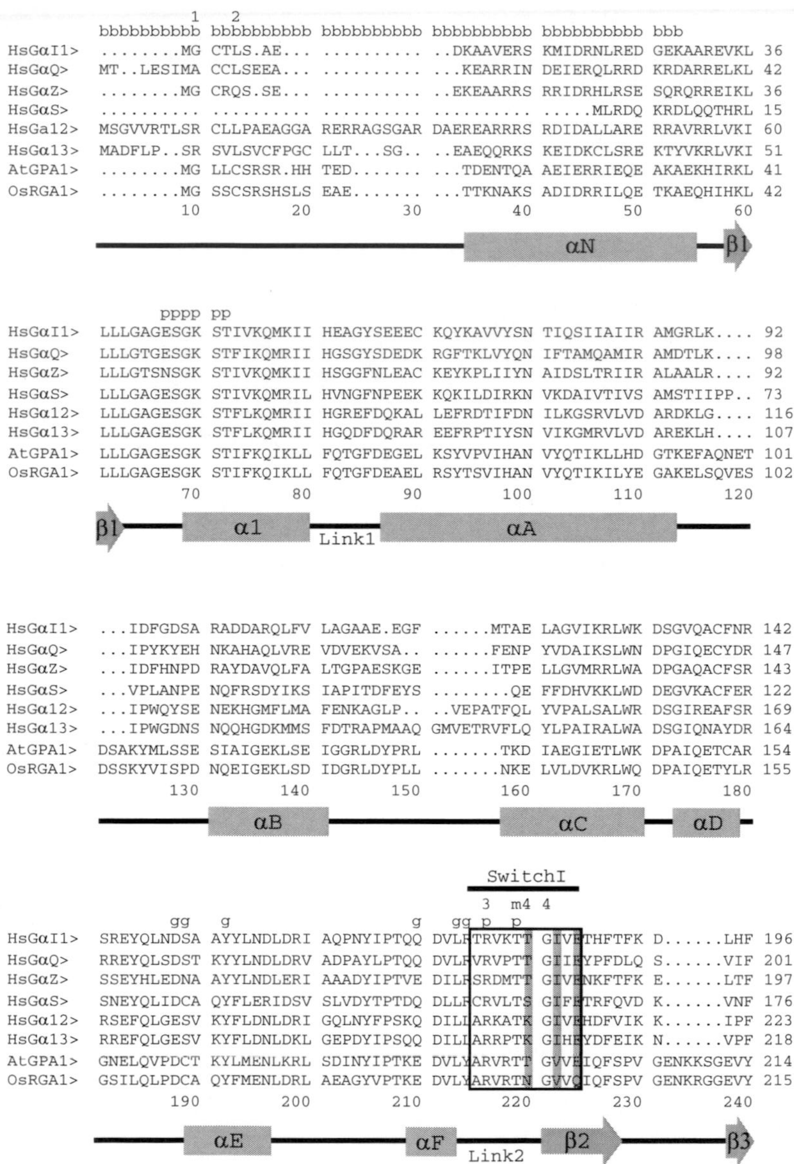

Fig. 2 (continued)

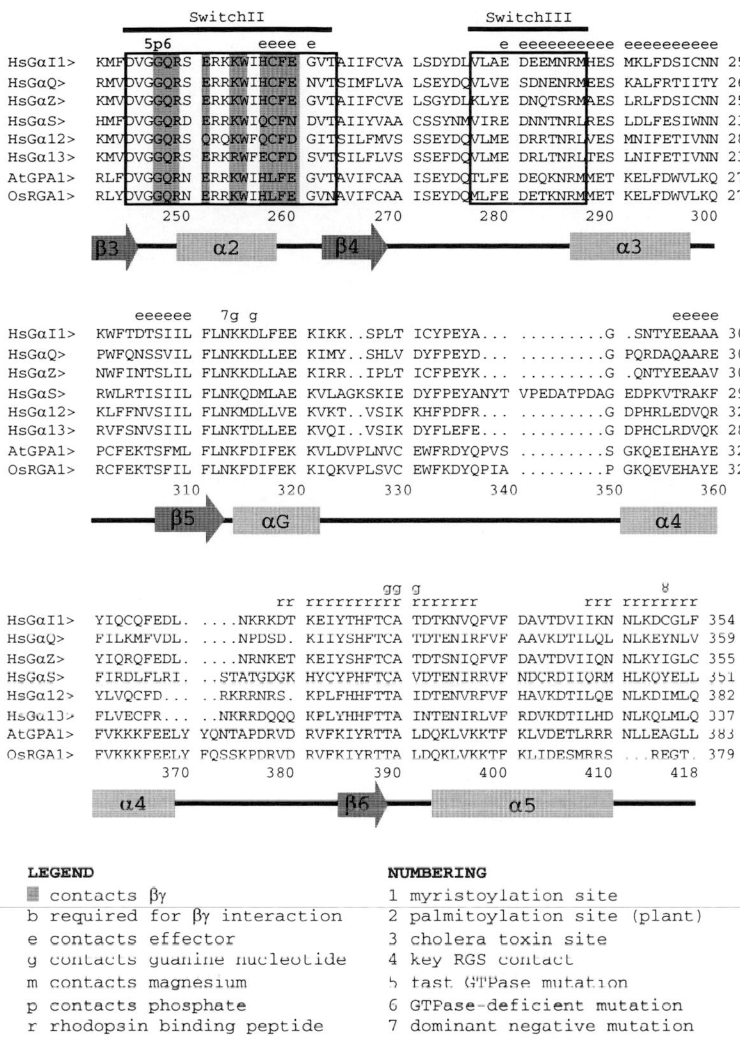

Fig. 2 *Multiple sequence alignment of representative Gα proteins from plants and animals. Homo sapien* $G\alpha_{i1}$ (NP_002060), *H. sapien* $G\alpha_o$ (AAH67850), *H. sapien* $G\alpha_z$ (NP_002064), *H. sapien* $G\alpha_s$ (P63092), *H. sapien* $G\alpha_{12}$ (NP_031379), *H. sapien* $G\alpha_{13}$ (NP_006563), *Arabidopsis thaliana* GPA1 (NP_180198), *Oryza sativa* (rice AAC41657). Sequence alignment was generated with ClustalW before small adjustments were made based on structural considerations. The inserts found in plant proteins were shifted away from secondary structural elements to loop regions. Legend describes annotations and numbering

In this regard, switches confer many of the core functions of Gα proteins, and the conformation of these domains determines whether or not the G protein dissociates from the Gβγ dimer and relays a signal to a downstream effector.

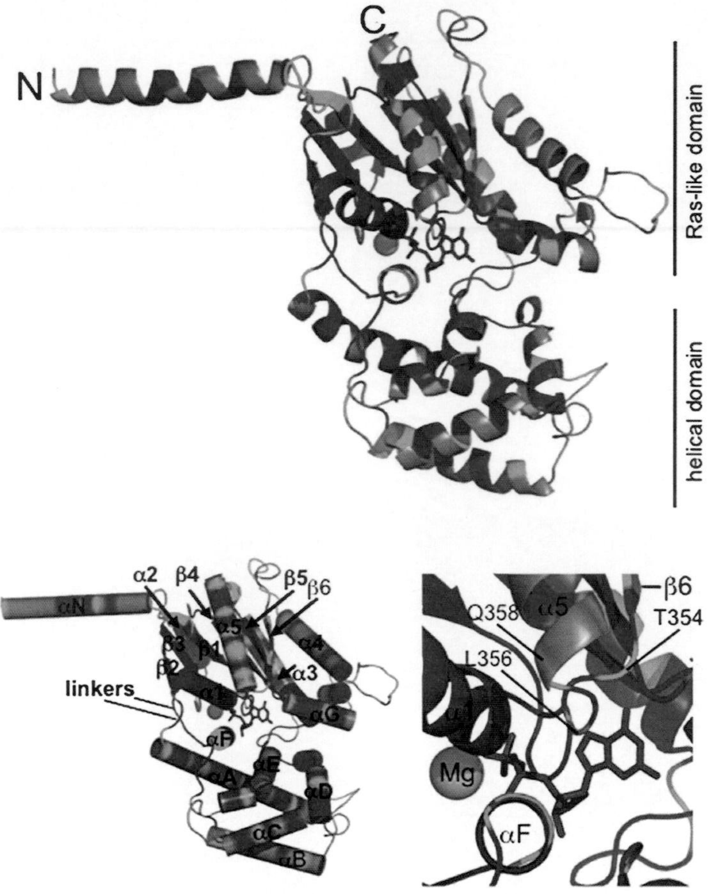


```
MGLLCSRSRHHTEDTDENTQAAEIERRIEQEAKAEKHIRKLLLLGAGESGKS
TIFKQIKLLFQTGFDEGELKSYVPVIHANVYQTIKLLHDGTKEFAQNETDSA
KYMLSSESIAIGEKLSEIGGRLDYPRLTKDIAEGIETLWKDPAIQETCARGN
ELQVPDCTKYLMENLKRLSDINYIPTKEDVLYARVRTTGVVEIQFSPVGENK
KSGEVYRLFDVGGQRNERRKWIHLFEGVTAVIFCAAISEYDQTLFEDEQKNR
MMETKELFDWVLKQPCFEKTSFMLFLNKFDIFEKKVLDVPLNVCEWFRDYQP
VSSGKQEIEHAYEFVKKKFEELYYQNTAPDRVDRVFKIYRTTALDQKLVKKT
FKLVDETLRRRNLLEAGLL*
```

Fig. 3 *Homology model of AtGPA1 bound to GDP*. (modified from (Temple and Jones 2007; Ullah et al. 2003). (*Top*) Residues in the homology model of AtGPA1 were colored based on a multiple sequence alignment, essentially as described by (Temple and Jones 2007; Ullah et al. 2003). Amino acids colored in blue are conserved throughout nearly all animal Gα protein classes and species. These residues include Switch regions and nucleotide-binding domains. Amino acids colored in gray are either divergent or class-specific. Amino acids colored in green are plant-specific. These residues are conserved throughout nearly all Gα protein classes besides plants. They likely confer the unique properties of plant Gα proteins. (*Middle left*) Labels correspond to secondary structures as marked in Fig. 1. (*Middle right*) Zoomed view of the GDP-binding pocket from the AtGPA1 homology model. Plant-specific residues in the GDP-binding loop and α5 helix

Given that switch regions confer basic Gα protein function, it is not surprising that these domains are nearly invariant across all classes of animal Gα proteins (G_s, $G_{i/o}$, G_q, $G_{12/13}$) and across species. Although plant Gα proteins share less than 40% overall similarity, all of the 9 residues in SwitchI, 18 of the 19 residues in SwitchII, and 6 of the 7 SwitchIII residues that are conserved across animal classes are also conserved in plants. Noteworthy is the substitution of an asparagine residue for a nearly invariant serine residue in the middle of SwitchII. The Gβ-interacting residues in the N-terminal helix are also conserved in plant Gα proteins. Switches are enclosed in a box and Gβ-interacting residues are shaded, in Fig. 2. More information about Gα-Gβγ interaction can be found in Section "Plant Gα Lipid Modification and Subcellular Localization."

2.2 Guanine Nucleotide-Binding Pocket

Crystal structures also reveal residues that interact with the guanine nucleotide. Like switch regions, these residues are nearly invariant across all families of Gα proteins and across species. Although the residues that interact with magnesium, ribose, and phosphate groups of the guanine nucleotide are nearly invariant in plants, two of the nine residues that interact with the guanine ring are not conserved in plants. These residues, located in a loop between the β6 strand and the α5 helix, are noteworthy because they are known determinants of protein stability and nucleotide exchange rates in mammalian G proteins. In mammals, the sequence is Cys–Ala–Thr, while it is Thr–Ala–Leu in plants. Mutation of the highly conserved cysteine within this loop (C325A in $G\alpha_o$) results in a 10-fold reduction in GDP affinity (Thomas et al. 1993). Similarly, mutation of the highly conserved alanine in this loop (A366S in $G\alpha_s$) results in rapid GDP release and high GTP occupancy (Iiri et al. 1994). These mutations map to residues that contact the N7 of the guanine ring, and this sequence divergence may confer some of the unique kinetic properties of plant Gα proteins (see Section "Kinetic Properties of the *Arabidopsis* Gα Protein"). Guanine nucleotide-interacting residues are marked with g (guanine ring), r (ribose), or p (phosphate) in Fig. 2, and the modeled nucleotide-binding pocket of a plant protein is shown in Fig. 3 (*bottom right*).

are labeled. (*Bottom*) Primary sequence of AtGPA1. Colored residues correspond to residue colors in the structural homology model. A detailed description of how the structural model was made can be found in (Temple and Jones 2007). Note that this analysis is based on a sequence alignment that contains multiple representatives of each mammalian class, whereas the alignment in Fig. 1 only has one representative from each class

2.3 Loop Insertions

One of the main differences between plant G proteins and animal G proteins is that plant G proteins contain several short amino acid insertions that are not found in the animal G proteins. First, plants have a six-residue insertion between the αA and αB helices, where affinity tags can be inserted without disrupting Gα function (Adjobo-Hermans et al. 2006; Bunemann et al. 2003; Gadella et al. 1999; Wang et al. 2008). Plant Gα proteins also have four- to six-residue insertions in loops between the αB and αC helices, the β2 and β3 strands (between SwitchI and SwitchII), and before and after the α4 helix. Although no function has been identified for these insertions, the residues that compose them are nearly invariant across plant species, suggesting selective pressure to maintain these elements in plants. The location of two of these insertions near where the Gα protein is thought to contact the plasma membrane and receptor could suggest a role in receptor coupling or membrane targeting. Otherwise, these plant-specific insertions may determine downstream targets of Gα proteins by forming an effector-binding interface (see Section "Candidate Plant Gα Effectors).

2.4 The α5 Helix

Overall, many of the variable Gα residues in plants reside near the C-terminus. In particular, the C-terminal α5 helix has received much attention as an element in animal Gα proteins that moves during nucleotide exchange (Natochin et al. 2001; Oldham et al. 2006). This ~17 residue helix spans the nearly 30 Å gap that is thought to lie between the receptor and the nucleotide-binding pocket. Insertion of a flexible linker between the α5 helix and the Gα C-terminus (where the receptor binds) decreases receptor-catalyzed nucleotide exchange (Natochin et al. 2001), possibly by reducing the ability of the activated receptor to translate a conformational change to the nucleotide-binding pocket. The α5 helix is thought to rotate away from the β2-β3 loop (toward the β6 strand) during nucleotide exchange (Oldham et al. 2006; Preininger et al. 2009).

Multiple sequence alignment (Fig. 2) identifies several plant-specific residues within the α5 helix. Interestingly, the amino acids that are unique to plants are spaced at an i to $i + 3$ distance, and therefore are expected to map to one face of the α-helix. Moreover, two plant-specific lysine residues in the α5 helix are potentially juxtaposed to two plant-specific positive residues (one lysine, one arginine) in the β6 strand. Electrostatic repulsion between these residues would rotate the α5 helix away from the β6 strand, and this movement could account for the unusual nucleotide exchange properties of some plant Gα proteins (see Section "Properties of Plant Gα Proteins"). In addition to its role in nucleotide exchange, the α5 helix of some Gα proteins also interacts with effectors (Lutz et al. 2007). Therefore, the divergence of residues in plant α5 helices may also confer effector specificity. The α5 helix is marked in Fig. 2 and is depicted in the model in Fig. 3.

2.5 RGS Interacting Interface

The Gα subunit of the heterotrimer has intrinsic GTPase activity, which serves to return the Gα protein to its inactive GDP-bound form. Proteins that contain RGS (regulator of G protein signaling) domains (Siderovski et al. 1996) bind to Gα proteins and accelerate their intrinsic rates of GTP hydrolysis (Berman et al. 1996b). RGS proteins are fairly ubiquitous throughout the plant kingdom, but plants typically only have a single RGS protein. The crystal structure of $G\alpha_i$ in a complex with an RGS protein (Tesmer et al. 1997a), along with mutational analysis (DiBello et al. 1998; Johnston et al. 2007a), revealed residues that are critical for the interaction between Gα proteins and RGS proteins. RGS-contacting residues are found in each of the three switch regions. With the exception of the rice Gα protein, these residues are conserved in plants. Residues critical for RGS interactions are marked in Fig. 2.

2.6 Receptor and Effector Coupling

Gα proteins physically associate with the intracellular loops of their cognate GPCRs. For animal Gα proteins, receptor-coupling specificity is conferred by the five C-terminal residues of the Gα protein (Conklin et al. 1996; Rasenick et al. 1994). Examination of this region in plant Gα proteins that have been sequenced shows that the C-termini of plant Gα proteins are nearly identical. The exception is OsRGA1, which is divergent from other plant species at its C-terminus. Candidate plant GPCRs are detailed in Section, "Candidate Plant GPCRs".

Residues that constitute the Gα–effector interface have been identified based on the Gα-effector cocrystal structures (Lutz et al. 2007; Slep et al. 2001; Tesmer et al. 1997b). Because different Gα classes signal to different effectors, these residues are class-specific. In other words, the effector interface in human $G\alpha_i$ is homologous to the same region in $G\alpha_i$ from *Drosophila*, but divergent from the same region in human $G\alpha_q$. Examination of the candidate effector-interacting residues in various plant Gα proteins shows high homology within this region among plant species, raising the possibility that these Gα proteins may interact with the same effector proteins. Effector-coupling residues are marked in Fig. 2, and candidate plant effectors are detailed in Sections "Candidate Plant Gα Effectors" and "Candidate Effectors in Plants Identified by Homology to Animal Effectors".

2.7 Cholera Toxin and Pertussis Toxin

Pertussis toxin and cholera toxin are commonly used as tools to study G protein signaling in living cells. These toxins covalently modify G proteins and alter their activation properties. Pertussis toxin ADP ribosylates a cysteine residue near the C-terminus of $G\alpha_i$ and $G\alpha_o$ (West et al. 1985). This modification disrupts coupling between the $G\alpha$ protein and receptor, and receptor-catalyzed activation of the G protein is greatly diminished (Cote et al. 1984). Plant G proteins lack the cysteine near the C-terminus that is found in $G\alpha_i$ and $G\alpha_o$, therefore it is unlikely that this toxin can be used as a tool for uncoupling plant G proteins from their cognate receptors.

Cholera toxin ADP ribosylates a SwitchI residue in $G\alpha_s$ proteins (Van Dop et al. 1984). This modification disrupts GTPase activity, resulting in constitutive activation of $G\alpha_s$ (Cassel and Pfeuffer 1978; Freissmuth and Gilman 1989). Although only $G\alpha_s$ is susceptible to constitutive activation by cholera toxin, the SwitchI arginine is conserved across G protein classes and in plants, indicating that this residue is not sufficient for modification (Freissmuth and Gilman 1989). One report suggests that OsRGA1, the $G\alpha$ protein from *Oryza sativa* (rice), is ADP ribosylated by cholera toxin (Seo et al. 1995). Toxin modification sites are marked with numbers in Fig. 2.

2.8 Contacts Between Ras-like and Helical Domains (Linkers 1 and 2)

As described earlier, $G\alpha$ proteins are composed of a helical domain and a Ras small GTPase-like domain. Two small linker segments connect these two domains, and the guanine nucleotide is buried between the two lobes. The helical and Ras-like domains must briefly separate for guanine nucleotide exchange to occur. It is thought that two conserved flexible glycine residues in the linkers give the conformational flexibility to allow nucleotide release since mutation to less flexible prolines increases nucleotide exchange and decreases GTP hydrolysis (Mello et al. 2002). The length and composition of linker 2, which is within SwitchI, is conserved in plant $G\alpha$ proteins, but the composition of linker 1 is divergent in plants. Linkers 1 and 2 are labeled in Figs. 2 and 3.

2.9 Other Residues of Interest: Tools for Studying G Protein Signaling

Along with crystal structures, point mutations in $G\alpha$ proteins have revealed residues that are critical for core G protein functions. Many of these mutant proteins have been used as tools to study G protein function in vivo. This section describes some of those residues that are conserved in plants. Plants have the SwitchII glutamine, which is commonly mutated to leucine to render Gα proteins GTPase-deficient and constitutively GTP-activated (Landis et al. 1989). The GTPase activity of this mutant is not susceptible to acceleration by RGS proteins. Plants also have the first arginine in SwitchI, which can be mutated to render Gα proteins GTPase-deficient yet susceptible to RGS activity (Berman et al. 1996b). Plants also have the Switch II glycine, which is mutated to confer accelerated GTPase activity (Thomas et al. 2004), and the SwitchI glycine residue, which is mutated to uncouple the Gα protein from the RGS protein (DiBello et al. 1998). Plants also have the guanine nucleotide contacting asparagine (NKxD), which is mutated to render a dominant negative, receptor-stabilized form of the Gα protein (Wu et al. 2004). Finally, plants have the SwitchII tryptophan, whose change in fluorescence is monitored to measure GTP binding rates in real-time in vitro (Higashijima et al. 1987a; Johnston et al. 2007a). Some of these tools have already been used for plant studies (Chen et al. 2003; Johnston et al. 2007a; Kato et al. 2004; Ullah et al. 2003), but more tools could likely be developed based on the sequence similarities outlined above. Residues discussed in this section are marked with numbers in Fig. 2.

2.10 Summary of Structural Comparison Between Plant and Mammalian Gα Proteins

In summary, the regions of Gα proteins that confer basic Gα function (nucleotide binding, GTP hydrolysis, Gβγ interaction, etc.) are mostly invariant between plant and animal proteins. Figs. 2 and 3 illustrate the similarities and differences between plant and animal G proteins. The next section highlights some of the properties of plant Gα proteins that differ from their mammalian homologs. The sequence alignment (Fig. 2) and structural model (Fig. 3) should aid in determining the structural features that confer unique functional properties to plant G proteins.

3 Properties of Plant Gα Proteins

The comparison of plant and animal Gα proteins in Section "Comparison of Plant Gα Proteins to Mammalian Gα Proteins" highlights structural differences between these families of proteins. Section "Properties of Plant Gα Proteins" expands this comparison to functional properties.

3.1 Kinetic Properties of the Arabidopsis Gα Protein

The G protein α subunit acts as a switch that turns cell signaling on and off in response to an extracellular signal. Signaling is turned on when the Gα protein binds GTP; signaling is turned off after the Gα protein hydrolyzes GTP to GDP (Fig. 1). For animal Gα proteins, GDP release is slow compared to GTP hydrolysis (Ferguson et al. 1986). Consequently, the Gα protein is not activated in a cell until a ligand-bound GPCR promotes fast GTP binding.

Studies with the *Arabidopsis* Gα protein, AtGPA1, have revealed some unusual activation properties. In contrast to typical G proteins, AtGPA1 does not require an activator because it has a fast spontaneous rate of nucleotide exchange and a relatively slow rate of GTP hydrolysis. In fact, exchange is nearly 100 times faster than hydrolysis for AtGPA1 in vitro (Johnston et al. 2007a). In a cell where GTP is in great excess over GDP, these properties would render AtGPA1 almost entirely GTP-bound and active. If AtGPA1 were constitutively active in vivo, the GTPase-deficient Q222L mutation would not alter GPA1-mediated processes. However, $GPA1^{Q222L}$ causes increased hypocotyl and primary root length (Chen et al. 2003) and decreased lateral root formation (Ullah et al. 2003) compared with plants expressing wild-type GPA1. These differences imply that AtGPA1 activation is negatively regulated in vivo.

For mammalian G proteins, Gα activation is known to be dampened by GTPase accelerating RGS proteins (Berman et al. 1996b; Dohlman et al. 1996), effectors (Cook et al. 2000; Mukhopadhyay and Ross 1999), and Gβγ proteins (Higashijima et al. 1987b). As detailed below in Section “Candidate Plant GPCRs: AtRGS1,” *Arabidopsis* has an RGS protein that accelerates GTP hydrolysis in vitro (Chen et al. 2003). The effect of Gβγ dimers and effectors on AtGPA1 activation remains to be determined, but together these proteins likely combine to diminish signaling in vivo. Future characterization of other plant Gα proteins will determine whether rapid nucleotide exchange and slow GTP hydrolysis are widespread in the plant kingdom or unique to the *Arabidopsis* Gα protein.

3.2 Kinetic Properties of Other Plant Gα Proteins

Although the *Arabidopsis* AtGPA1 protein is the best-characterized plant G protein to date, RGA1 from *Oryza sativa* (rice) and TGα1 from *Lycopersicon esculentum* (tomato) have also been purified and characterized. Two groups reported kinetic properties of OsRGA1. One group reported RGA1 kinetics reminiscent of slow nucleotide exchange by mammalian Gα proteins (Seo et al. 1995), whereas another group reported faster nucleotide exchange (Iwasaki et al. 1997). A third group measured the kinetic properties of OsRGA1 and found a fast nucleotide exchange rate which is more consistent with the findings of (Iwasaki et al. 1997) (M. Grosso, unpublished data). In contrast to AtGPA1, which has a slow rate of GTP hydrolysis, OsRGA1 has a GTP hydrolysis rate that is faster than its exchange rate (Iwasaki et al. 1997). Therefore, like mammalian G proteins, nucleotide exchange remains rate-limiting for this protein. Notably, OsRGA1 lacks the class-specific threonine in SwitchI that is critical for RGS interaction, and no canonical GAP has been identified in rice. $G\alpha_{12/13}$ and $G\alpha_s$ proteins also lack this residue, and in accordance no RGS protein has been identified that has activity toward these proteins (Fig. 2, residue #4).

GTP hydrolysis kinetics for Gα purified from tomato have also been reported (Aharon et al. 1998). The k_{cat} for this protein is reported as 0.08–0.13 min. This rate is determined from a steady-state GTP hydrolysis experiment (with multiple cycles of GTP binding, GTP hydrolysis, and GDP release), and the rate-limiting step is not identified, so it is unclear if this rate reflects the nucleotide exchange or GTP hydrolysis rate.

3.3 Possible Structural Determinants of Rapid Nucleotide Exchange

A detailed comparison of slowly exchanging mammalian Gα proteins and rapidly exchanging plant proteins may reveal structural constraints on nucleotide exchange rates. The sequence comparison from Section "Comparison of Plant Gα Proteins to Mammalian Gα Proteins" (along with other literature described above) points to a few candidate regions where plants have divergent sequences that may confer their unusual kinetic properties, namely the α5 helix and linker 1.

Typical Gα proteins require a GPCR to stabilize the nucleotide-free protein, an unstable intermediate between GDP release and GTP binding. This is how GPCRs catalyze nucleotide exchange. The mechanism of nucleotide exchange remains one of the most poorly understood processes in G protein signaling, partly due to the heroic effort that would be required to obtain a co-crystal structure of a nucleotide-free Gα protein in complex with Gβγ and receptor. The *Arabidopsis* G protein does not require a GPCR for rapid exchange, suggesting that this protein may be independently stable in the nucleotide-free form. Perhaps, nucleotide-free AtGPA1 could be characterized structurally as a surrogate for the transition state of receptor-catalyzed nucleotide exchange.

3.4 Plant Gα Lipid Modification and Subcellular Localization

Like animal G proteins, plant G proteins are predominantly found in the plasma membrane, with some localization at inner membranes (Adjobo-Hermans et al. 2006; Iwasaki et al. 1997; Kato et al. 2004; Wang et al. 2008). Plasma membrane localization facilitates interaction with GPCRs, effectors, and Gβγ proteins. Membrane targeting of animal Gα proteins requires a lipid anchor such as a myristate group or a palmitate group (Mumby et al. 1990). Moreover, animal Gα proteins require a second membrane-targeting feature such as another lipid anchor or association with the Gβγ protein, which is anchored at the membrane by a prenyl group at the C-terminus of the Gγ protein (Evanko et al. 2001; Morales et al. 1998;

Muntz et al. 1992; Sternweis 1986). Plant Gα subunits contain the conserved site for myristoylation and a potential site for palmitoylation near their N-termini (Fig. 2, residues 1 and 2). Mutation of either of these residues reduces plasma membrane localization of fluorescently tagged AtGPA1, even when a lipidated Gγ protein is present as a second membrane-targeting feature (Adjobo-Hermans et al. 2006). Unlike animal Gβγ proteins which require the lipid-modified Gα protein for plasma membrane targeting (Takida and Wedegaertner 2003), the *Arabidopsis* Gβγ dimer is membrane-localized without the Gα protein (Zeng et al. 2007). Together these data show that plants and animals rely on similar strategies for membrane-targeting, although the plant heterotrimer may form at the plasma membrane (Adjobo-Hermans et al. 2006), whereas the animal heterotrimer is thought to assemble before the complex reaches the plasma membrane (Fishburn et al. 2000; Takida and Wedegaertner 2003).

3.5 Candidate Plant GPCRs

GPCRs couple extracellular cues to intracellular responses by activating the Gα subunit of the heterotrimeric G protein. When activated by a ligand, GPCRs in turn activate G proteins by binding to and stabilizing the nucleotide-free form of the Gα protein, which facilitates GDP release and GTP binding. Accordingly, GPCRs have three main characteristics: (1) they have membrane-spanning domains (2) they are activated by ligands, and (3) they physically associate with and activate a G protein. Typically GPCRs are composed of seven transmembrane (7TM) domains, but GPCRs are difficult to identify through sequence alignment due to sequence divergence. Although predictive algorithms have been used to identify divergent GPCR candidates from plants (Devoto et al. 1999; Gookin et al. 2008; Moriyama et al. 2006), this review will only focus on candidate plant GPCRs that have been confirmed to physically interact with their cognate Gα protein. These proteins are considered "candidates" because to date no plant protein has been confirmed to possess all three of the GPCR properties listed above. Other candidate receptors are discussed in chapters 13 and 11 (two chapters, one by Etsuko Moriyama and one by Ralph Panstruga).

3.5.1 Candidate Plant GPCR: GCR1

Of the candidate plant GPCRs, GCR1 from *Arabidopsis* is one of the best characterized. GCR1 has a predicted 7TM topology and shares the highest sequence similarity (25%) with the CR1A GPCR from *Dictyostelium discoideum* (Josefsson and Rask 1997; Plakidou-Dymock et al. 1998). Physical interaction with the *Arabidopsis* Gα protein, GPA1, has been detected in a split ubiquitin assay in yeast and by immunoprecipitation from plants, and this interaction requires a free GCR1 C-terminus as expected for a GPCR-Gα interaction (Pandey and Assmann

2004). Although GCR1 physically interacts with GPA1, it also has Gα-independent function: *gcr1* and *gpa1* single deletion plants are less sensitive to the positive effects of giberrellins and brassinosteroids on seed germination, but *gcr1 gpa1* double null plants are even less sensitive than the single null mutant, indicating that GCR1 also functions outside of the AtGPA1 signaling pathway (Chen et al. 2004).

3.5.2 Candidate Plant GPCR: AtRGS1

AtRGS1 is also a well-characterized candidate plant GPCR. This protein has an unusual domain architecture (Chen et al. 2003), composed of an N-terminal 7TM domain and a C-terminal RGS box. This architecture suggests that, unlike typical GPCRs that activate Gα proteins, AtRGS1 may function as a receptor that accelerates GTPase activity. The *Arabidopsis* Gα protein is well suited to have regulated GTPase activity because of its unusual kinetic properties (see Section "Kinetic Properties of the *Arabidopsis* Gα Protein"). AtGPA1 spontaneously binds GTP rapidly (Johnston et al. 2007a), and therefore does not require a GPCR for activation. This property renders AtGPA1 predominantly GTP-bound and consequently subject to regulation of inactivation by an RGS protein.

Several studies support the hypothesis that AtRGS1 is a receptor that accelerates GTPase activity. First, RGS1 has predicted transmembrane domains, and while typical RGS proteins are cytoplasmic, AtRGS1 localizes to the plasma membrane (Chen et al. 2003). Second, *rgs1* deletion phenotypes are mimicked by Gα constitutive activation, indicating that AtRGS1 functions primarily as a GTPase accelerating protein instead of a guanine nucleotide exchange factor (Chen et al. 2003). Moreover, AtRGS1 physically interacts with the *Arabidopsis* Gα protein: the RGS box of AtRGS1 binds directly to AtGPA1 in vitro in the presence of GDP and aluminum fluoride (GTP hydrolysis transition state mimic), and the RGS box has GTPase accelerating activity toward AtGPA1 in vitro. Finally, genetic experiments reveal D-glucose (or a metabolic product) as a possible ligand for AtRGS1 (Chen and Jones 2004), and in vivo interaction between AtRGS1 and AtGPA1 is modulated by D-Glucose (Johnston et al. 2007a).

Other RGS proteins with predicted transmembrane domains have been identified in some fungi, plant pathogens, and other plant species including cotton, lettuce, potato, and grape (Johnston et al. 2007a). Although animals lack 7TM-RGS proteins, receptor-RGS interactions have been described for all animal RGS families (reviewed in Abramow-Newerly et al. 2006). In some cases, the receptor (and the receptor ligand) can increase or decrease RGS activity toward a Gα protein (Garzon et al. 2005; Ingi et al. 1998; Saitoh et al. 2002). Receptor-RGS interaction is also found in yeast, where the pheromone pathway RGS protein (Sst2) binds to the pheromone receptor. This interaction with the receptor is essential for Sst2 activity in vivo, and the receptor-interacting domain from Sst2 is sufficient to target other RGS proteins to the receptor wherein they compensate for Sst2 deletion (Ballon et al. 2006). In other words, RGS activity and specificity for this Gα are conferred by RGS interaction with the receptor that activates the Gα.

3.5.3 GPCR Type G Proteins (GTG)

A recent report suggests that *Arabidopsis* plants have a novel type of G protein called a GPCR-type G protein (Pandey et al. 2009). These proteins (GTG1 and GTG2) interact with GPA1 in a yeast split ubiquitin assay and coprecipitate with AtGPA1 from plants. They share 68% similarity (45% identity) to human orphan receptor GPR89. In addition to nine predicted transmembrane domains and a cell surface localization, the authors also report that GTG1 and GTG2 can bind and hydrolyze GTP. Plants lacking GTG1 and GTG2 are less sensitive to the effects of the plant hormone, abscisic acid (ABA), and the authors report that these GTG proteins bind ABA with a nanomolar affinity. Overall, the findings raise the possibility of a fascinating new complexity to plant G protein signaling, although the impact of the data is greatly limited by the very low specific activity of the purified GTG proteins. The stoichiometry of ABA binding was reported as only 1% (0.01 mol ABA/mol protein), and the stoichiometry of GTP binding was not reported.

3.5.4 GCR2

GCR2 from *Arabidopsis* was also proposed as a plant GPCR. The original report suggests that GCR2 is a transmembrane protein that binds to AtGPA1 and the plant hormone ABA (Liu et al. 2007). This report was challenged on many fronts (Guo et al. 2008; Illingworth et al. 2008; Johnston et al. 2007b). First, the algorithm that predicted GCR2 as a 7TM domain protein has a high false-positive rate for predicting transmembrane domains, and the GCR2 purification method from *E. coli* is entirely inconsistent with 7TM proteins. Alternatively, GCR2 is more likely an intracellular receptor for ABA. Finally, *gcr2* null mutants retain ABA responsiveness. The authors proposed that *gcr2* null mutant plants are only slightly less sensitive to ABA because of compensation by GCL1 and GCL1 (GCR2-like genes) (Liu et al. 2007). However, *gcr2 gcl1 gcl2* triple null mutant plants have wild-type sensitivity to the inhibitory effect of ABA on seed germination and plant development (Guo et al. 2008).

3.6 Candidate Plant Gα Effectors

Once Gα proteins are activated by a GPCR, both the Gα protein and the Gβγ dimer activate downstream effectors including enzymes and regulators. Effectors include protein kinases, ion channels, and enzymes that generate second messengers. These effectors begin to convert the signal initiated by an extracellular ligand into an intracellular response. One well-characterized example of how G protein activation of an effector leads to a cellular response lies in $G\alpha_s$ activation of adenylate cyclase in response to stress. Epinephrine is a hormone that is secreted from adrenal glands in response to acute stress to prepare muscles for action. Muscle cells have β-adrenergic GPCRs that bind epinephrine and activate $G\alpha_s$. $G\alpha_s$ activates adenylate cyclase to produce a second messenger molecule (cAMP). This second messenger activates cAMP-dependent protein kinase (PKA), an effector that phosphorylates Phosphorylase B kinase. Phosphorylase B kinase activates Phosphorylase A, an enzyme that converts glycogen to glucose-1-phosphate, which can enter glycolysis to be used by the cell for energy. This pathway has been rigorously defined through decades of research (reviewed in Gilman Nobel Lecture, 1984).

Although G proteins in plants were implicated in a range of functions (reviewed in Perfus-Barbeoch et al. 2004), very few candidate effectors were identified from plants. Studies in *Arabidopsis* and rice showed that the Gα proteins from these species associate with very large multiprotein complexes in vivo that dissociate when GTP is added (Kato et al. 2004; Wang et al. 2008), but no effectors have been identified from these complexes. Moreover, yeast two-hybrid interaction screens revealed only a few candidate effectors (Huang et al. 2006; Lapik and Kaufman 2003). Section "Candidate Plant Gα Effectors" highlights known Gα-interacting proteins that may be effectors. Section "Candidate Effectors in plants Identified by Homology to Animal Effectors" describes plant proteins that are homologous to known effectors for animal Gα proteins. Although plants have homologs of known Gβγ effectors from animals, including ion channels (Aharon et al. 1998; Wang et al. 2001), the following sections will focus mainly on Gα effectors.

3.6.1 Phospholipase Dα (PLDα)

Phospholipase Dα proteins from *Arabidopsis* (Zhao and Wang 2004) and tobacco (Lein and Saalbach 2001) were identified as possible effectors for plant Gα proteins. This enzyme hydrolyzes phospholipids into phosphatidic acid (PA) and a head group. Purified *Arabidopsis* PLDα1 binds to purified AtGPA1 in vitro, and these proteins coprecipitate in plants (Zhao and Wang 2004). GDP–AtGPA1 slightly inhibits in vitro production of PA by PLDα1, whereas GTP–GPA1 slightly stimulates this activity. Similarly, GDP-Gα from tobacco inhibits PLDα activity (Lein and Saalbach 2001). Further supporting PLDα as a candidate Gα effector in plants, both proteins have been implicated in ABA-mediated regulation of stomatal aperture (Mishra et al. 2006). The stimulatory effect of GTP-AtGPA1 on AtPLDα1 activity is at odds with the fact that GTP analogs decrease the association between AtGPA1 and AtPLDα1.

3.6.2 Other Gα-Interacting Proteins

The *Arabidopsis* Pirin protein was identified as a Gα-interacting protein in a yeast two-hybrid screen (Lapik and Kaufman 2003). Like AtGPA1, AtPirin is linked to seed germination (Lapik and Kaufman 2003). AtPirin may link GPA1 to transcription, since animal pirin proteins are nuclear proteins known to interact with transcription factors (Wendler et al. 1997). AtGPA1 also interacts with THF1 in vivo and in vitro (Huang et al. 2006). Unlike typical effectors, which interact preferentially with activated Gα proteins, THF1 interacts with both inactive and active GPA1 in vitro.

3.7 Candidate Effectors in Plants Identified by Homology to Animal Effectors

The 23 mammalian G proteins have been subdivided into four classes based on sequence similarity and function. Members of the $G\alpha_s$ family stimulate adenylate cyclase (see Section "$G\alpha_s$ and $G\alpha_i$ Effector: Adenylate Cyclase"), leading to increased cyclic AMP (cAMP); $G\alpha_i$ family members inhibit adenylate cyclase. $G\alpha_q$ proteins activate phospholipase Cα (see Section "$G\alpha_q$ Effector: Phospholipase Cβ1 (PLCβ1)"), leading to increased diacylglycerol (DAG) and inositol triphosphate (IP3) second messengers. $G\alpha_{12/13}$ proteins activate RhoGEFs (see Section "$G\alpha_{12/13}$ Effector: RhoGEF"), which in turn activate Rho GTPases. Plant Gα proteins are divergent from each class of animal Gα proteins in their effector-interacting regions, so no prediction can be made from sequence analysis as to which effectors may be activated by plant Gα proteins. As detailed below, some homologs of animal effectors are possible targets of G protein signaling in plants; some of these targets are not found in plants.

3.7.1 $G\alpha_s$ and $G\alpha_i$ Effector: Adenylate Cyclase

In animals, transmembrane adenylate cyclase is activated by $G\alpha_s$ and inhibited by $G\alpha_i$ family members. Adenylate cyclase converts ATP to the second messenger (cAMP). Plants lack a canonical membrane-associated adenylate cyclase protein. However, there is evidence for a cAMP system in plants (Lomovatskaya et al. 2008; Roef et al. 1996), and cAMP has been loosely tied to some of the same processes as plant G proteins (reviewed in Lomovatskaya et al. 2008). Plants also lack cGMP phosphodiesterase, which are downstream effectors of the $G\alpha_s$ class member, transducin.

3.7.2 $G\alpha_q$ Effector: Phospholipase Cβ1 (PLCβ1)

Phospholipases relay signals by generating lipid second messengers. In animals, $G\alpha_q$ directly activates PLCβ proteins to produce DAG and IP3 second messengers. Proteins that contain lipid-binding motifs are then recruited to sites where these second messengers accumulate, and they can directly activate enzymes and ion channels (Kishimoto et al. 1980; Nishizuka 1984; Slessareva et al. 2006; Streb et al.

1983; Takai et al. 1979). Lipid-binding domains, including PH, FYVE, C2, and PX domains, have been identified in plant proteins, and lipid second messengers have been connected to known to be G protein-mediated processes (Xiong et al. 2001). Although plants lack a canonical PLCβ, they possess very small PLC δ-like genes (Hartweck et al. 1997). These components leave the possibility of G protein-PLC signaling in plants, although the two have not been firmly connected so far. For complete review of lipid signaling in plants, see (Meijer and Munnik 2003).

3.7.3 $G\alpha_{12/13}$ Effector: RhoGEF

In animals, $G\alpha_{12/13}$ proteins activate Rho guanine nucleotide exchange factors (GEFs) (Hart et al. 1998), which in turn activate Rho small GTPases. Rho GTPases (called Rop GTPases) are found in plants where they regulate diverse processes. Two groups simultaneously identified a 14-member family of *Arabidopsis* Rop-GEFs with a yeast-two hybrid screen (Berken et al. 2005; Gu et al. 2006), and BLAST searches show RopGEF sequences in a number of plant species. Animal RhoGEFs typically contain catalytic DH domains and a PH domain (reviewed in (Cerione and Zheng 1996), which also form the Gα-interacting interface (Lutz et al. 2007). Plant RopGEFs instead contain a catalytic DUF315 domain. In plants, Gα proteins have not been connected to Rho GTPases yet, although genetic studies in rice show that the Gα protein acts upstream of a small GTPase in defense signaling (Suharsono et al. 2002).

4 Conclusions: Plant Gα Proteins are like Animal Gα Proteins, but Different

Overall plant Gα proteins are nearly identical to animal Gα proteins in regions of the protein that confer core function. Accordingly, basic Gα properties including plasma membrane localization, nucleotide binding, GTP hydrolysis, receptor coupling, and Gβγ sequestration have been confirmed in plant Gα proteins. The unusual kinetic properties of the *Arabidopsis* Gα protein show that animal characteristics do not universally apply to plant Gα proteins. Moreover, plants lack some of the well-characterized effector enzymes from animals, and plants have unusual signaling components, including a receptor-RGS hybrid protein (Chen et al. 2003).

One of the main unanswered questions in plant signaling is how a small repertoire of G protein signaling components can regulate such a large and diverse group of cellular processes. Plants typically only have one or two Gα proteins, and a few Gβγ dimer combinations. Despite this simplicity, plant G proteins are tied to seed germination, cell division, plant morphology, pathogen defense, stress response (damage, drought, oxidative, etc.), hormone sensitivity, gene expression,

ion channels, root growth, stomatal aperture, flowering, lipid signaling, and sugar/nutrient sensing (Perfus-Barbeoch et al. 2004). Identification of GPCRs, ligands, and downstream effectors of plant Gα and Gβγ will begin to shed light on how G proteins mediate these diverse cellular responses. It is possible that response diversity is achieved through assembly of the G protein into large Gβγ- and receptor-dependent multiprotein complexes at the plasma membrane (Wang et al. 2008). Alternatively, it is also possible that the plant G protein only serves as a nutrient gauge (Chen and Jones 2004), and that all of the other deficiencies in plants that lack functional G proteins are secondary to improper nutrient-sensing. Undoubtedly, elucidation of plant signaling pathways will inform how to engineer plants with more desirable characteristics including improved drought tolerance and pathogen defense.

References

Abramow-Newerly M, Roy AA, Nunn C, Chidiac P (2006) RGS proteins have a signalling complex: interactions between RGS proteins and GPCRs, effectors, and auxiliary proteins. Cell Signal 18:579–591

Adjobo-Hermans MJ, Goedhart J, Gadella TW Jr (2006) Plant G protein heterotrimers require dual lipidation motifs of Galpha and Ggamma and do not dissociate upon activation. J Cell Sci 119:5087–5097

Aharon GS, Gelli A, Snedden WA, Blumwald E (1998) Activation of a plant plasma membrane Ca^{2+} channel by TGalpha1, a heterotrimeric G protein alpha-subunit homologue. FEBS Lett 424:17–21

Ballon D, Flanary P, Gladue D, Konopka J, Dohlman H, Thorner J (2006) DEP domains link a regulator-of-G-protein-signaling protein to its cognate G-protein-coupled receptor. Cell 126(6):1079–1093

Berken A, Thomas C, Wittinghofer A (2005) A new family of RhoGEFs activates the Rop molecular switch in plants. Nature 436:1176–1180

Berman DM, Kozasa T, Gilman AG (1996a) The GTPase-activating protein RGS4 stabilizes the transition state for nucleotide hydrolysis. J Biol Chem 271:27209–27212

Berman DM, Wilkie TM, Gilman AG (1996b) GAIP and RGS4 are GTPase-activating proteins for the Gi subfamily of G protein alpha subunits. Cell 86:445–452

Bunemann M, Frank M, Lohse MJ (2003) Gi protein activation in intact cells involves subunit rearrangement rather than dissociation. Proc Natl Acad Sci USA 100:16077–16082

Cassel D, Pfeuffer T (1978) Mechanism of cholera toxin action: covalent modification of the guanyl nucleotide-binding protein of the adenylate cyclase system. Proc Natl Acad Sci USA 75:2669–2673

Cerione RA, Zheng Y (1996) The Dbl family of oncogenes. Curr Opin Cell Biol 8:216–222

Chen JG, Jones AM (2004) AtRGS1 function in *Arabidopsis thaliana*. Methods Enzymol 389:338–350

Chen JG, Pandey S, Huang J, Alonso JM, Ecker JR, Assmann SM, Jones AM (2004) GCR1 can act independently of heterotrimeric G-protein in response to brassinosteroids and gibberellins in Arabidopsis seed germination. Plant Physiol 135:907–915

Chen JG, Willard FS, Huang J, Liang J, Chasse SA, Jones AM, Siderovski DP (2003) A seven-transmembrane RGS protein that modulates plant cell proliferation. Science 301:1728–1731

Coleman DE, Berghuis AM, Lee E, Linder ME, Gilman AG, Sprang SR (1994) Structures of active conformations of Gi alpha 1 and the mechanism of GTP hydrolysis. Science 265:1405–1412

Conklin BR, Herzmark P, Ishida S, Voyno-Yasenetskaya TA, Sun Y, Farfel Z, Bourne HR (1996) Carboxyl-terminal mutations of Gq alpha and Gs alpha that alter the fidelity of receptor activation. Mol Pharmacol 50:885–890

Cook B, Bar-Yaacov M, Cohen Ben-Ami H, Goldstein RE, Paroush Z, Selinger Z, Minke B (2000) Phospholipase C and termination of G-protein-mediated signalling in vivo. Nat Cell Biol 2:296–301

Cote TE, Frey EA, Sekura RD (1984) Altered activity of the inhibitory guanyl nucleotide-binding component (Ni) induced by pertussis toxin. Uncoupling of Ni from receptor with continued coupling of Ni to the catalytic unit. J Biol Chem 259:8693–8698

Devoto A, Piffanelli P, Nilsson I, Wallin E, Panstruga R, von Heijne G, Schulze-Lefert P (1999) Topology, subcellular localization, and sequence diversity of the Mlo family in plants. J Biol Chem 274:34993–35004

DiBello PR, Garrison TR, Apanovitch DM, Hoffman G, Shuey DJ, Mason K, Cockett MI, Dohlman HG (1998) Selective uncoupling of RGS action by a single point mutation in the G protein alpha-subunit. J Biol Chem 273:5780–5784

Dohlman HG, Song J, Ma D, Courchesne WE, Thorner J (1996) Sst2, a negative regulator of pheromone signaling in the yeast *Saccharomyces cerevisiae*: expression, localization, and genetic interaction and physical association with Gpa1 (the G-protein alpha subunit). Mol Cell Biol 16:5194–5209

Evanko DS, Thiyagarajan MM, Siderovski DP, Wedegaertner PB (2001) Gbeta gamma isoforms selectively rescue plasma membrane localization and palmitoylation of mutant Galphas and Galphaq. J Biol Chem 276:23945–23953

Ferguson KM, Higashijima T, Smigel MD, Gilman AG (1986) The influence of bound GDP on the kinetics of guanine nucleotide binding to G proteins. J Biol Chem 261:7393–7399

Fishburn CS, Pollitt SK, Bourne HR (2000) Localization of a peripheral membrane protein: Gbetagamma targets Galpha(Z). Proc Natl Acad Sci USA 97:1085–1090

Freissmuth M, Gilman AG (1989) Mutations of GS alpha designed to alter the reactivity of the protein with bacterial toxins. Substitutions at ARG187 result in loss of GTPase activity. J Biol Chem 264:21907–21914

Gadella TW Jr, van der Krogt GN, Bisseling T (1999) GFP-based FRET microscopy in living plant cells. Trends Plant Sci 4:287–291

Garzon J, Rodriguez-Munoz M, Sanchez-Blazquez P (2005) Morphine alters the selective association between mu-opioid receptors and specific RGS proteins in mouse periaqueductal gray matter. Neuropharmacology 48:853–868

Gookin TE, Kim J, Assmann SM (2008) Whole proteome identification of plant candidate G-protein coupled receptors in Arabidopsis, rice, and poplar: computational prediction and in vivo protein coupling. Genome Biol 9:R120

Gu Y, Li S, Lord EM, Yang Z (2006) Members of a novel class of Arabidopsis Rho guanine nucleotide exchange factors control Rho GTPase-dependent polar growth. Plant Cell 18:366–381

Guo J, Zeng Q, Emami M, Ellis BE, Chen JG (2008) The GCR2 gene family is not required for ABA control of seed germination and early seedling development in Arabidopsis. PLoS ONE 3:e2982

Hart MJ, Jiang X, Kozasa T, Roscoe W, Singer WD, Gilman AG, Sternweis PC, Bollag G (1998) Direct stimulation of the guanine nucleotide exchange activity of p115 RhoGEF by Galpha13. Science 280:2112–2114

Hartweck LM, Llewellyn DJ, Dennis ES (1997) The Arabidopsis thaliana genome has multiple divergent forms of phosphoinositol-specific phospholipase C1. Gene 202:151–156

Higashijima T, Ferguson KM, Sternweis PC, Ross EM, Smigel MD, Gilman AG (1987a) The effect of activating ligands on the intrinsic fluorescence of guanine nucleotide-binding regulatory proteins. J Biol Chem 262:752–756

Higashijima T, Ferguson KM, Sternweis PC, Smigel MD, Gilman AG (1987b) Effects of Mg^{2+} and the beta gamma-subunit complex on the interactions of guanine nucleotides with G proteins. J Biol Chem 262:762–766

Huang J, Taylor JP, Chen JG, Uhrig JF, Schnell DJ, Nakagawa T, Korth KL, Jones AM (2006) The plastid protein THYLAKOID FORMATION1 and the plasma membrane G-protein GPA1 interact in a novel sugar-signaling mechanism in Arabidopsis. Plant Cell 18:1226–1238

Iiri T, Herzmark P, Nakamoto JM, van Dop C, Bourne HR (1994) Rapid GDP release from Gs alpha in patients with gain and loss of endocrine function. Nature 371:164–168

Illingworth CJ, Parkes KE, Snell CR, Mullineaux PM, Reynolds CA (2008) Criteria for confirming sequence periodicity identified by Fourier transform analysis: application to GCR2, a candidate plant GPCR? Biophys Chem 133:28–35

Ingi T, Krumins AM, Chidiac P, Brothers GM, Chung S, Snow BE, Barnes CA, Lanahan AA, Siderovski DP, Ross EM, Gilman AG, Worley PF (1998) Dynamic regulation of RGS2 suggests a novel mechanism in G-protein signaling and neuronal plasticity. J Neurosci 18:7178–7188

Iwasaki Y, Kato T, Kaidoh T, Ishikawa A, Asahi T (1997) Characterization of the putative alpha subunit of a heterotrimeric G protein in rice. Plant Mol Biol 34:563–572

Johnston CA, Taylor JP, Gao Y, Kimple AJ, Grigston JC, Chen JG, Siderovski DP, Jones AM, Willard FS (2007a) GTPase acceleration as the rate-limiting step in Arabidopsis G protein-coupled sugar signaling. Proc Natl Acad Sci USA 104:17317–17322

Johnston CA, Temple BR, Chen JG, Gao Y, Moriyama EN, Jones AM, Siderovski DP, Willard FS (2007b) Comment on "A G protein coupled receptor is a plasma membrane receptor for the plant hormone abscisic acid". Science 318:914

Josefsson LG, Rask L (1997) Cloning of a putative G-protein-coupled receptor from *Arabidopsis thaliana*. Eur J Biochem 249:415–420

Kato C, Mizutani T, Tamaki H, Kumagai H, Kamiya T, Hirobe A, Fujisawa Y, Kato H, Iwasaki Y (2004) Characterization of heterotrimeric G protein complexes in rice plasma membrane. Plant J 38:320–331

Kishimoto A, Takai Y, Mori T, Kikkawa U, Nishizuka Y (1980) Activation of calcium and phospholipid-dependent protein kinase by diacylglycerol, its possible relation to phosphatidylinositol turnover. J Biol Chem 255:2273–2276

Lambright DG, Noel JP, Hamm HE, Sigler PB (1994) Structural determinants for activation of the alpha-subunit of a heterotrimeric G protein. Nature 369:621–628

Lambright DG, Sondek J, Bohm A, Skiba NP, Hamm HE, Sigler PB (1996) The 2.0 A crystal structure of a heterotrimeric G protein. Nature 379:311–319

Landis CA, Masters SB, Spada A, Pace AM, Bourne HR, Vallar L (1989) GTPase inhibiting mutations activate the alpha chain of Gs and stimulate adenylyl cyclase in human pituitary tumours. Nature 340:692–696

Lapik YR, Kaufman LS (2003) The Arabidopsis cupin domain protein AtPirin1 interacts with the G protein alpha-subunit GPA1 and regulates seed germination and early seedling development. Plant Cell 15:1578–1590

Lein W, Saalbach G (2001) Cloning and direct G-protein regulation of phospholipase D from tobacco. Biochim Biophys Acta 1530:172–183

Liu X, Yue Y, Li B, Nie Y, Li W, Wu WH, Ma L (2007) A G protein-coupled receptor is a plasma membrane receptor for the plant hormone abscisic acid. Science 315:1712–1716

Lomovatskaya LA, Romanenko AS, Filinova NV (2008) Plant adenylate cyclases. J Recept Signal Transduct Res 28:531–542

Lutz S, Shankaranarayanan A, Coco C, Ridilla M, Nance MR, Vettel C, Baltus D, Evelyn CR, Neubig RR, Wieland T, Tesmer JJ (2007) Structure of Galphaq-p63RhoGEF-RhoA complex reveals a pathway for the activation of RhoA by GPCRs. Science 318:1923–1927

Meijer HJ, Munnik T (2003) Phospholipid-based signaling in plants. Annu Rev Plant Biol 54:265–306
Mello LV, Millner PA, Findlay JB (2002) Biochemical characteristics of guanine nucleotide binding protein alpha-subunit recombinant protein and three mutants: investigation of a domain motion involved in GDP-GTP exchange. J Protein Chem 21:29–34
Mishra G, Zhang W, Deng F, Zhao J, Wang X (2006) A bifurcating pathway directs abscisic acid effects on stomatal closure and opening in Arabidopsis. Science 312:264–266
Mixon MB, Lee E, Coleman DE, Berghuis AM, Gilman AG, Sprang SR (1995) Tertiary and quaternary structural changes in Gi alpha 1 induced by GTP hydrolysis. Science 270:954–960
Morales J, Fishburn CS, Wilson PT, Bourne HR (1998) Plasma membrane localization of G alpha z requires two signals. Mol Biol Cell 9:1–14
Moriyama EN, Strope PK, Opiyo SO, Chen Z, Jones AM (2006) Mining the *Arabidopsis thaliana* genome for highly-divergent seven transmembrane receptors. Genome Biol 7:R96
Mukhopadhyay S, Ross EM (1999) Rapid GTP binding and hydrolysis by G(q) promoted by receptor and GTPase-activating proteins. Proc Natl Acad Sci USA 96:9539–9544
Mumby SM, Heukeroth RO, Gordon JI, Gilman AG (1990) G-protein alpha-subunit expression, myristoylation, and membrane association in COS cells. Proc Natl Acad Sci USA 87:728–732
Muntz KH, Sternweis PC, Gilman AG, Mumby SM (1992) Influence of gamma subunit prenylation on association of guanine nucleotide-binding regulatory proteins with membranes. Mol Biol Cell 3:49–61
Natochin M, Moussaif M, Artemyev NO (2001) Probing the mechanism of rhodopsin-catalyzed transducin activation. J Neurochem 77:202–210
Nishizuka Y (1984) The role of protein kinase C in cell surface signal transduction and tumour promotion. Nature 308:693–698
Noel JP, Hamm HE, Sigler PB (1993) The 2.2 A crystal structure of transducin-alpha complexed with GTP gamma S. Nature 366:654–663
Oldham WM, Van Eps N, Preininger AM, Hubbell WL, Hamm HE (2006) Mechanism of the receptor-catalyzed activation of heterotrimeric G proteins. Nat Struct Mol Biol 13:772–777
Pandey S, Assmann SM (2004) The Arabidopsis putative G protein-coupled receptor GCR1 interacts with the G protein alpha subunit GPA1 and regulates abscisic acid signaling. Plant Cell 16:1616–1632
Pandey S, Nelson DC, Assmann SM (2009) Two novel GPCR-type G proteins are abscisic acid receptors in Arabidopsis. Cell 136:136–148
Perfus-Barbeoch L, Jones AM, Assmann SM (2004) Plant heterotrimeric G protein function: insights from Arabidopsis and rice mutants. Curr Opin Plant Biol 7:719–731
Plakidou-Dymock S, Dymock D, Hooley R (1998) A higher plant seven-transmembrane receptor that influences sensitivity to cytokinins. Curr Biol 8:315–324
Preininger A, Funk M, Meier S, Oldham W, Johnston C, Adhikary S, Kimple A, Siderovski D, Hamm H, Iverson T (2009) Helix dipole movement and conformational variability contribute to allosteric GDP release in Gi subunits. Biochemistry (In press)
Rasenick MM, Watanabe M, Lazarevic MB, Hatta S, Hamm HE (1994) Synthetic peptides as probes for G protein function. Carboxyl-terminal G alpha s peptides mimic Gs and evoke high affinity agonist binding to beta-adrenergic receptors. J Biol Chem 269:21519–21525
Roef L, Witters E, Gadeyne J, Marcussen J, Newton RP, Van Onckelen HA (1996) Analysis of 3', 5'-cAMP and adenylyl cyclase activity in higher plants using polyclonal chicken egg yolk antibodies. Anal Biochem 233:188–196
Saitoh O, Murata Y, Odagiri M, Itoh M, Itoh H, Misaka T, Kubo Y (2002) Alternative splicing of RGS8 gene determines inhibitory function of receptor type-specific Gq signaling. Proc Natl Acad Sci USA 99:10138–10143
Seo HS, Kim HY, Jeong JY, Lee SY, Cho MJ, Bahk JD (1995) Molecular cloning and characterization of RGA1 encoding a G protein alpha subunit from rice (*Oryza sativa* L. IR-36). Plant Mol Biol 27:1119–1131

Siderovski DP, Hessel A, Chung S, Mak TW, Tyers M (1996) A new family of regulators of G-protein-coupled receptors? Curr Biol 6:211–212
Slep KC, Kercher MA, He W, Cowan CW, Wensel TG, Sigler PB (2001) Structural determinants for regulation of phosphodiesterase by a G protein at 2.0 A. Nature 409:1071–1077
Slessareva JE, Routt SM, Temple B, Bankaitis VA, Dohlman HG (2006) Activation of the phosphatidylinositol 3-kinase Vps34 by a G protein alpha subunit at the endosome. Cell 126:191–203
Sternweis PC (1986) The purified alpha subunits of Go and Gi from bovine brain require beta gamma for association with phospholipid vesicles. J Biol Chem 261:631–637
Streb H, Irvine RF, Berridge MJ, Schulz I (1983) Release of Ca^{2+} from a nonmitochondrial intracellular store in pancreatic acinar cells by inositol-1, 4, 5-trisphosphate. Nature 306:67–69
Suharsono U, Fujisawa Y, Kawasaki T, Iwasaki Y, Satoh H, Shimamoto K (2002) The heterotrimeric G protein alpha subunit acts upstream of the small GTPase Rac in disease resistance of rice. Proc Natl Acad Sci USA 99:13307–13312
Takai Y, Kishimoto A, Iwasa Y, Kawahara Y, Mori T, Nishizuka Y, Tamura A, Fujii T (1979) A role of membranes in the activation of a new multifunctional protein kinase system. J Biochem 86:575–578
Takida S, Wedegaertner PB (2003) Heterotrimer formation, together with isoprenylation, is required for plasma membrane targeting of Gbetagamma. J Biol Chem 278:17284–17290
Temple BR, Jones AM (2007) The plant heterotrimeric G-protein complex. Annu Rev Plant Biol 58:249–266
Tesmer JJ, Berman DM, Gilman AG, Sprang SR (1997a) Structure of RGS4 bound to AlF4–activated G(i alpha1): stabilization of the transition state for GTP hydrolysis. Cell 89:251–261
Tesmer JJ, Sunahara RK, Gilman AG, Sprang SR (1997b) Crystal structure of the catalytic domains of adenylyl cyclase in a complex with Gsalpha.GTPgammaS. Science 278:1907–1916
Thomas CJ, Du X, Li P, Wang Y, Ross EM, Sprang SR (2004) Uncoupling conformational change from GTP hydrolysis in a heterotrimeric G protein alpha-subunit. Proc Natl Acad Sci USA 101:7560–7565
Thomas TC, Schmidt CJ, Neer EJ (1993) G-protein alpha o subunit: mutation of conserved cysteines identifies a subunit contact surface and alters GDP affinity. Proc Natl Acad Sci USA 90:10295–10298
Ullah H, Chen JG, Temple B, Boyes DC, Alonso JM, Davis KR, Ecker JR, Jones AM (2003) The beta-subunit of the Arabidopsis G protein negatively regulates auxin-induced cell division and affects multiple developmental processes. Plant Cell 15:393–409
Van Dop C, Tsubokawa M, Bourne HR, Ramachandran J (1984) Amino acid sequence of retinal transducin at the site ADP-ribosylated by cholera toxin. J Biol Chem 259:696–698
Wall MA, Coleman DE, Lee E, Iniguez-Lluhi JA, Posner BA, Gilman AG, Sprang SR (1995) The structure of the G protein heterotrimer Gi alpha 1 beta 1 gamma 2. Cell 83:1047–1058
Wall MA, Posner BA, Sprang SR (1998) Structural basis of activity and subunit recognition in G protein heterotrimers. Structure 6:1169–1183
Wang S, Assmann SM, Fedoroff NV (2008) Characterization of the Arabidopsis heterotrimeric G protein. J Biol Chem 283:13913–13922
Wang XQ, Ullah H, Jones AM, Assmann SM (2001) G protein regulation of ion channels and abscisic acid signaling in Arabidopsis guard cells. Science 292:2070–2072
Wendler WM, Kremmer E, Forster R, Winnacker EL (1997) Identification of pirin, a novel highly conserved nuclear protein. J Biol Chem 272:8482–8489
West RE Jr, Moss J, Vaughan M, Liu T, Liu TY (1985) Pertussis toxin-catalyzed ADP-ribosylation of transducin. Cysteine 347 is the ADP-ribose acceptor site. J Biol Chem 260:14428–14430
Wu YL, Hooks SB, Harden TK, Dohlman HG (2004) Dominant-negative inhibition of pheromone receptor signaling by a single point mutation in the G protein alpha subunit. J Biol Chem 279:35287–35297

Xiong L, Lee B, Ishitani M, Lee H, Zhang C, Zhu JK (2001) FIERY1 encoding an inositol polyphosphate 1-phosphatase is a negative regulator of abscisic acid and stress signaling in Arabidopsis. Genes Dev 15:1971–1984

Zeng Q, Wang X, Running MP (2007) Dual lipid modification of Arabidopsis Ggamma-subunits is required for efficient plasma membrane targeting. Plant Physiol 143:1119–1131

Zhao J, Wang X (2004) Arabidopsis phospholipase Dalpha1 interacts with the heterotrimeric G-protein alpha-subunit through a motif analogous to the DRY motif in G-protein-coupled receptors. J Biol Chem 279:1794–1800

Regulatory and Cellular Functions of Plant RhoGAPs and RhoGDIs

Benedikt Kost

Abstract Rho GTPases play central roles in the regulation of essential cellular processes, such as directional expansion, motility, and division. RhoGEFs (Guanine Nucleotide Exchange Factors) have key functions in the stimulus-induced spatio-temporal control of Rho GTPase activity. RhoGAPs (GTPase activating proteins) and RhoGDIs (Guanine nucleotide dissociation inhibitors) have long been seen as less important regulators of Rho GTPase activity, with functions largely restricted to the constitutive attenuation of Rho signaling. Extended families of diverse RhoGAPs, as well as small families of structurally similar RhoGDIs, have been identified in yeast, animals, and plants. Recent research has established that members of these protein families play much more important and complex roles than previously anticipated in the regulation of Rho GTPase activity and cellular processes. Non-plant RhoGAPs and RhoGDIs were shown to be tightly regulated by upstream signaling, and the same is likely to be true for their plant homologs as well. The recent functional characterization of plant RhoGAPs and RhoGDIs has allowed exciting and universally important insights into the molecular mechanisms underlying the control of Rho GTPase activity by these proteins.

1 Introduction

Rho family small GTPases are important eukaryotic regulators of signaling lipid metabolism, ROS production, transcription, cytoskeletal dynamics, and membrane trafficking. Through the coordination of these processes, Rho GTPases play key roles in the control of cell motility, division, and growth. In plants, Rho signaling also regulates hormone and stress responses (Jaffe and Hall 2005; Nibau et al. 2006,

B. Kost
Uppsala BioCenter, Plant Biology and Forest Genetics, Swedish University of Agricultural Sciences, P.O. Box 7080, 75007 Uppsala, Sweden
e-mail: benedikt.kost@vbsg.slu.se

S. Yalovsky et al. (eds.), *Integrated G Proteins Signaling in Plants*,
Signaling and Communication in Plants,
DOI 10.1007/978-3-642-03524-1_2, © Springer-Verlag Berlin Heidelberg 2010

see chapter "RAC/ROP GTPases in the Regulation of Polarity and Polar Cell Growth").

The diverse functions of Rho GTPases all depend on tight spatial and/or temporal control of their activity. In many cell types, spatially restricted Rho signaling orchestrates pronounced polarization (Etienne-Manneville and Hall 2002; Kost 2008). Most Rho GTPases are associated with the plasma membrane, often at specific sites of activation, depending on posttranslational prenylation and, at least in some cases, on activation-trigged acylation (Wennerberg and Der 2004; Yalovsky et al. 2008, see chapter "ROPs, Vesicle Trafficking and Lipid Modifications"). Rho GTPases stimulate downstream signaling when bound to GTP, and are inactive with respect to signaling in the GDP bound conformation after GTP hydrolysis (Fig. 1). RhoGEFs (guanine nucleotide exchange factors), a large and diverse family of typically membrane-associated factors, activate Rho GTPases by promoting GDP for GTP exchange and play a key role in the control of Rho signaling in response to upstream regulators (Fig. 1; Berken et al. 2005; Rossman et al. 2005; see chapter "Structure and Function of ROPs and Their GEFs").

RhoGAPs (GTPase activating proteins) and RhoGDIs (guanine nucleotide dissociation inhibitors) have long been seen as less important, in comparison to RhoGEFs, regulators of Rho activity, with functions largely restricted to constitutive signal attenuation. RhoGAPs can inactivate the signaling functions of Rho GTPases by increasing their low intrinsic GTPase activity, which promotes conversion to the GDP-bound conformation (Fig. 1; Tcherkezian and Lamarche-Vane 2007). RhoGDIs contain a hydrophobic binding pocket capable of accommodating the prenyl tail of Rho GTPases. These proteins can transfer Rho GTPases from the plasma membrane to the cytoplasm, where they are thought to form inactive heterodimers with them (Fig. 1; Hoffman et al. 2000). Activation-dependent acylation of Rho GTPase is likely to inhibit RhoGDI binding (Yalovsky et al. 2008, see chapter "ROPs, Vesicle Trafficking and Lipid Modifications"). Some Rho GTPases appear to interact with RhoGDIs, specifically, in the inactive GDP-bound form (Ueda et al. 1990; Klahre et al. 2006; see section "NtRhoGDI2: Maintenance of Polarized Rho GTPase Activation at the Tip of Tobacco Pollen Tubes), whereas others seem to bind to these proteins independently of their activation status (Nomanbhoy and Cerione 1996). Accordingly, RhoGDIs have been proposed to generally act as negative regulators of Rho signaling (Etienne-Manneville and Hall 2002), and to contribute to the control of the intracellular targeting of activated forms of some Rho GTPases (Del Pozo et al. 2002). Dissociation of Rho GTPase/RhoGDI complexes and subsequent reassociation of Rho GTPases with the plasma membrane, a prerequisite for RhoGEF-dependent activation, appear to be promoted by membrane-associated proteins or lipids acting as RhoGDFs (RhoGDI dissociation factors; Fig. 1; DerMardirossian and Bokoch 2005).

Recent work in non-plant systems has shown that upstream regulators control Rho signaling not only via RhoGEFs, but also by directly regulating RhoGAP activity and Rho GTPase/RhoGDI interaction. A number of signaling pathways that control essential cellular functions alter the activity of RhoGAPs based on different mechanisms including direct binding of lipid or protein factors, proteolytic

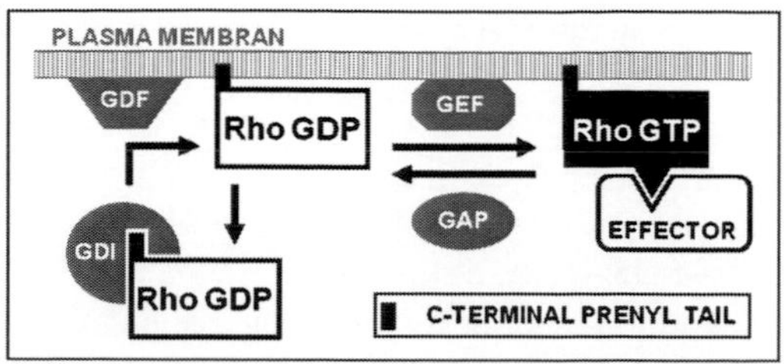

Fig. 1 *Control of Rho GTPase activity by regulatory proteins.* Most Rho GTPases associate with the plasma membrane via posttranslationally attached C-terminal prenyl tails. They interact with effectors to activate downstream signaling when bound to GTP, and are inactive in the GDP-bound form. RhoGAPs (GTPase activating proteins) enhance the low intrinsic GTPase activity of Rho GTPases, and inactivate their signaling function. RhoGEFs (Guanine nucleotide exchange factors) activate Rho GTPases by promoting GDP for GTP exchange. RhoGDIs (Guanine nucleotide dissociation inhibitors) transfer GDP-bound Rho GTPases to the cytoplasm, where they form inactive heterodimers with them. They can also translocate GTP-bound Rho GTPases between different membrane domains. RhoGDFs (GDI displacement factors) destabilize Rho GTPase/RhoGDI complexes and promote reassociation of Rho GTPases with the plasma membrane, which is required for RhoGEF-mediated activation

degradation, and phosphorylation (Bernards and Settleman 2004; Tcherkezian and Lamarche-Vane 2007; Yoshida and Pellman 2008). Rho GTPase/RhoGDI interactions are also modulated by stimulus-induced phosphorylation either of Rho GTPases or of RhoGDIs by a range of different protein kinases (DerMardirossian and Bokoch 2005; DerMardirossian et al. 2006; Knezevic et al. 2007; Qiao et al. 2008). Much less is known about regulatory mechanisms controlling RhoGAP and RhoGDI activity in plants, although membrane association of a RhoGAP in tobacco pollen tubes was proposed to be modulated by phosphorylation-dependent interaction with a 14-3-3 protein (Klahre and Kost 2006; see section "NtRhoGAP1: Polarity Maintenance at the Tip of Tobacco Pollen Tubes").

Consistent with the regulation of RhoGAP and RhoGDI activity by elaborate signaling mechanisms, recent research has shown that members of these two protein families play much more important and complex roles in the control of Rho GTPase activity and essential cellular functions than previously anticipated (Bernards and Settleman 2004). Work in plants has been at the forefront of this exciting discovery. The purpose of this chapter is to discuss recent advances in our understanding of RhoGAP and RhoGDI functions in plants in the light of a comparison of the plant RhoGAP and RhoGDI families with those of non-plant organisms.

2 RhoGAP Protein Families

Interestingly, yeast and animal RhoGAP families generally appear to be two to three larger than the corresponding Rho GTPase families. In budding yeast (*Saccharomyces cerevisiae*), five Rho GTPases and eight RhoGAPs have been characterized, with two additional predicted RhoGAP genes identified in the genome. Mammalian genomes encode 22 Rho GTPases and 59–70 predicted RhoGAPs, of which about half have been characterized. Fly (*Drosophila melanogaster*) and worm (*Caenorhabditits elegans*) genomes are predicted to contain 11 and 10 Rho GTPase genes, respectively, along with about 20 RhoGAP genes (Tcherkezian and Lamarche-Vane 2007).

All characterized and predicted RhoGAP proteins contain a homologous RhoGAP domain (Lamarche and Hall 1994) with a conserved arginine residue required for catalytic activity (Graham et al. 1999; Rittinger et al. 1997). Outside this domain, non-plant RhoGAPs are highly diverse and contain a bewildering variety of additional domains with many different confirmed or predicted functions, which include protein–protein interaction (e.g., SH3, IQ, RA, etc.), lipid/membrane binding (e.g., PH, C2, PX, etc.) and enzymatic activity (RhoGEF, S/T kinase, myosin motor, etc.) (Bernards and Settleman 2004; Tcherkezian and Lamarche-Vane 2007). Up to 11 different functional domains have been identified in individual non-plant RhoGAPs (Tcherkezian and Lamarche-Vane 2007). An important function of multidomain RhoGAPs was proposed to be the integration of different signaling pathways (Bernards and Settleman 2004; Tcherkezian and Lamarche-Vane 2007). Consistent with this hypothesis, the domain organization of complex RhoGAPs appears to be conserved between different organisms (Bernards 2003).

Despite an extensive discussion in the literature, the significance of the striking diversity within non-plant RhoGAP families has remained a bit of mystery. Although a few members of these families are differentially expressed in distinct cell types or tissues, and/or display RhoGAP activity selectively towards single Rho GTPases, most of them are ubiquitous and can attenuate a range of Rho signaling pathways (Tcherkezian and Lamarche-Vane 2007). Several proteins have been found to contain inactive RhoGAP domains, which bind to Rho GTPases without promoting GTP hydrolysis. These proteins modulate Rho signaling by acting as a scaffold for complex formation (Chiang et al. 2003). A plausible hypothesis suggests that each RhoGAP selectively regulates a single signaling pathway, which is embedded in a complex Rho signaling network, and controls a specific cellular function. Consistent with this idea, knock-out or knock-down of individual RhoGAPs can cause highly specific cellular or developmental defects in mammals and in flies (Tcherkezian and Lamarche-Vane 2007).

The situation in plants is considerably less complex. The *Arabidopsis* genome encodes 11 highly similar Rho GTPases referred to as AtROPs (Rho of plant; Vernoud et al. 2003), of which most have been characterized, at least to some extent (Gu et al. 2004; Nibau, et al. 2006; Yalovsky et al. 2008, see chapter "ROP Evolution and ROPs in Grasses"). Only nine *Arabidopsis* genes coding for predicted AtROPGAP proteins with a RhoGAP domain have been identified (Wu et al. 2000; Hwang et al. 2008). These proteins contain a single recognizable functional domain in addition to the RhoGAP domain, and can be divided into two subfamilies each consisting of structurally very similar proteins, which share a high degree of sequence identity (Fig. 2). One subfamily consists of six relatively small proteins (331–466 amino acids) called AtROPGAP1-6 (Wu et al. 2000), which all contain a

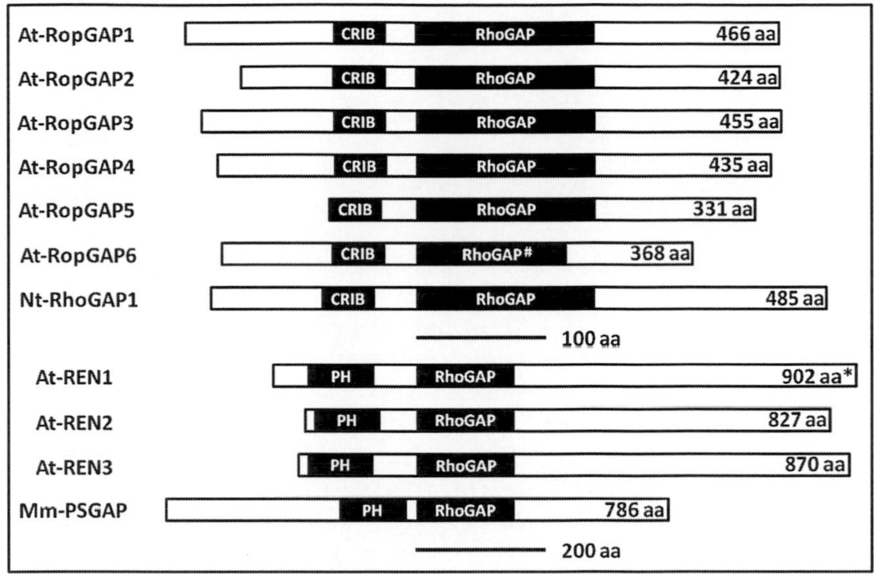

Fig. 2 *Domain structure of AtROPGAPs and related proteins*. Structures of AtROPGAP1-6 (At5G22400, At4G03100, At2G46710, At3G11490, At1G08340, At2G27440), NtRhoGAP1 (DQ813657), AtREN1-3 (At4G24580, At5G12150, At5g19390), and MmPSGAP (AF297030) are shown drawn to scale (CRIB domain proteins > scale bar: 100 amino acids; PH domain proteins > scale bar: 200 amino acids). All structures are aligned such that the first amino acid of the RhoGAP domain is positioned on a vertical line. #AtROPGAP6 is missing a fragment of the conserved RhoGAP domain. *Displayed is the structure of AtREN1 according to TAIR (www.arabidopsis.org). Full-length cDNA sequencing by Hwang et al. (2008) has shown that the part of the protein between the RhoGAP domain and the C-terminus contains 18 additional amino acids

CRIB (CDC42/Rac interactive binding) domain adjacent to the N-terminal end of the RhoGAP domain (Fig. 2). The gene encoding AtROPGAP6 has been annotated as a pseudo gene, presumably because corresponding EST/cDNA sequences remain to be identified (http://www.arabidopsis.org/). This gene has a normal structure, contains an intact ORF and appears to be expressed according to the GeneVestigator database, although at somewhat lower levels as compared to other subfamily members (https://www.genevestigator.com). However, the predicted AtROPGAP6 protein is lacking a part of the conserved RhoGAP domain (Fig. 2). The second *Arabidopsis* RhoGAP subfamily consists of three larger (870–902 amino acids) proteins called AtREN1-3 (Hwang et al. 2008), which contain a PH domain near the N-terminus (Fig. 2). Rho GTPase and RhoGAP families similar to those of *Arabidopsis* also appear to be present in other plant species (Wu et al. 2000; Klahre and Kost 2006; Hwang et al. 2008). Members of both plant RhoGAP subfamilies display RhoGAP activity towards ROP GTPases in vitro and in vivo (Klahre and Kost 2006; Hwang et al. 2008). Plant RhoGAPs with CRIB domains were also shown to stimulate the GTPase activity of mammalian Rho GTPases in vitro (Wu et al. 2000; Klahre and Kost 2006; see section "NtRhoGAP1: Polarity Maintenance at the Tip of Tobacco Pollen Tubes").

PH domain containing RhoGAPs with similar structure as AtREN1-3 are also found in non-plant systems (Ren et al. 2001; Tcherkezian and Lamarche-Vane 2007; Fig. 2). Consistent with the ability of PH domains to bind to phospholipids, the PH domain of the mammalian RhoGAP PSGAP (Ren et al. 2001) has been shown to be essential for membrane association and correct intracellular targeting. The presence of a CRIB domain in the RhoGAPs of one of the plant subfamilies is more surprising, as such domains are not found in any of the non-plant RhoGAPs identified to date (Tcherkezian and Lamarche-Vane 2007). CRIB domains mediate the interaction of many plant and non-plant Rho effectors, specifically with activated forms of Rho GTPases (Pirone et al. 2001). Because RhoGAP domains on their own are sufficient for specific binding to activated Rho GTPases (Wu et al. 2000; Klahre and Kost 2006), the functions of the CRIB domain of plant RhoGAPs is not entirely clear. In any case, CRIB domains of plant RhoGAPs have been shown to modulate interactions of these proteins with Rho GTPases, as well as their RhoGAP activity (Wu et al. 2000; Klahre and Kost 2006; see section "NtRhoGAP1: Polarity Maintenance at the Tip of Tobacco Pollen Tubes").

GeneVestigator data supported by some experimental verification (Baxter-Burrell et al. 2002; Klahre and Kost 2006) suggest that plant RhoGAPs with a CRIB domain are constitutively expressed at very low levels, whereas genes coding for two of the three PH domain containing *Arabidopsis* ROPGAPs proteins (AtREN1+2, no expression data are available to date for AtREN3) seem to be differentially expressed at a much higher levels, similar to genes encoding AtROPs. One possible explanation for these observations, which clearly need further experimental confirmations, could be that CRIB domains confer a particularly high affinity for active ROP GTPases to plant RhoGAPs. Together with the catalytic nature of RhoGAP activity, this may allow RhoGAPs with CRIB domains to function at low expression levels.

Results of the recent functional characterization of members of both RhoGAP subfamilies found in plants (Baxter-Burrell et al. 2002; Klahre and Kost 2006; Hwang et al. 2008) are summarized in the following sections of this chapter. These results demonstrate that despite the comparably low complexity of these protein families, RhoGAPs play the key roles in the control of Rho GTPase activity and cellular processes also in plants. Furthermore, they have allowed exciting new insights into RhoGAP-dependent signaling mechanisms, which are likely to have important functions also in non-plant systems.

2.1 Plant RhoGAP SubFamily I: CRIB domain proteins

2.1.1 AtROPGAP4: Regulation of Oxygen Deprivation Tolerance

To maintain energy-dependent metabolism under flooding-induced oxygen deprivation, plant roots activate ethanolic fermentation and induce alcohol dehydrogenase (ADH) expression. In *Arabidopsis* roots, oxygen deprivation induces increased

ADH expression via ROP activation, which stimulates NADPH oxidase-dependent production of reactive oxygen species (ROS) that act as second messengers. Rho GTPase-dependent stimulation of ROS production also has important functions in the control of tip growth in root hairs (Takeda et al. 2008; see section "AtROPGDI1: Maintenance of Cellular Polarity Required for Root Hair Initiation and Growth") and in pollen tubes (Potocky et al. 2007), as well as of pathogen defense in plants (Ono et al. 2001) and in animals (Bokoch 1994). Constitutive overexpression of dominant negative AtROP2 (DN-AtROP2) in *Arabidopsis* roots prevents ADH activity from increasing under low oxygen conditions, whereas ADH activity is enhanced under nonstress conditions in the presence of constitutively active forms of this protein (CA-AtROP2). After the transfer of wild-type *Arabidopsis* roots to oxygen deprived conditions, levels of activated GTP-bound ROP, which can be pulled down from extracts using a CRIB domain containing effector protein, increase for 12 h, and then start dropping. By contrast, ADH transcript levels and activity keep rising for at least 24 h.

A mutant in which AtROPGAP4 expression is disrupted displays increased levels of GTP-bound ROP, as well as slightly enhanced ADH expression and activity, under nonstress conditions. During the first 12 h of oxygen deprivation, levels of GTP-bound ROP increase and ADH expression and activity in mutant roots rise much steeper than in wild-type roots. However, in striking contrast to what happens in wild-type roots, within the next 12 h under low oxygen conditions, ADH expression and activity in mutant roots massively drop, whereas levels of GTP-bound ROP remain constant. Consistent with these unexpected observations, mutant roots display reduced resistance to oxygen deprivation, and are compromised in their ability to recover from this condition.

Within 12 h, oxygen deprivation also results in the accumulation of about three times higher ROS levels in mutant roots than in wild-type roots, presumably because of increased NADPH oxidase stimulation by ROP overactivation in the absence of AtROPGAP4. Oxidative stress caused by the accumulation of excessive amounts of ROS is likely to be responsible for the collapse of ADH activity in mutant roots after 12 h of oxygen deprivation, as well for reduced resistance to this condition.

On the basis of these observations, AtROPGAP4 appears to have an essential function in restraining ROP activation, and consequently NADPH oxidase stimulation, during oxygen deprivation. Interestingly, AtROPGAP4 transcription is induced by low oxygen conditions, CA-AtROP2 overexpression, and by treatments resulting in increased ROS levels. Furthermore, the stimulation of AtROPGAP4 transcription by oxygen deprivation can be blocked by DN-AtROP2 overexpression and by treatments preventing ROS accumulation. These observations strongly suggest negative feedback regulation of the ROP/NADPH oxidase/ROS signaling pathway in *Arabidopsis* roots via the stimulation of AtROPGAP4 expression. A delicate balance between ROP activation and AtROPGAP4 expression maintained by this feedback loop appears to be required for an effective response of *Arabidopsis* roots to oxygen deprivation. Consistent with this hypothesis, constitutive overexpression of CA-AtROP2, which carries a mutation that disrupts GTPase activity even in the presence of RhoGAPs (Klahre and Kost 2006), also results in reduced resistance to low oxygen conditions.

2.1.2 NtRhoGAP1: Polarity Maintenance at the Tip of Tobacco Pollen Tubes

Vegetative pollen tubes cells rapidly elongate in a strictly polarized manner based on F-actin-dependent tip-directed targeted secretion of cell wall material (Hepler et al. 2001). Rho GTPases, such AtROP1 and the closely related tobacco NtRac5, accumulate at the plasma membrane selectively at the tip of growing pollen tubes (Lin et al. 1996; Kost et al. 1999, Klahre et al. 2006). Specific activation of these Rho GTPases at the pollen tube apex (Hwang et al. 2005) is thought to play a central role in the control of pollen tube tip growth through the coordination of F-actin dynamics and membrane trafficking (Yalovsky et al. 2008). Consistent with this view, overexpression of wild-type or CA Rho GTPases depolarizes pollen tube growth and causes massive tip ballooning, presumably as it results in an extension of the area of the apical plasma membrane-associated with Rho GTPase activity. By contrast, pollen tube elongation is strongly inhibited by the overexpression of DN Rho GTPases (Li et al. 1999; Kost et al. 1999; Klahre et al. 2006). These observations strongly suggest that tight spatial control of Rho GTPase activity is essential for the maintenance of the polarity of pollen tube tip growth.

AtROPGAP1 and its close tobacco homolog NtRhoGAP1 were identified in yeast-two hybrid screens using CA-AtROP1 and CA-NtRac5, respectively, as bait. Pull-down and yeast-two hybrid assays established that full-length AtROPGAP1 and NtRhoGAP1, as well as their isolated CRIB and RhoGAP domains, preferentially interact with active GTP-bound forms of their target Rho GTPases (AtROP1 and NtRac5, respectively). The CRIB domain of AtROPGAP1, but not the one of NtRhoGAP1, also displays affinity to the nucleotide-free transition state of its target Rho GTPase (Wu et al. 2000; Klahre and Kost 2006).

AtROPGAP1 and Nt-RhoGAP1 dramatically enhance the GTPase activity of their target Rho GTPases in vitro. Removal of the CRIB domain strongly reduces the in vitro RhoGAP activity of both proteins towards these targets. AtROPGAP1 and NtRhoGAP1 also show in vitro RhoGAP activity towards mammalian Cdc42 and Rac1, respectively, although in the case of AtROPGAP1, this activity is relatively weak. Surprisingly, removal of the CRIB domain enhances A-ROPGAP1 activity towards Cdc42. These observations are in agreement with the hypothesis that the function of the CRIB domains of plant RhoGAP is to modulate the strength of target Rho GTPase binding. RhoGAP activity may be reduced not only by decreased Rho GTPase affinity, but also by excessively strong target binding, which interferes with substrate turnover (Wu et al. 2000; Klahre and Kost 2006).

Consistent with the ability of AtROPGAP1 and NtRhoGAP1 to inactivate Rho signaling, overexpression of these proteins strongly inhibits pollen tube growth (Hwang et al. 2005; Klahre and Kost 2006). However, the intracellular distribution of N- and C-terminal NtRhoGAP1 YFP (Yellow Fluorescent Protein) fusion

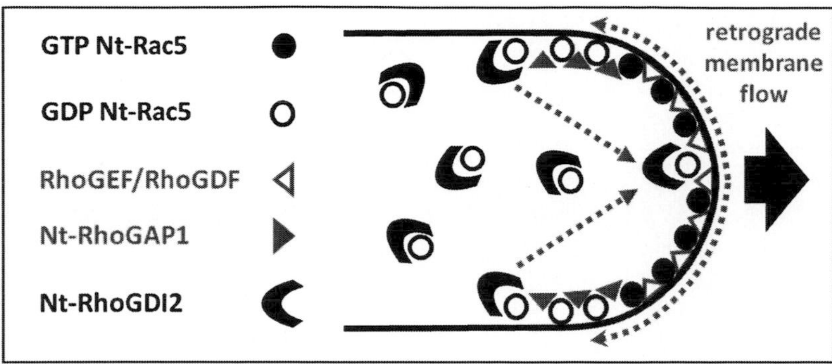

Fig. 3 *Maintenance of apical NtRAC5 activity at the tip of tobacco pollen tubes.* This model of the molecular mechanisms responsible for the polarization of NtRAC5 activity at pollen tube tip predicts that the stimulation of the fusion of secretory vesicles with the apical plasma membrane by active NtRAC5 generates a constant retrograde flow of plasma membrane material, which laterally displaces this protein. NtRAC5 inactivation by NtRhoGAP1 at the flanks of the tip is required to prevent a depolarization of NtRAC5 activity. NtRhoGDI2 extracts inactive NtRAC5 from the plasma membrane at the flanks of the tip and escorts it through the cytoplasm back to apex. To complete NtRhoGDI2-mediated recycling at the apex, RhoGDFs (such as perhaps the membrane lipid PIP_2) promote NtRAC5 reassociation with the plasma membrane by destabilizing NtRAC5/NtRhoGDI2 complexes, and RhoGEFs subsequently reactivate NtRAC5 by promoting GDP for GTP exchange

proteins can be visualized in normally elongating tobacco pollen tubes based on low-level transient expression. Interestingly, both YFP fusion proteins associate with the plasma membrane at the flanks of the pollen tube tip, but not at the apex where active NtRAC5 presumably accumulates (Klahre and Kost 2006). These findings strongly suggest that NtRhoGAP1 plays a central role in the maintenance of the polarized distribution of NtRAC5 activity at the pollen tube tip, which controls directional cell expansion. The stimulation of the fusion of secretory vesicles with the apical plasma membrane by activated NtRAC5 presumably causes constant lateral displacement of this protein. The resulting depolarization of NtRAC5 activity and cell growth can be prevented by NtRhoGAP1-mediated stimulation of GTPase activity at the flanks of the tip. RhoGDI-mediated recycling of inactivated NtRAC5 from this location back to the apex appears to be required for the maintenance of apical NtRAC5 activity (see section "NtRhoGDI2: Maintenance of Polarized Rho GTPase Activation at the Tip of Tobacco Pollen Tubes"; Fig. 3).

A key function of NtRhoGAP1 in the spatial control of NtRAC5 signaling is supported by the observation that overexpression of a dominant negative (DN) mutant form of this protein in tobacco pollen tubes induces ballooning at the tip (Klahre and Kost 2006). DN-NtRhoGAP1 is missing an essential arginine residue in the catalytic domain and does not stimulate NtRAC5 GTPase activity in vitro. As expected, based on the fact that NtRhoGAP1 preferentially associates with GTP-bound NtRAC5 (see *above*), DN-NtRhoGAP1, which fails to promote GTP hydrolysis, displays enhanced interaction with NtRac5 in yeast two-hybrid assays. When overexpressed as a YFP fusion protein, DN-NtRhoGAP1, particularly, strongly accumulates at plasma membrane of swollen pollen tube tips, which is consistent with increased NtRac5 activation (Klahre and Kost 2006). T-DNA insertions disrupting the expression of the *Arabidopsis* NtRhoGAP1 homolog AtROPGAP1 do not affect pollen tube growth (Klahre and Kost 2006; Hwang et al. 2008). GeneVestigator data indicate that at least four additional AtROPGAPs with a CRIB domain are expressed in *Arabidopsis* pollen tubes, two of them at higher levels. This suggests redundant functions of multiple CRIB domain AtROP-GAPs in the control of Rho GTPase activity in pollen tubes, which underscores the importance of the role this protein family plays in the control of tip growth.

Essential functions in cell polarity establishment or maintenance of RhoGAPs, which are associated with distinct plasma membrane domains and confine Rho activity to others, have recently also been identified in animal cells (Simoes et al. 2006; Anderson et al. 2008) and were proposed to underlie bud formation in yeast (Knaus et al. 2007). To further advance our understanding of the spatial control of Rho signaling during cellular polarization, it is essential to investigate the molecular mechanisms responsible for RhoGAP targeting to specific membrane domains. The analysis of the intracellular distributions of truncated forms NtRhoGAP1 fused to YFP (Klahre et al. 2006) has established that a large C-terminal CRIB/RhoGAP-domain-containing fragment, as well as the CRIB domain alone, associates with the plasma membrane at the apex of normally elongating tobacco pollen tubes, where active Rho GTPases accumulate. Interestingly, a short (95 amino acid) N-terminal fragment, which is complementary to the C-terminal CRIB/RhoGAP domain containing fragment (see above) and displays an even cytoplasm distribution on its own, is required for the subapical association of full-length NtRhoGAP1 with the plasma membrane at the flanks of the pollen tube tip.

A yeast-two hybrid screen for proteins interacting with the N-terminal 95 amino acid NtRhoGAP1 fragment resulted in the identification of Nt14-3-3b-1 (Klahre et al. 2006), a member of a protein family implicated in the relocation of target proteins between cellular compartments (Aitken 2002). Nt14-3-3b-1 is specifically expressed at high levels in tobacco pollen and pollen tubes, has no effect on tobacco pollen tube growth when overexpressed on its own, and displays an even distribution throughout the cytoplasm of these cells when fused to YFP. However, Nt14-3-3b-1 coexpression strongly alleviates the inhibition of tobacco pollen tube growth induced by NtRhoGAP1 overexpression, and almost completely prevents the accumulation of NtRhoGAP1 YFP fusion proteins at the plasma membrane. NtRho-GAP1 contains a consensus motif predicted to confer phosphorylation-sensitive binding to 14-3-3 proteins. A point mutation mimicking the phosphorylated state of this consensus motif, which is predicted to enhance interaction with 14-3-3 proteins, reduces membrane association of NtRhoGAP1 YFP fusion proteins, whereas mutations preventing phosphorylation of this motif has the opposite effect (Klahre et al. 2006). Altogether, these observations establish an important function of Nt14-3-3b-1 in the control of NtRhoGAP1 targeting, although the exact molecular mechanism of this process remains to be determined. Recently, the mammalian

RhoGAP DLC1 was also shown to be inactivated by the interaction with a 14-3-3 protein (Scholz et al. 2009).

2.2 *Plant RhoGAP Subfamily II: PH Domain Proteins*

2.2.1 AtREN1: Polarity Maintenance at the Tip of *Arabidopsis* Pollen Tubes

AtREN1 was identified in a screen for *Arabidopsis* mutants showing enhanced effects of low-level overexpression of a GFP AtROP1 fusion protein. The disruption of AtREN1 expression results in massive ballooning at the tip of pollen tubes expressing this fusion protein. In a wild-type background, the absence of AtREN1 expression has the same effect in a less severe form, and also causes precocious pollen germination and results in male sterility. AtREN1 overexpression under the control of the strong pollen-specific Lat52 promoter (Twell et al. 1991) restores normal fertility and in vitro pollen tube growth, whereas an AtREN1 GFP fusion protein complements the *ren-1* mutant phenotype only partially under the same conditions. Interestingly, overexpression of CA-AtROP1 causes similar defects in pollen tubes as the disruption of AtREN1 expression. These data show that AtREN1 is required for the maintenance of polar pollen tube growth, and indicate that it may participate in the restriction of Rho GTPase activity.

GeneVestigator data and RT-PCR suggest preferential expression of AtREN1 at high level in mature pollen. Full-length AtREN1 and an isolated N-terminal fragment of this protein containing the PH and RhoGAP domains selectively bind to activated AtROP1 in vitro. The in vitro GTPase activity of AtROP1 is strongly enhanced in the presence of full-length AtREN1, but not in the presence of mutant AtREN1 missing a conserved arginine residue in the catalytic domain. These observations establish that AtREN1 has RhoGAP activity towards AtROP1.

By contrast to the overexpression of AtROPGAP1 or NtRhoGAP1, which strongly inhibits pollen tube growth (see 2.1.1.2.), overexpression of AtREN1 on its own does not seem to affect this process (see *above*). However, AtREN1 cooverexpression suppresses the depolarization of tobacco pollen tubes induced by the overexpression of a GFP AtROP1 fusion protein, whereas cooverexpression of mutant AtREN1 without RhoGAP activity has the opposite effect, presumably because the mutant protein dominant negatively inhibits endogenous ROPGAPs. Swollen tips of *Atren1* mutant pollen tubes display enhanced plasma membrane association of ROP GTPases, and of the CRIB domain containing AtROP effector AtRIC4. All these data are consistent with a role of AtREN1 in the downregulation of pollen tube ROP GTPase activity.

Immunolabeling and GFP tagging show that AtREN1 accumulates in the cytoplasm at the pollen tube tip, where it extensively overlaps with the styryl dye FM4-64 and with a YFP AtRabA4D fusion protein, which are thought to be associated with subapical endocytic and/or post-Golgi secretory vesicles (Parton et al. 2001; Szumlanski and Nielsen 2009). The intracellular distribution of full-length AtREN1 was compared to that of an N-terminal fragment, which contains the PH and RhoGAP domains, specifically interacts with active AtROP1 (see above), and activates its GTPase activity, but lacks the C-terminus with two coiled-coil domains. This fragment does not accumulate at the pollen tube tip when expressed as a GFP fusion protein. It also fails to suppress the pollen tube phenotype of *Atren1* mutants when expressed as a free protein under the control of the Lat52 promoter. These results demonstrate (1) that the C-terminus is more important than the PH for the intracellular targeting of AtREN1, and (2) that the accumulation of this protein at the pollen tube tip is essential for its function.

Vesicle transport and dynamics at the pollen tube apex are thought to be highly sensitive to latrunculin B (LatB), which at low concentrations appears to specifically disrupt fine apical F-actin structures (Vidali et al. 2001), and to brefeldin A (BFA), which blocks secretion and causes endocytic and post-Golgi organelles to aggregate to "BFA compartments". These compartments have been shown to trap FM4-64 and other plasma membrane markers undergoing endocytic recycling (Nebenführ et al. 2002; Helling et al. 2006). Treatment with LatB or BFA reduces the rate of pollen tube growth and results in loss of tip-specific accumulation of an AtREN1-GFP fusion protein. Unlike FM4-64, this fusion protein does not accumulate in BFA compartments, suggesting that it may be associated with secretory, rather than with endocytic, vesicles. It will be interesting to test whether LatB and BFA have similar effects on the intracellular distribution of an AtRabA4D-YFP fusion protein, which is also thought to be associated with post-Golgi vesicles. In any case, the intracellular distribution of an AtREN1 GFP fusion protein is severely disrupted in mutant *Atraba4d* pollen tubes, which display depolarized growth resulting from defects in the targeted of secretory vesicles (Szumlanski and Nielsen 2009). Furthermore, the introduction of a weak *Atren1* allele into the *Atraba4d* background results in a synergistically enhanced phenotype.

Additional support for an import role of vesicle transport in the function of AtREN1 is provided by the observation that mutant *Atren1* pollen tubes with a phenotype partially complemented by the overexpression of an AtREN1-GFP fusion protein (see *above*) are hypersensitive to LatB and BFA. Treatment of such pollen tubes with these drugs at concentrations that affect wild-type pollen tube only weakly causes pronounced growth depolarization and tip ballooning. Cultured pollen tubes can show growth oscillations. A quantitative correlation analysis was performed (1) of the fluorescence emitted by an AtREN1-GFP fusion protein directly adjacent to the apical plasma membrane of oscillating pollen tubes, and (2) of the growth rate of these cells. Fluorescence levels were found to peak immediately before growth rate maxima both in *Arabidopsis* and in tobacco pollen tubes, although the AtREN1-GFP fusion protein did not display a strong accumulation near the tip of the latter. On the basis of the established phasal relationship between peak growth rate and maximal ROP GTPase activity at the apex of tobacco pollen tubes (Hwang et al. 2005), it was concluded that maximal ROP GTPase activity is followed first by a peak in GFP fluorescence near the apical plasma membrane, which may indicate delivery of secretory vesicles associated with GFP-tagged AtREN1, and then by a peak in growth rate.

Together, all these observation have led to the proposal that global inhibition of AtROP1 by AtREN1 is required for the maintenance of the polarity of pollen tube tip growth, and that delivery of secretory vesicles, with which AtREN1 is associated, to the plasma membrane is essential for the function of this protein. Conceivably, the maintenance of polarized Rho signaling during tip growth depends on two proteins with RhoGAP activity, one that associates subapically with the plasma membrane where it spatially restricts ROP GTPase activity to the apex (NtRhoGAP1, see section "NtRhoGAP1: Polarity Maintenance at the Tip of Tobacco Pollen Tubes"), and one that globally attenuates this activity after delivery to this site by secretory vesicle (AtREN1).

3 RhoGDI Protein Families

By contrast to the large families of diverse RhoGAPs found in eukaryotic organisms, in each of these organisms, only a few RhoGDIs have been identified, which share a high degree of sequence identity and are structurally very similar. The human genome appears to encode three RhoGDIs: HsRhoGDI, HsLy/D4GDI, and HsRhoGDIγ. HsRhoGDI is ubiquitously expressed, while the two other proteins are selectively present in a few cell types and tissues. While HsRhoGDI and HsLy/D4GDI, like most other RhoGDIs, are localized in the cytoplasm, HsRhoGDIγ is associated with vesicular membranes (DerMardirossian and Bokoch 2005). In budding yeast, a single cytoplasmic RhoGDI seems to be expressed (Masuda et al. 1994).

RhoGDIs are small proteins with two highly conserved functional domains (Fig. 4). A C-terminal immunoglobulin-like domain (Fig. 4; IG-like) contains a hydrophobic binding pocket, which can accommodate prenyl tails anchoring Rho GTPases in the plasma membrane. This domain is responsible for the ability of

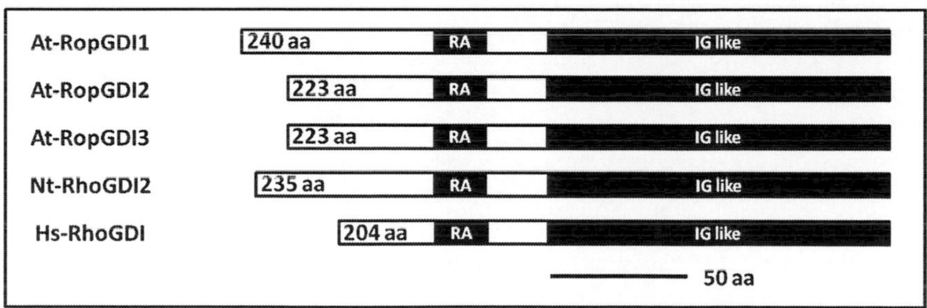

Fig. 4 *Domain structure of AtROPGDIs and related proteins*. Structures of AtROPGDI 1-3 (At3g07880, At1g12070, At1g62450), NtRhoGDI2 (DQ416769), and HsRhoGDI (AAP35530) are shown drawn to scale (scale bar: 50 amino acids). All structures are aligned such that the first amino acid of the IG-like domain is positioned on a vertical line. *RA* regulatory arm, *IG-like* immunoglobulin like domain

RhoGDIs to transfer Rho GTPases from the plasma membrane to the cytoplasm. RhoGDIs typically seem to preferentially translocate inactive Rho GTPase to the cytoplasm (Ueda et al. 1990; Klahre et al. 2006), although some Rho GTPases have been found to interact equally well in the GDP and in the GTP bound with RhoGDIs (Nomanbhoy and Cerione 1996). Once a cytoplasmic Rho GTPase/RhoGDI heterodimer has been formed, the regulatory arm (Fig. 4; RA) located between the N-terminus and the IG-like domain of RhoGDIs interacts with Rho GTPase regions involved in guanine nucleotide, GEF, and effector binding. This interaction prevents GTP hydrolysis even in the presence of RhoGAP activity, GDP dissociation, GEF-mediated nucleotide exchange, and the activation of downstream signaling pathways. In agreement with the biological inertness of Rho GTPase/RhoGDI heterodimers, RhoGDIs are generally thought to act as negative regulators of Rho signaling (DerMardirossian and Bokoch 2005).

In a range of mammalian cell types, the molar RhoGDI amount is roughly equal to the combined molar amount of all major Rho GTPases (Michaelson et al. 2001). In some of these cell types, Rho GTPases are largely present in cytoplasmic heterodimers with RhoGDIs, whereas in others variable levels of free Rho GTPases are detected (Chuang et al. 1993; Fritz et al. 1994). Consistent with these observations, mammalian Rho GTPases differ widely in their affinities to RhoGDIs, suggesting that they may be subject to regulation by these proteins to a variable extent (DerMardirossian and Bokoch 2005).

RhoGDI overexpression has been shown to downregulate Rho signaling, and to interfere with Rho regulated process in many different cell types and organisms. However, disruption of the genes encoding the single RhoGDI identified in budding yeast (Masuda et al. 1994), or the ubiquitously expressed mouse homolog of HsRhoGDI (MmRhoGDI), causes surprisingly mild phenotypes. Because mice in which additional RhoGDI genes have been disrupted are not yet available, it is possible that stronger defects in the development of MmRhoGDI knock-out mice were masked by the upregulation of other RhoGDI genes (DerMardirossian and Bokoch 2005).

The *Arabidopsis* genome contains three genes encoding proteins with a high degree of sequence identity, which are very similar to characterized mammalian and yeast RhoGDIs (Fig. 4). These proteins are called AtROPGDI1-3 and, as their homologs identified in other plants species, contain short, highly divergent N-terminal extensions, which remain to be functionally characterized (Carol et al. 2005; Klahre et al. 2006). GeneVestigator data suggest that transcripts of all three AtROPGDI genes reach high levels in mature pollen, and are ubiquitously present at much lower levels in other organs throughout plant development. AtROPGDI1 transcripts seem to be the most abundant in all cell types and tissues. Interestingly, the GeneVestigater expression patterns displayed by all three AtROPGDI genes nicely overlap with that of the AtROP gene family as a whole. Also, the combined expression levels of all AtROPGDIs and AtROPs in each tissue and cell type seem to be similar. This suggests that, as in mammalian cells, roughly equal molar amounts of AtROPGDIs and AtROPs may be present in *Arabidopsis* cells. The recent functional characterization of AtROPGDI1, and of a tobacco pollen tube homolog of this protein, has established that these proteins are not simply down-regulating ROP signaling, but are essential for the establishment and maintenance of spatially restricted ROP activity, which controls polarized cell expansion (Carol et al. 2005; Klahre et al. 2006). The results of these studies are summarized in the sections below.

3.1 AtROPGDI1: Maintenance of Cellular Polarity Required for Root Hair Initiation and Growth

Root hairs are single, uniaxial, highly elongated protrusions growing out from the basal end of root epidermal cells (trichoblasts). They elongate by tip growth much in the same way as pollen tubes (see section "NtRhoGAP1: Polarity Maintenance at the Tip of Tobacco Pollen Tubes"). AtROP2 and other AtROP GTPases accumulate at the trichoblast plasma membrane selectively at sites from which root hairs will emerge, and remain associated specifically with the apical plasma membrane of elongating root hairs. Root hair tip growth, like the expansion of pollen tubes, is depolarized upon expression of CA-AtROP GTPases, and is blocked by DN forms of these proteins (Jones et al. 2002; Molendijk et al. 2001). The stimulation of ROS production by NADPH oxidases appears to play a key role in the control of root hair elongation downstream of ROP GTPase activation, as it does in the regulation of other processes by ROP GTPases, including the response of *Arabidopsis* roots to oxygen deprivation (see section "AtROPGAP4: Regulation of Oxygen Deprivation Tolerance"). ROS accumulate in trichoblasts at sites of future root hair outgrowth, as well as at the apex of elongating root hairs, depending on the activity of the NADPH oxidase RHD2/AtrbohC. Trichoblast of *Arabidopsis* mutants defective in the gene coding for this protein form correctly positioned but highly stunted root hairs, which fail to accumulate ROS at the tip (Foreman et al. 2003).

By contrast, trichoblasts of the *Arabidopsis supercentipede1* (*scn1*) mutant initiate multiple (about three in average) root hairs at random positions, which remain very short and often split at the tip to form multiple growth sites. An AtROP2 GFP fusion protein displays strongly enhanced association with the plasma membrane of mutant trichoblasts, and is mislocalized to all sites of cell expansion, where ROS also rise to high levels. Interestingly, the *scn1* phenotype is caused by defects in the gene that encodes AtROPGDI1, which either disrupt gene expression, or result in the production of a mutant protein in which a highly conserved glutamate residue at position 181 is replaced by glycine. Consistent with an important function of this residue in Rho GTPase binding predicated based on modeling of an AtROP GTPase/AtROPGDI1 complex, recombinant mutant AtROPGDI1^{Glu188Gly} displays a strongly reduced ability to pull-down ROP GTPases from cauliflower extracts (Carol et al. 2005).

These observations demonstrate that in trichoblasts, AtROPGDI1 is essential for the establishment and the maintenance of cellular polarity. Interaction of

AtROPGDI1 with AtROP GTPases, which spatially restricts ROP GTPase activity and RHD2/AtrbohC-mediated ROS production, is required for the determination of a single site of cell expansion during root hair formation.

3.2 *NtRhoGDI2: Maintenance of Polarized Rho GTPase Activation at the Tip of Tobacco Pollen Tubes*

Like NtRhoGAP1, NtRhoGDI2 was identified in a yeast two-hybrid screen for proteins that interact with the Rho GTPase NtRac5, a key regulator of tobacco pollen tube tip growth (see section "NtRhoGAP1: Polarity Maintenance at the Tip of Tobacco Pollen Tubes"). NtRhoGDI2 is highly similar to *Arabidopsis* and mammalian RhoGDIs (Fig. 4). Northern analysis has shown that transcripts encoding this protein accumulate to high levels specifically in tobacco pollen and pollen tubes. In yeast two-hybrid assays, NtRhoGDI2 shows strong interaction with NtRAC5, weaker interaction with CA-NtRAC5, and no interaction with DN-NtRAC5. These observations suggest that NtRhoGDI2, like other RhoGDIs (see Section "RhoGDI Protein Families"), preferentially interacts with its target Rho GTPase in the inactive GDP bound form (Klahre et al. 2006). RhoGDIs generally do not interact well with DN forms of Rho GTPases, which are considered nucleotide-free because of their low affinity for both GDP and GTP (Strassheim et al. 2000). Interestingly, a point mutation preventing the prenylation of NtRAC5, which is expected to promote nuclear targeting required for the detection of yeast two-hybrid interactions, weakens the interaction of NtRAC5 with NtRhoGDI2 in yeast two-hybrid assays. This is in agreement with an important function of interactions between the prenyl tail of NtRAC5 and the hydrophobic pocket in the IG-like domain of NtRhoGDI2 in the binding of the two proteins to each other (Scheffzek et al. 2000).

Cell fractionation, as well as C- and N-terminal YFP tagging, has established that NtRhoGDI2 accumulates in the cytoplasm of tobacco pollen tubes. NtRhoGDI2 overexpression strongly inhibits the growth of these cells and effectively suppresses the accumulation of a coexpressed NtRAC5 YFP fusion protein at the apical plasma membrane. Cooverexpression of NtRhoGDI2 and NtRAC5 at different relative levels has shown that excess NtRhoGDI2 activity inhibits the elongation without inducing tip swelling, whereas excess NtRAC5 activity depolarizes the growth and results in ballooning at the tip. Interestingly, pollen tubes overexpressing NtRhoGDI2 and NtRAC5 at similar levels can grow normally, suggesting that the two proteins can neutralize each other. NtRhoGDI2 was also cooverexpressed at similar levels with YFP tagged or free CA-NtRAC5. In these experiments, plasma membrane accumulation of CA-NtRAC5 fused to YFP was only partially suppressed and growth depolarization by free CA-NtRAC5 was not inhibited at all. These results provide further evidence for the preferential interaction of NtRhoGDI2 with inactive GDP-bound NtRac5. Together, the observations summarized in this section are consistent with the view that NtRhoGDI2, like other

RhoGDIs (see section "RhoGDI Protein Families"), transfers GDP-bound NtRAC5 from the plasma membrane to the cytoplasm, where the two proteins form inactive heterodimers (Klahre et al. 2006).

Replacement of the arginine at position 69 by alanine specifically disrupts the ability of NtRAC5 to interact with NtRhoGDI2. The in vitro GTPase activity of NtRAC5 is not affected by this mutation. In yeast two-hybrid assays, Ntrac5^{arg69ala} does not interact with NtRhoGDI2, but shows normal interaction with NtRhoGAP1 as well as with NtRAC5 effectors. Interestingly, an Ntrac5^{arg69ala} YFP fusion protein accumulates strongly at the plasma membrane at the flanks of tobacco pollen tube tips, but is almost completely absent from the apex, where NtRAC5 accumulates (see section "NtRhoGAP1: Polarity Maintenance at the Tip of Tobacco Pollen Tubes"). YFP-tagged DN-Ntrac5, which does not interact with NtRhoGDI2 either, shows the same aberrant intracellular distribution. Furthermore, in contrast to NtRAC5, Ntrac5^{arg69ala} fails to depolarize pollen tube growth when overexpressed and does not block the inhibition of this process by cooverexpressed NtRhoGDI2. Together, these results strongly suggest that interaction with NtRhoGDI2 is required for the accumulation of NtRAC5 at the pollen tube apex, as well as for the activation of this protein at this location (Klahre et al. 2006).

On the basis of the observations summarized above, NtRhoGDI2 has been proposed (1) to extract NtRac5 from the plasma membrane at the flanks of the pollen tube tip after its inactivation by NtRhoGAP1 (see section "NtRhoGAP1: Polarity Maintenance at the Tip of Tobacco Pollen Tubes"), and (2) to subsequently escort this protein in the GDP-bound form through the cytoplasm back to the apex (Fig. 3). RhoGDF activity, which destabilizes NtRAC5/NtRhoGDI2 complexes and promotes NtRAC5 reassociation with the plasma membrane, together with Rho-GEF activity, which promotes GDP for GTP exchange, is thought to reactivate NtRAC5 at the apex (Fig. 3). Consistent with this model, an *Arabidopsis* protein with RhoGEF activity accumulates at the plasma membrane at the tip of tobacco pollen tubes when expressed in these cells (Gu et al. 2006). The membrane lipid phosphatidylinositol 4,5-bisphosphate (PIP_2) also accumulates specifically at the apex of tobacco pollen tubes, is generated by a lipid kinase activity pulled down form extracts of these cells together with ROP GTPases (Kost et al. 1999), and has GDF activity in animal cells (Fauré et al. 1999). The stimulation of PIP_2 production by active NtRAC5, together with the promotion of NtRAC5 activation by PIP_2-mediated destabilization of NtRAC5/NtRhoGDI2 complexes, could potentially create a positive feedback loop that helps focusing NtRAC5 activity at the pollen tube apex (Klahre et al. 2006).

The results of the functional characterization of NtRhoGDI2 suggest that the primary role of this protein is not to downregulate NtRac5 signaling. Rather, NtRhoGDI2-mediated recycling from the flanks of the pollen tube tip to the apex appears to be required for NtRac5 activation specifically at this location (Kost 2008). Consistent with this view, the growth of tobacco pollen tubes containing constructs designed to silence NtRhoGDI2 expression is strongly reduced, but not depolarized (Fig. 5). AtROPGDI1 in *Arabidopsis* trichoblasts and root hairs (see section "AtROPGDI1: Maintenance of Cellular Polarity Required for Root Hair

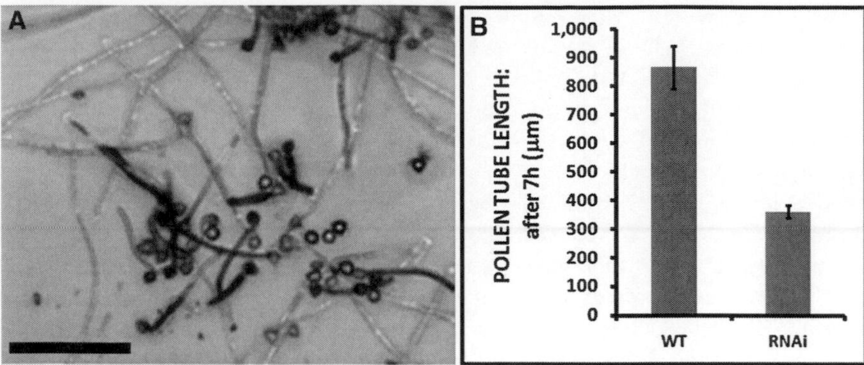

Fig. 5 *Constructs designed to silence NtRhoGDI2 expression inhibit tobacco pollen tube growth without depolarizing this process*. Pollen was collected from heterozygous transgenic tobacco plants containing a single T-DNA inserts that confers pollen-specific expression of (1) an RNAi construct (Wesley et al. 2001) designed to downregulate NtRhoGDI2 transcript levels (Lat52 promoter; Twell et al. 1991), and (2) a GUS (β-glucuronidase) gene (AtProfilin4 promoter; Christensen et al. 1996). (**a**) Seven hours after pollen plating on solid culture medium (Read et al. 1993), a histochemical GUS assay was performed, which resulted in the selective blue (dark) staining of transgenic pollen tubes, whereas nontransgenic pollen tubes remained transparent. As expected for pollen produced by heterozygous transformants, 50% of the analyzed pollen tubes were transgenic and displayed GUS activity. These pollen tubes were clearly shorter that non-transformed pollen tubes, but did not display ballooning at the tip. (**b**) Statistical analysis of pollen tube length showed that within the first 7 h after germination the presence of an NtRhoGDI2 RNAi construct reduced pollen tube growth by roughly 50%. Essentially, the same results were obtained with 3 independent transgenic tobacco lines. Scale bar: 200 µm; error bars: 95% confidence interval

Initiation and Growth"), as well as non-plant RhoGDIs at least in some systems (Lin et al. 2003), are likely to function in a similar manner as NtRhoGDI2.

4 Conclusions

Recent research has established that RhoGAPs and RhoGDIs are not simply attenuating Rho GTPase signaling in a constitutive manner. Members of both protein families play important and highly complex roles in the control of Rho GTPase activity and of cellular processes in animals, yeast, and plants. The functions of non-plant RhoGAPs and RhoGDIs are tightly regulated by upstream signaling and there are indications in the literature that the same is also true for plant homologs of these proteins. The functional characterization of plant RhoGAPs and RhoGDIs during the past few years has allowed exciting insights in the molecular mechanism underlying the spatio-temporal control of Rho GTPase activity by these proteins, which are likely to be relevant also for non-plant systems.

Acknowledgements BK thanks DFG, BBSRC, VR, and FORMAS for funding.

References

Aitken A (2002) Functional specificity in 14–3-3 isoform interactions through dimer formation and phosphorylation. Chromosome location of mammalian isoforms and variants. Plant Mol Biol 50:993–1010

Anderson DC, Gill JS, Cinalli RM, Nance J (2008) Polarization of the C. elegans embryo by RhoGAP-mediated exclusion of PAR-6 from cell contacts. Science 320:1771–1774

Baxter-Burrell A, Yang Z, Springer PS, Bailey-Serres J (2002) ROPGAP4-dependent ROP GTPase rheostat control of Arabidopsis oxygen deprivation tolerance. Science 296:2026–2028

Berken A, Thomas C, Wittinghofer A (2005) A new family of RhoGEFs activates the ROP molecular switch in plants. Nature 436:1176–1180

Bernards A (2003) GAPs galore! A survey of putative Ras superfamily GTPase activating proteins in man and Drosophila. Biochim Biophys Acta 1603:47–82

Bernards A, Settleman J (2004) GAP control: regulating the regulators of small GTPases. Trends Cell Biol 14:377–385

Bokoch GM (1994) Regulation of the human neutrophil NADPH oxidase by the Rac GTP-binding proteins. Curr Opin Cell Biol 6:212–218

Carol RJ, Takeda S, Linstead P, Durrant MC, Kakesova H, Derbyshire P, Drea S, Zarsky V, Dolan L (2005) A RhoGDP dissociation inhibitor spatially regulates growth in root hair cells. Nature 438:1013–1016

Chiang SH, Hwang J, Legendre M, Zhang M, Kimura A, Saltiel AR (2003) TCGAP, a multi-domain Rho GTPase-activating protein involved in insulin-stimulated glucose transport. EMBO J 22:2679–2691

Christensen HEM, Ramachandran S, Tan C-T, Surana U, Dong C-H, Chua N-H (1996) Arabidopsis profilins are functionally similar to yeast profilins: identification of a vascular bundle-specific profilin and a pollen-specific profilin. Plant J 10:269–279

Chuang TH, Bohl BP, Bokoch GM (1993) Biologically active lipids are regulators of Rac.GDI complexation. J Biol Chem 268:26206–26211

Del Pozo MA, Kiosses WB, Alderson NB, Meller N, Hahn KM, Schwartz MA (2002) Integrins regulate GTP-Rac localized effector interactions through dissociation of Rho-GDI. Nat Cell Biol 4:232–239

DerMardirossian C, Bokoch GM (2005) GDIs: central regulatory molecules in Rho GTPase activation. Trends Cell Biol 15:356–363

DerMardirossian C, Rocklin G, Seo JY, Bokoch GM (2006) Phosphorylation of RhoGDI by Src regulates Rho GTPase binding and cytosol-membrane cycling. Mol Biol Cell 17:4760–4768

Etienne-Manneville S, Hall A (2002) Rho GTPases in cell biology. Nature 420:629–635

Fauré J, Vignais PV, Dagher MC (1999) Phosphoinositide-dependent activation of Rho A involves partial opening of the RhoA/Rho-GDI complex. Eur J Biochem 262:879–889

Foreman J, Demidchik V, Bothwell JH, Mylona P, Miedema H, Torres MA, Linstead P, Costa S, Brownlee C, Jones JD, Davies JM, Dolan L (2003) Reactive oxygen species produced by NADPH oxidase regulate plant cell growth. Nature 422:442–446

Fritz G, Lang P, Just I (1994) Tissue-specific variations in the expression and regulation of the small GTP-binding protein Rho. Biochim Biophys Acta 1222:331–338

Graham DL, Eccleston JF, Lowe PN (1999) The conserved arginine in rho-GTPase-activating protein is essential for efficient catalysis but not for complex formation with Rho.GDP and aluminum fluoride. Biochemistry 38:985–991

Gu Y, Li S, Lord EM, Yang Z (2006) Members of a novel class of Arabidopsis Rho Guanine Nucleotide Exchange Factors control Rho GTPase-dependent polar growth. Plant Cell 18:366–381

Gu Y, Wang Z, Yang Z (2004) ROP/RAC GTPase: an old new master regulator for plant signaling. Curr Opin Plant Biol 7:527–536

Helling D, Possart A, Cottier S, Klahre U, Kost B (2006) Pollen tube tip growth depends on plasma membrane polarization mediated by tobacco PLC3 activity and endocytic membrane recycling. Plant Cell 18:3519–3534

Hepler PK, Vidali L, Cheung AY (2001) Polarized cell growth in higher plants. Annu Rev Cell Dev Biol 17:159–187

Hoffman GR, Nassar N, Cerione RA (2000) Structure of the Rho family GTP-binding protein Cdc42 in complex with the multifunctional regulator RhoGDI. Cell 100:345–356

Hwang J-U, Vernoud V, Szumlanski A, Nielsen E, Yang Z (2008) A tip-localized RhoGAP controls cell polarity by globally inhibiting Rho GTPase at the cell apex. Curr Biol 18:1907–1916

Hwang JU, Gu Y, Lee YJ, Yang Z (2005) Oscillatory ROP GTPase activation leads the oscillatory polarized growth of pollen tubes. Mol Biol Cell 16:5385–5399

Jaffe AB, Hall A (2005) Rho GTPases: biochemistry and biology. Annu Rev Cell Dev Biol 21:247–269

Jones MA, Shen JJ, Fu Y, Li H, Yang Z, Grierson CS (2002) The Arabidopsis ROP2 GTPase is a positive regulator of both root hair initiation and tip growth. Plant Cell 14:763–776

Klahre U, Becker C, Schmitt AC, Kost B (2006) Nt-RhoGDI2 regulates Rac/ROP signaling and polar cell growth in tobacco pollen tubes. Plant J 46:1018–1031

Klahre U, Kost B (2006) Tobacco RhoGTPase ACTIVATING PROTEIN1 spatially restricts signaling of RAC/ROP to the apex of pollen tubes. Plant Cell 18:3033–3046

Knaus M, Pelli-Gulli MP, van Drogen F, Springer S, Jaquenoud M, Peter M (2007) Phosphorylation of Bem2p and Bem3p may contribute to local activation of Cdc42p at bud emergence. EMBO J 26:4501–4513

Knezevic N, Roy A, Timblin B, Konstantoulaki M, Sharma T, Malik AB, Mehta D (2007) GDI-1 phosphorylation switch at serine 96 induces RhoA activation and increased endothelial permeability. Mol Cell Biol 27:6323–6333

Kost B (2008) Spatial control of Rho (Rac-ROP) signaling in tip-growing plant cells. Trends Cell Biol 18:119–127

Kost B, Lemichez E, Spielhofer P, Hong Y, Tolias K, Carpenter C, Chua N-H (1999) Rac homologues and compartmentalized phosphatidylinositol 4, 5-bisphosphate act in a common pathway to regulate polar pollen tube growth. J Cell Biol 145:317–330

Lamarche N, Hall A (1994) GAPs for rho-related GTPases. Trends Genet 10:436–440

Li H, Lin YK, Heath RM, Zhu MX, Yang ZB (1999) Control of pollen tube tip growth by a ROP GTPase-dependent pathway that leads to tip-localized calcium influx. Plant Cell 11:1731–1742

Lin Q, Fuji RN, Yang W, Cerione RA (2003) RhoGDI is required for Cdc42-mediated cellular transformation. Curr Biol 13:1469–1479

Lin Y, Wang Y, Zhu J-K, Yang Z (1996) Localization of a Rho GTPase implies a role in tip growth and movement of the generative cell in pollen tubes. Plant Cell 8:293–303

Masuda T, Tanaka K, Nonaka H, Yamochi W, Maeda A, Takai Y (1994) Molecular cloning and characterization of yeast rho GDP dissociation inhibitor. J Biol Chem 269:19713–19718

Michaelson D, Silletti J, Murphy G, D'Eustachio P, Rush M, Philips MR (2001) Differential localization of Rho GTPases in live cells: regulation by hypervariable regions and RhoGDI binding. J Cell Biol 152:111–126

Molendijk AJ, Bischoff F, Rajendrakumar CSV, Friml J, Braun M, Gilroy S, Palme K (2001) Arabidopsis thaliana ROP GTPases are localized to tips of root hairs and control polar growth. EMBO J 20:2779–2788

Nebenführ A, Ritzenthaler C, Robinson DG (2002) Brefeldin A: deciphering an enigmatic inhibitor of secretion. Plant Physiol 130:1102–1108

Nibau C, Wu HM, Cheung AY (2006) RAC/ROP GTPases: 'hubs' for signal integration and diversification in plants. Trends Plant Sci 11:309–315

Nomanbhoy TK, Cerione RA (1996) Characterization of the interaction between RhoGDI and Cdc42Hs using fluorescence spectroscopy. J Biol Chem 271:10004–10009

Ono E, Wong HL, Kawasaki T, Hasegawa M, Kodama O, Shimamoto K (2001) Essential role of the small GTPase Rac in disease resistance of rice. Proc Natl Acad Sci USA 98:759–764

Parton RM, Fischer-Parton S, Watahiki MK, Trewavas AJ (2001) Dynamics of the apical vesicle accumulation and the rate of growth are related in individual pollen tubes. J Cell Sci 114:2685–2695

Pirone DM, Carter DE, Burbelo PD (2001) Evolutionary expansion of CRIB-containing Cdc42 effector proteins. Trends Genet 17:370–373

Potocky M, Jones MA, Bezvoda R, Smirnoff N, Zarsky V (2007) Reactive oxygen species produced by NADPH oxidase are involved in pollen tube growth. New Phytol 174:742–751

Qiao J, Holian O, Lee BS, Huang F, Zhang J, Lum H (2008) Phosphorylation of GTP dissociation inhibitor by PKA negatively regulates RhoA. Am J Physiol Cell Physiol 295:C1161–1168

Read SM, Clarke AE, Bacic A (1993) Stimulation of growth of cultured *Nicotiana tabacum* W38 pollen tubes by poly(ethylene glycol) and $Cu_{(II)}$ salts. Protoplasma 177:1–14

Ren XR, Du QS, Huang YZ, Ao SZ, Mei L, Xiong WC (2001) Regulation of CDC42 GTPase by proline-rich tyrosine kinase 2 interacting with PSGAP, a novel pleckstrin homology and Src homology 3 domain containing rhoGAP protein. J Cell Biol 152:971–984

Rittinger K, Walker PA, Eccleston JF, Smerdon SJ, Gamblin SJ (1997) Structure at 1.65 A of RhoA and its GTPase-activating protein in complex with a transition-state analogue. Nature 389:758–762

Rossman KL, Der CJ, Sondek J (2005) GEF means go: turning on RHO GTPases with guanine nucleotide-exchange factors. Nat Rev Mol Cell Biol 6:167–180

Scheffzek K, Stephan I, Jensen ON, Illenberger D, Gierschik P (2000) The Rac-RhoGDI complex and the structural basis for the regulation of Rho proteins by RhoGDI. Nat Struct Biol 7:122–126

Scholz RP, Regner J, Theil A, Erlmann P, Holeiter G, Jahne R, Schmid S, Hausser A, Olayioye MA (2009) DLC1 interacts with 14–3-3 proteins to inhibit RhoGAP activity and block nucleocytoplasmic shuttling. J Cell Sci 122:92–102

Simoes S, Denholm B, Azevedo D, Sotillos S, Martin P, Skaer H, Hombria JC, Jacinto A (2006) Compartmentalisation of Rho regulators directs cell invagination during tissue morphogenesis. Development 133:4257–4267

Strassheim D, Porter RA, Phelps SH, Williams CL (2000) Unique in vivo associations with SmgGDS and RhoGDI and different Guanine Nucleotide Exchange activities exhibited by RhoA, dominant negative RhoAAsn-19, and activated RhoAVal-14. J Biol Chem 275:6699–6702

Szumlanski AL, Nielsen E (2009) The Rab GTPase RabA4d regulates pollen tube tip growth in *Arabidopsis thaliana*. Plant Cell 21:526–544

Takeda S, Gapper C, Kaya H, Bell E, Kuchitsu K, Dolan L (2008) Local positive feedback regulation determines cell shape in root hair cells. Science 319:1241–1244

Tcherkezian J, Lamarche-Vane N (2007) Current knowledge of the large RhoGAP family of proteins. Biol Cell 99:67–86

Twell D, Yamaguchi J, Wing RA, Ushiba J, McCormick S (1991) Promoter analysis of genes that are coordinately expressed during pollen development reveals pollen-specific enhancer sequences and shared regulatory elements. Genes Dev 5:496–507

Ueda T, Kikuchi A, Ohga N, Yamamoto J, Takai Y (1990) Purification and characterization from bovine brain cytosol of a novel regulatory protein inhibiting the dissociation of GDP from and the subsequent binding of GTP to rhoB p20, a ras p21-like GTP-binding protein. J Biol Chem 265:9373–9380

Vernoud V, Horton AC, Yang Z, Nielsen E (2003) Analysis of the small GTPase gene superfamily of Arabidopsis. Plant Physiol 131:1191–1208

Vidali L, McKenna ST, Hepler PK (2001) Actin polymerization is essential for pollen tube growth. Mol Biol Cell 12:2534–2545

Wennerberg K, Der CJ (2004) Rho-family GTPases: it's not only Rac and Rho (and I like it). J Cell Sci 117:1301–1312

Wesley SV, Helliwell CA, Smith NA, Wang MB, Rouse DT, Liu Q, Gooding PS, Singh SP, Abbott D, Stoutjesdijk PA, Robinson SP, Gleave AP, Green AG, Waterhouse PM (2001) Construct design for efficient, effective and high-throughput gene silencing in plants. Plant J 27:581–590

Wu G, Li H, Yang ZB (2000) Arabidopsis ROPGAPs are a novel family of Rho GTPase-activating proteins that require the Cdc42/Rac-interactive binding motif for ROP-specific GTPase stimulation. Plant Physiol 124:1625–1636

Yalovsky S, Bloch D, Sorek N, Kost B (2008) Regulation of membrane trafficking, cytoskeleton dynamics, and cell polarity by ROP/RAC GTPases. Plant Physiol 147:1527–1543

Yoshida S, Pellman D (2008) Plugging the GAP between cell polarity and cell cycle. EMBO Rep 9:39–41

Structure and Function of ROPs and their GEFs

Christoph Thomas and Antje Berken

Abstract Rho-related small guanine nucleotide binding proteins, termed ROP, are important molecular switches that cycle between a GDP-bound "off" and a GTP-loaded "on" state in order to regulate vital signaling pathways in plants. They constitute a unique class within the Rho family with distinctive features, and we are just beginning to understand the mode of action, the specificities, and molecular mechanisms of signal transduction by taking advantage of three-dimensional structure data. Here, we give an overview about the structure–function relationships in ROPs in terms of nucleotide binding, GTP hydrolysis, and membrane association. Moreover, we emphasize their plant- and isoform-specific properties regarding distinct interactions with regulators or effectors of the switch. We further address the reaction mechanism of ROP activation by novel RopGEFs and discuss the structural basis for the function and interaction of those unique regulators in the physiological context of the plant cell.

1 Introduction

Rho proteins of plants, known as ROPs or RACs, are key regulatory components of G protein-mediated signal transduction in fundamental processes such as polar growth and differentiation (see the Chapters "ROP GTPases and the Cytoskeleton" and "RAC/ROP GTPases in the Regulation of Polarity and Polar Cell Growth"), interactions with microbes (see the Chapter "The Role of Seven Transmembrane Domain MLO Proteins, Heterotrimeric G Proteins and Monomeric RAC/ROPs in

A. Berken (✉)
Max Planck Institute of Molecular Physiology, Otto Hahn Strasse 11, 44227, Dortmund, Germany
e-mail: antje.berken@mpi-dortmund.mpg.de

C. Thomas
Departments of Molecular and Cellular Physiology and Structural Biology, Stanford University School of Medicine, Stanford, CA 94305, USA
e-mail: cthomas3@stanford.edu

S. Yalovsky, et al. (eds.), *Integrated G Proteins Signaling in Plants*,
Signaling and Communication in Plants,
DOI 10.1007/978-3-642-03524-1_3, © Springer-Verlag Berlin Heidelberg 2010

Plant Defense"), hormone signaling (see Chapter "Auxin, Brassinosteroids and G Protein Signaling"), and adaptation to abiotic stress situations (Yang 2002; Berken 2006; Nibau et al. 2006; Yang and Fu 2007). ROPs belong to the Ras superfamily of small, monomeric guanine nucleotide binding proteins that act as molecular switches cycling between an inactive GDP- and an active GTP-bound conformation (Bourne et al. 1991; Vetter and Wittinghofer 2001). This usually occurs in association with membranes where most small G proteins are anchored via a *C*-terminal prenyl modification, i.e., farnesyl or geranylgeranyl chains, or via acylation with saturated fatty acids (Paduch et al. 2001) (for lipid modifications in ROPs see Chapter "Protein-Lipid Modifications and Targeting of ROP/RAC and Heterotrimeric G Protein" and Section "Structural Characteristics of ROP Proteins"). In the GTP-bound "on" state, the G proteins recognize their targets, the so-called effectors, which generate a cellular response until GTP hydrolysis returns the switch back to the "off" state. Both guanine nucleotides are bound with high affinities in presence of an associated Mg^{2+} ion. Therefore, the intrinsic nucleotide exchange is pretty slow and requires catalysis by guanine nucleotide exchange factors (GEFs) to ensure a fast reaction to incoming stimuli. GEFs function to accelerate nucleotide release (see the Section. "Insights into the Catalytic Mechanism of RopGEFs"), which ultimately leads to G protein activation after reassociation with GTP. While there is generally no preference of GTP over GDP *in vitro*, GTP binding is favored in cells because of the higher concentration of the triphosphate form. The intrinsic GTP hydrolysis reaction is also very slow with turnover numbers (k_{cat}) of 10^{-3} to 10^{-1} min^{-1} so that the G proteins as such cannot be regarded as true GTPase enzymes. Still, they manage GTP cleavage in the presence of GTPase-activating proteins, short GAPs (see the Chapter "Regulatory and Cellular Functions of Plant RhoGAPs and RhoGDIs"). Those enzymes stabilize a hydrolysis competent G protein conformation and usually supply an essential catalytic group in *trans* to promote the GTPase reaction by up to five orders of magnitude (Scheffzek and Ahmadian 2005; Bos et al. 2007). Another level of regulation for some small G proteins, including the ROPs, is provided by guanine nucleotide dissociation inhibitors (GDIs) that control the equilibrium between the cytosolic and the membrane-associated G protein pool (see the Chapter "Regulatory and Cellular Functions of Plant RhoGAPs and RhoGDIs"). They bind the prenylated forms with high affinities, thereby extracting their substrates from the membrane disturbing the proper localization and function in the cellular context (Seabra and Wasmeier 2004; DerMardirossian and Bokoch 2005).

Five principal families of small G proteins with overall structural and functional similarities are known in animals and fungi (Takai et al. 2001; Wennerberg et al. 2005): the Ran, Rab, and Arf proteins are the main regulators of intracellular trafficking, while the members of the Ras and the Rho family are rather considered as "signaling G proteins," which relay extracellular signals to the cytosol and the nucleus. The involved pathways affect mainly gene expression, cell cycle progression, and cytoskeletal dynamics in the opisthokonts, and equivalent functions are essential for normal plant development and physiology, too. Thus, it is surprising that true homologues of the important Ras proteins and the major Rho subfamilies

Rho, Rac, and Cdc42 are absent in the plant kingdom. Yet, the lack of those usually conserved protein classes is apparently compensated in plants by the evolution of multiple isoforms of the plant-specific ROP proteins (see the Chapter "ROP Evolution and ROPs in Grasses"), which are mostly related to Rac but still constitute an extra branch within the Rho family with unique structural and functional features and novel regulatory traits.

2 Structure and Function of ROPs

As *bona fide* members of the Rho family, the ROP proteins are in general quite similar to their animal and fungal counterparts Rho, Rac, and Cdc42 and fulfill some equivalent cellular functions via conserved effector targets (Berken 2006; Nibau et al. 2006, Brembu et al. 2006). However, the repertoire of ROP functions appears to be more diversified and possibly includes certain Ras-related signal transduction events. Additional roles are specifically associated with the plant life style and, accordingly, require plant-type effectors, such as the Cinnamoyl-CoA reductase that plays a role in lignin biosynthesis (Kawasaki et al. 2006). On the other hand, conserved functions of the Rho family, for example in cytoskeletal regulation or during exocytosis, are partly mediated by novel effectors (RICs: see Chapter "ROP GTPases and the Cytoskeleton"; ICR: see Chapter "RAC/ROP GTPases in the Regulation of Polarity and Polar Cell Growth") in plants that otherwise find no match in animals and fungi. Moreover, unconventional regulators (RopGAPs: see the Chapter "Regulatory and Cellular Functions of Plant RhoGAPs and RhoGDIs"; RopGEFs: see the Section "RopGEFs: Novel Activators for Rho Proteins in Plants") have been identified in the recent years (Wu et al. 2000; Berken et al. 2005; Gu et al. 2006), making the ROP system quite unique within the Rho family. Yet, we are just beginning to understand the mode of action, the specificities, and molecular mechanisms of this exciting system taking advantage of detailed structural analyses.

2.1 Structural Characteristics of ROP Proteins

Like animal and fungal Rho proteins, the ROPs comprise about 200 amino acids with a molecular weight of 21–24 kDa. The best studied ROP family members are the 11 isoforms from the model plant *Arabidopsis thaliana* (ROP1–ROP11). Four crystal structures of ROP proteins from *Arabidopsis* are currently available. Two of those are high-resolution structures of the free G proteins, ROP9 and ROP5, in the GDP-bound conformation (Sørmo et al. 2006; Fricke and Berken 2009), while the others, ROP4 and ROP7 (Thomas et al. 2007 and 2009), are from complexes with a catalytic GEF domain (see the Section "Architecture of RopGEFs and Mode of Substrate Binding"). Their conserved structural and functional element is the globular G domain, which folds into a central β-sheet surrounded on both sides by α-helices (Fig. 1a). A shallow pocket accommodates the guanine nucleotide and the associated Mg^{2+} ion. Five highly conserved sequence motifs, designated G-box motifs (G1–G5), are mainly located in the interconnecting loop regions of the G domain. They contain essential amino acids for the binding of the nucleotide, the coordination of the Mg^{2+} ion, and for GTP hydrolysis (see the Section "Insights into the Catalytic Mechanism of RopGEFs"). Two loops, known as the switch regions (switch I and switch II) in G proteins, are particularly flexible and change their conformation upon GTP binding through interactions with the γ-phosphate (Vetter and Wittinghofer 2001). Those switches are the main contact sites in Rho complexes with effectors or regulatory molecules (Dvorsky and Ahmadian 2004). In addition to the G domain, ROPs and other Rho proteins display a helical insertion (αi) followed by a short loop structure known as the "Rho insert." The sequence composition of this insertion is rather diverse among and even within distinct Rho subfamilies, including the ROPs, making it especially suited to define subfamily- and isoform-specific interactions with effectors and regulatory molecules (see the Section "Substrate Specificity of RopGEFs"). Another quite unconserved portion is present at the *C*-terminus of the proteins. This hypervariable region (HVR) contains a polybasic amino acid stretch that is believed to facilitate membrane association both through interactions with the head groups of the inner leaflet of the lipid bilayer and enhancement of the prenylation reaction (Yalovsky et al. 2008, see also Chapter "ROPs, Vesicle Trafficking, and Lipid Modifications"). The extreme carboxyl ends of the polypeptide chain differ among the multiple ROP isoforms, and two ROP subtypes have been described based on those regions and their lipid modifications (Winge et al. 2000; Ivanchenko et al. 2000; Lavy et al. 2002). Type-I ROPs, including ROP1–ROP8 from *Arabidopsis,* end with a "CaaL" motif in which the prenylatable cysteine (C) is followed by two aliphatic residues (a) and a leucine (L). Prenylation of the cysteine is usually accompanied by proteolytic cleavage of the following tripeptide and subsequent carboxymethylation (Yalovsky et al. 2008, see also Chapter "Protein-Lipid Modifications and Targeting of ROP/RAC and Heterotrimeric G Protein"). Type-II ROPs, e.g., ROP9–ROP11 from *A. thaliana,* lack the characteristic "CaaL" motif and, instead, undergo stable *S*-acylation at a unique sequence element known as the "GCCG" box in which the two cysteines for posttranslational modification are flanked by glycines (G) and separated by five to six mostly aliphatic amino acids (Lavy et al. 2002; Lavy and Yalovsky 2006). Transient S-acylation also occurs in type-I ROPs at a conserved cysteine following the last β-strand (β6) in the G domain, and this seems to be associated with ROP activation and the presence of ROP in detergent-resistant membrane fractions (DRM) reminiscent of lipid rafts (Sorek et al. 2007). The available type-I ROP structures, however, reveal that the thiol group of the involved Cys residue (C156) is buried and, therefore, not readily accessible for acylating enzymes (Fig. 1a). Acylation of this residue, thus, would imply a conformational change, probably involving a movement of the *C*-terminal helix α5. Although no structural information is currently available for the active, GTP-bound state of a type-I ROP, it can be assumed from comparisons of active and inactive conformations of related small G proteins that this region is not affected through GTP binding. Still, it is possible

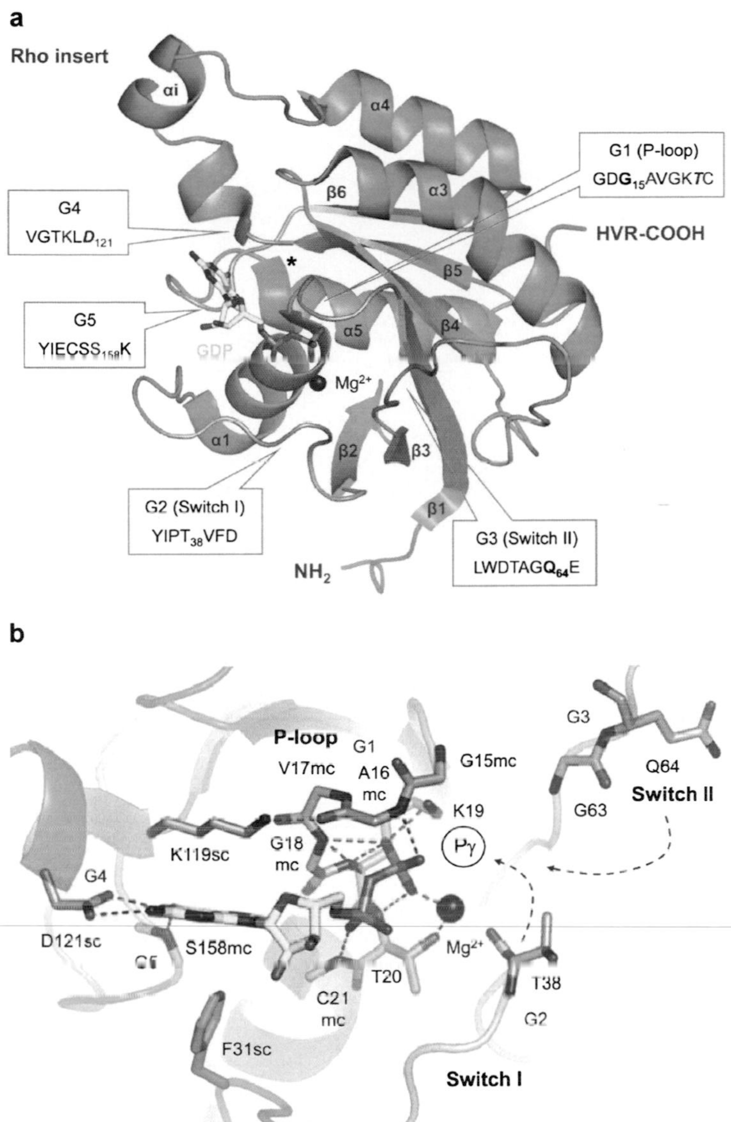

Fig. 1 Three-dimensional structure of GDP-bound ROP9 (PDB-code: 2J0V). (**a**) Overall structure as ribbon plot with GDP as sticks (element colors: C = yellow, P = orange, N = blue, O = red) and Mg^{2+} as *blue sphere*. α-helices and β-strands are numbered consecutively (Note: no α2 is present in switch II of this structure; αi is the insert helix). The G-box motifs (G1–G5) are highlighted in *beige* in the structure and sequences in one-letter code are given in *boxes*. Amino acids interacting with GDP or Mg^{2+} (see b) are marked *red*. Essential sites for GTP hydrolysis which are altered in constitutive-active mutants are shown in **bold**, residues mutated in dominant-negative versions are depicted ***bold in italics***. *NH_2, COOH* amino- and carboxy-termini; *HVR* hypervariable region; *indicates the homologous position of the side chain of an acylatable

that the required rearrangement for S-acylation at the internal cysteine occurs in association with the membrane or depending on the presence of ROPs in DRMs.

2.2 Nucleotide Binding, GTPase Activity, and Commonly Used Mutants

The high-resolution crystal structures of the GDP-bound ROPs reveal that nucleotide binding is essentially the same as in other small G proteins wherein the five conserved G-box motifs line up along the nucleotide binding site (Vetter and Wittinghofer 2001). GDP is bound by a large number of polar contacts and hydrophobic interactions (Fig. 1b), reflecting the high affinity for the nucleotide with equilibrium dissociation constants (K_D) in the low nanomolar range (e.g., 2 nM for the fluorescent GDP analog *N*-methylanthraniloyl-GDP and ROP4; Berken, 2006). The guanine moiety sits in a hydrophobic pocket where it is stabilized via hydrogen bonding and nonpolar interactions: the carboxylate group of an invariant aspartate residue (D121) in the G4 region forms two essential hydrogen bonds to the base, and another polar contact is achieved via the main-chain amide of serine 158 in G5. The guanine base interaction with the aspartate in G4 is believed to be particularly important for tight nucleotide binding and for the high specificity of the G proteins for guanine nucleotides. Mutation of the homologous residue (D119N) in Ras results in drastically reduced nucleotide affinity, increased nucleotide dissociation rates, and higher GEF affinity (Schmidt et al. 1996; Cool et al. 1999). The same can be assumed for a corresponding ROP mutant which was successfully used to isolate RopGEFs through a strong interaction in the yeast two-hybrid system (Berken et al. 2005; Gu et al. 2006; see the Section "Identification of RopGEFs"). Mutation of the G4-aspartate is considered to have a dominant-negative (DN) effect *in vivo* by competing with the normal G protein for the interaction with the GEF. Yet, the mutant is still able to bind to effectors, and the strongly reduced nucleotide affinity together with the higher GTP concentration in cells can also lead to GTP loading in a GEF independent manner (Cool et al. 1999). Binding of the guanine base in ROP is further supported by an aromatic–aromatic interaction with a phenylalanine (F31) at the beginning of switch I and a hydrophobic contact to the alkyl part of the long side chain of lysine 119 in G4. The latter links the G4 region to the phosphate binding P-loop by contacting the main-chain carbonyl of alanine 16. The P-loop embraces the phosphate moieties of the nucleotide like a kinked hand involving interactions of the conserved G1 region:

Fig. 1 (continued) cysteine in type-I ROPs. (**b**) Closeup view of the nucleotide binding site showing polar interactions of amino acids with GDP and Mg^{2+} at 3.0 Å and less as *grey* dashes. Selected amino acids are depicted as *sticks* [element colors: C = beige, N and O as in (a)] with their position number in the sequence. Water molecules were excluded for this figure. The approximate position of the γ-phosphate (Pγ) in a GTP-bound state is indicated, the expected movement of the switches is shown by *dashed arrows*. *sc* side chain; *mc* main chain

main-chain amide hydrogens of several amino acids form crucial bonds with the α- and β-phosphates of GDP (Fig. 1b). A further interaction occurs between the ε-amino group of an invariant P-loop lysine (K19) and the β-phosphate, and this is believed to be particularly important for high-affinity nucleotide binding (Sigal et al. 1986; John et al. 1988). The side chain of threonine 20 also contributes to coordinate the Mg^{2+} ion, which supports the tight binding of GDP. Mutations in the P-loop usually affect the proper function of ROP, leading to constitutive-active (CA) or DN versions when G15 or T20 are changed, respectively. The CA effect is believed to involve a disturbed interaction with a conserved arginine (arginine finger) of the catalytic GAP due to sterical hindrance when the small glycine is replaced by any other amino acid (Scheffzek et al. 1997). The DN mutant T20N is assumed to sequester the exchange factor because homologous mutations, for example S17N in Ras, have a strongly reduced nucleotide affinity and bind tightly to GEFs, but their interaction with effectors is impaired (Feig 1999). While the nucleotide interactions with the P-loop and the base binding sites are virtually the same in active and inactive G proteins, the switch regions provide further contacts in the GTP-bound form but the extent of the conformational change is different for different proteins (Milburn et al. 1990; Bourne et al. 1991, Vetter and Wittinghofer 2001). Thus, a detailed description of the GTP-state of ROP would require a 3D structure of the activated protein. However, important ROP interactions with the γ-phosphate can still be deduced from homologous G proteins based on the conservation of involved key amino acids in the G2 and G3 regions. Those interactions include threonine 38 and glycine 63 in the switches that most likely form hydrogen bonds via their main-chain amides to the γ-phosphate oxygens, and thereby foster the restructuring of the ROP switch regions. The shifted switch II would bring an invariant glutamine residue (Q64 in ROP) close to the γ-phosphate where it is required to position a catalytic water molecule during GTP hydrolysis (Milburn et al. 1990; Vetter and Wittinghofer 2001). This crucial function is impaired when the Gln is mutated in small G proteins so that a Q64 exchange in ROP is another effective way to generate a constitutive-active protein that is stuck in the "on"-state.

2.3 *Plant- and Isoform-Specific Structural Features*

Although the overall structure and the basic function of the ROPs as molecular switches align well with other Rho family members, the plant proteins reveal some unique structural features which are believed to be involved in specific interactions with novel regulators or effectors (Berken and Wittinghofer 2008; Fricke and Berken 2009; see the Section "Substrate Specificity of RopGEFs"). Those plant-specific protein features mainly apply to the switch II loop and the insert region in ROPs. While the former is usually well conserved throughout the Rho proteins, it contains two strikingly different amino acids in ROPs (Fig. 2): an invariant arginine (R76 in ROP5/ROP9), which replaces a conserved proline in animal and fungal Rho proteins, protrudes from the ROP5 and ROP9 structures, and therefore is amenable

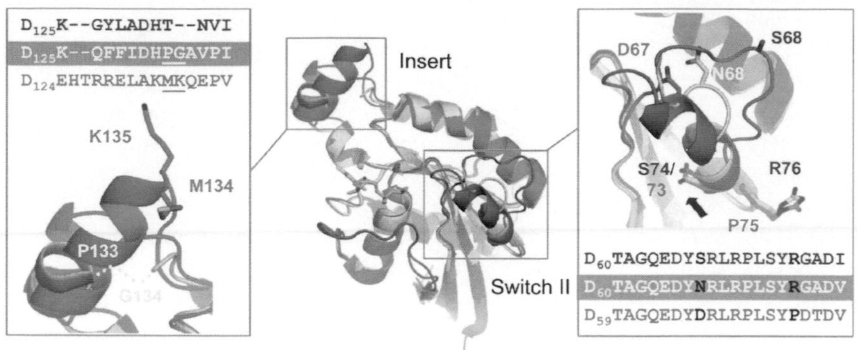

Fig. 2 Structure and sequence comparisons of plant and animal Rho proteins. Three-dimensional structure overlay of ROP9 (*green*), ROP5 (PDB-code: 3BWD; *yellow*) and RhoA (PDB-code: 1FTN; *blue*) with closeup views and amino acid alignments of switch II and the insert region. Important residues are shown as *sticks* in the structure. Significantly different switch II sites are depicted in *red* in the sequence alignment, the *red arrow* marks a putative phosphorylation site for serine/threonine kinases. Underlined residues in the insert region participate in determining substrate specificity of RopGEFs. *Dotted line*: region with undefined electron density in the structure

to binding partners. It is part of a putative recognition site for serine/threonine kinases which may phosphorylate the proximal serine 74 as a possible means of regulation. This serine is conserved in Rho proteins, for example in Rac1, where it is modified by protein kinase B (Akt), which in turn inhibits the GTP-binding activity (Kwon et al. 2000). However, it remains to be established if and how a respective phosphorylation may affect the function or interaction of the ROPs. Another peculiar variation in switch II is an accessible serine or asparagine residue in the ROPs (S68 in ROP9/N68 in ROP5) instead of a conserved aspartate in nonplant Rho proteins implying a different behavior or interaction due to distinct residue size and charge. Switch II usually comprises a short helical element which is also present in ROP5 but not in the crystal structure of ROP9. Still, it is unclear if this conformational feature of ROP9 is relevant in solution because switch II was also involved in packing interactions within the crystal used for structure determination (Sørmo et al. 2006). A possible helix disruption in ROP9 could be explained by the presence of the serine at position 68, which may prevent the formation of a stable secondary structure in switch II. The resulting higher flexibility of the exclusively loop-shaped region could, in turn, be responsible for the faster intrinsic nucleotide dissociation rate of this ROP isoform (Fricke and Berken 2009) and might as well enable specific interactions (Sørmo et al. 2006). Further specificity toward distinct interaction partners is likely achieved via pronounced structural differences among the insert regions of the different Rho family members. Apart from the highly variable sequence composition, the insert is shorter and spatially displaced in ROPs as opposed to other Rho proteins such as RhoA (Fig. 2). Those differences together with the exposed position at the protein surface make the insert region highly suited to participate in the recognition of Rho subfamily-specific targets.

Moreover, an involvement of the insert in isoform-specific interactions can be assumed from the variability of this structural element even within the ROP subfamily. In this context, ROP9 is again quite exceptional because the insert is even shorter as compared to ROP5 and resides in a distinct position in 3D structure comparisons. A similar role in defining the specificity for interaction partners is also assumed for the insert regions of different animal Rho proteins (Freeman et al. 1996; Karnoub et al. 2001; Thapar et al. 2002; Walker and Brown 2002), but there is still only sparse experimental evidence to univocally claim this function. On the other hand, the recent structural and biochemical data on the ROP system clearly demonstrate that the insert is, indeed, a major element in determining the specific recognition by interaction partners, such as the ROP-activating GEFs in plants (Fricke and Berken 2009; see the Section "Substrate Specificity of RopGEFs").

3 RopGEFs: Novel Activators for Rho Proteins in Plants

Like all Rho proteins, the ROPs need to be activated to fulfill their physiological function in cells, but the responsible GEFs had long been a matter of mystery in plants. The main activators of Rho proteins in opisthokonts (Metazoa and fungi) instead have been known for several years and are well characterized in terms of structure and function (Zheng 2001; Schmidt and Hall 2002; Rossman et al. 2005). The catalytic core of these so-called Dbl-type GEFs (Dbl: Diffuse B cell lymphoma) consists of a Dbl-homology (DH) domain encompassing about 180 amino acids that occurs in tandem with a pleckstrin-homology (PH) domain. Currently, 69 different Dbl-type GEFs are known in humans (Rossman et al. 2005). Another, distinct family of RhoGEFs in opisthokonts, with 11 members in mammals, is represented by the recently identified Dock180-related proteins (Meller et al. 2005; Côté and Vuori 2007). Like Dbl-type GEFs, the Dock180-related proteins harbor a tandem module of domains: while the Dock-homology region (DHR)-1 domain seems to be responsible for localizing the exchange factors to sites of signaling (Côté and Vuori 2007), the *C*-terminal DHR-2 domain spanning 450–550 residues is responsible for catalysis of nucleotide exchange. Plants completely lack the family of Dbl-type RhoGEFs and only a single member of the Dock180-related proteins, named SPIKE1, has been described in plants (Qiu et al. 2002; Basu et al. 2008). SPIKE1 displays features of a true GEF for ROP but it seems to be dispensable for plant survival and reproduction suggesting the presence of further RopGEFs in plants.

3.1 *Identification of RopGEFs*

The essential plant RopGEFs were first identified in a yeast two-hybrid screen with the special ROP mutant ROP4(D121N) as bait (Berken et al. 2005), which was expected to have a significantly reduced affinity for guanine nucleotides, an increased nucleotide dissociation rate, and a higher affinity for exchange factors. The same strategy was also successfully used later to isolate RopGEFs via a yeast two-hybrid interaction with the corresponding D121 mutant of ROP1 (Gu et al. 2006). As G proteins and GEFs only form stable complexes in the absence of nucleotide, a decrease in nucleotide affinity promotes a stable association between G protein and exchange factor. Thus, ROP4(D121N) and ROP1(D121N) were ideal baits for catching RopGEFs from cDNA libraries of *A. thaliana*. Among other targets, two homologous clones were identified as effective binding partners of the mutants. They shared 40% protein sequence identity and contained a conserved central domain of unknown function (DUF315). This domain was only found in plant proteins and had homology neither to Dbl-type GEFs nor to Dock180-related proteins nor to any other known nucleotide exchange factor for G proteins. When tested in a GEF activity assay *in vitro*, the identified proteins, termed RopGEF1 and RopGEF2, accelerated specifically the spontaneous nucleotide exchange of ROP with catalytic properties comparable to those of Dbl-type GEFs (Berken et al. 2005). Moreover, several lines of evidence also support the ROP-activating function of the novel RopGEFs *in vivo* (Gu et al. 2006; Zhang and McCormick 2007). The characterized proteins turned out to be members of a large plant-specific protein family found in Chloroplastida and Rhodophyta (Berken et al. 2005; Elias 2008), including 14 different isoforms in *A. thaliana*. This complexity involving presumably redundant proteins further supports the crucial role of the new RopGEFs as the main ROP activators in plants.

3.2 Architecture of RopGEFs and Mode of Substrate Binding

The new plant GEFs are characterized by a novel catalytic domain dubbed PRONE (plant-specific ROP nucleotide exchanger), which comprises about 380 amino acids and is only found in RopGEFs (Berken et al. 2005). This PRONE domain is flanked by *N*- and *C*-terminal regions variable in sequence and length. The structure of the PRONE domain of RopGEF8 (PRONE8) from *A. thaliana* has been solved by X-ray crystallography, both alone and in complex with two different substrates, ROP4 and ROP7 from *A. thaliana* (Thomas et al. 2007; Thomas et al. 2009). PRONE exhibits an almost purely α-helical fold and is divided into two subdomains (Fig. 3a). The larger subdomain 1 consists of helices α1–5 plus α13, and subdomain 2 comprises the remaining helices α6–12. The only β-structural element is a β-hairpin called β-arm which protrudes from the main body of the molecule. Another remarkable feature of PRONE8 is an extended loop structure between helices α4 and α5 spanning more than 40 residues. Because of two invariant, consecutive tryptophans, this loop has been named WW-loop. PRONE forms a butterfly-shaped constitutive dimer via its *N*-terminal helix α1 and parts of the succeeding loop. Dimerization of PRONE proved to be essential for catalysis (Thomas et al. 2007) while most other GEF domains function as monomers. The only other GEF domain for which dimerization seems to be catalytically

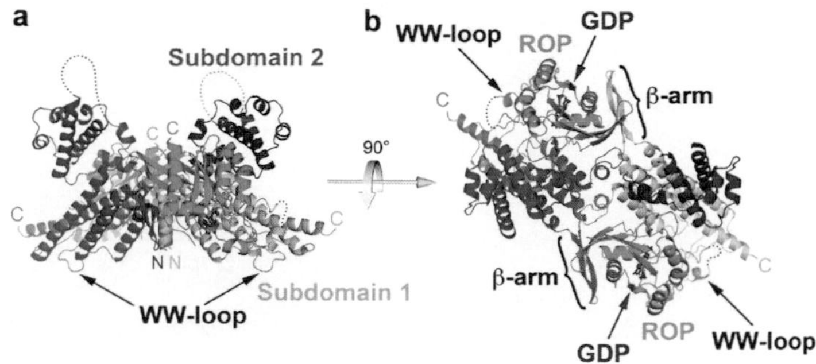

Fig. 3 Three dimensional structure of the ROP4 GDP PRONE8 complex (PDB code: 2NTY). Side view (**a**) and top view (**b**) of the ternary complex in ribbon representation with GDP depicted as *sticks*. The two PRONE protomers are shown in *blue* and *magenta/beige*, respectively. Important structural elements are labeled. The regions of PRONE that could not be modeled due to ill-defined electron density are indicated by *dotted lines*. N, C: *N*- and *C*-termini

essential is that of Sec2p which forms a parallel coiled coil (Dong et al. 2007; Sato et al. 2007a, b).

The structures of PRONE in complex with its substrate ROP show a hetero-tetrameric complex of two ROP molecules interacting with one PRONE dimer (Fig. 3b). Each ROP molecule is contacted by both PRONE protomers: one protomer hugs ROP via its β-arm, the other protomer interacts with the same ROP molecule primarily through subdomain 1. This binding mode of ROP captured between both PRONE protomers nicely explains why dimerization of the exchange factor is a prerequisite for catalytic activity. The ROP–PRONE interface is unusually large for a protein–protein interaction (Lo Conte et al. 1999), with 4,390 Å^2 of surface area buried upon complex formation. However, interfaces in G protein–GEF complexes tend to be larger than the average interface of protein complexes. The unusually large interaction surface is thought to stabilize the unstable nucleotide-free state of the G protein and the large enthalpy contribution compensates for the entropy loss caused by the stabilization of switch II observed in G protein–GEF complexes (Cherfils and Chardin 1999). An extended interaction surface is also observed in the switch II region of the ROP–PRONE complexes. Further contacts of ROP to its exchange factor are primarily mediated by residues of the P-loop, switch I, the first β-strand of the interswitch region, and the end of the Rho insert, and most of those interactions are functionally important for the catalytic mechanism of nucleotide exchange.

3.3 Insights into the Catalytic Mechanism of RopGEFs

GEF-catalyzed nucleotide exchange on small G proteins can be envisioned as a cyclic reaction cascade involving stable binary and transient ternary complexes of

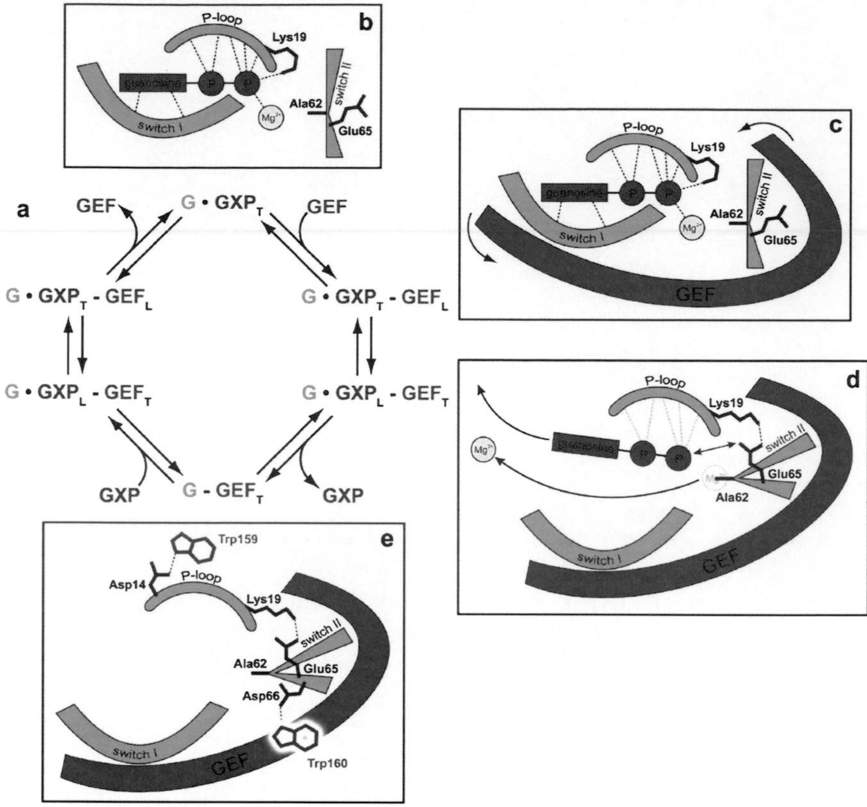

Fig. 4 Mechanism of PRONE-catalyzed nucleotide exchange. (**a**) The catalytic cycle involves different intermediates of G protein, nucleotide, and exchange factor. The structural rearrangements in ROP induced by PRONE are shown schematically in (**b**)–(**e**). *G* G protein; *GEF* exchange factor; *GXP* GDP or GTP; *L* loosely-bound; *T* tightly-bound; *P* phosphate

G protein, nucleotide, and exchange factor (Fig. 4) (Wittinghofer 2000). The encounter of GEF and nucleotide-bound G protein results in a transient ternary complex in which the nucleotide is still tightly bound and the exchange factor is loosely associated. Through conformational changes, this initial complex is converted into another short-lived complex in which the affinities of nucleotide and exchange factor are reversed. Nucleotide release from this trimeric complex leads to the formation of a stable binary G protein–GEF complex. Rebinding of nucleotide finally regenerates the nucleotide-bound G protein after formation of the transient intermediates and dissociation of the exchange factor. As in cells the concentration of GTP is much higher than that of GDP, the last steps of the reaction cycle lead to the GTP-loaded, activated G protein *in vivo*. While the structures of several stable binary complexes of G proteins and their cognate exchange factors are known, the trimeric intermediates are scarcely characterized.

The first X-ray structure of a ROP-PRONE complex (ROP4-GDP-PRONE8) showed GDP molecules in the nucleotide binding pocket of both ROP molecules (Thomas et al. 2007). This complex thus represents one of the ternary intermediates of the GEF catalytic cycle depicted in the Fig. 4. The structure shows PRONE-bound ROP molecules in which the nucleotide affinity is decreased. This reduction of nucleotide affinity is achieved by several PRONE-induced conformational perturbations in ROP (Fig. 4). The interactions of the guanine base with the G protein are virtually unchanged compared with uncomplexed GDP-bound Rho/ROP protein structures (Wei et al. 1997; Sørmo et al. 2006). The same is true for the phosphate-coordinating P-loop, except for the invariant lysine (K19), whose ε-amino group no longer contacts the β-phosphate, but is now further oriented toward a conserved glutamate (E65) in the dramatically remodeled switch II region. The interaction between the P-loop lysine and the nucleotide phosphate moiety is known to be essential for tight nucleotide binding in small G proteins (Sigal et al. 1986; see the Section "Nucleotide Binding, GTPase Activity, and Commonly Used Mutants"), so that the reorientation of its side chain, as seen in the ROP4-GDP-PRONE8 complex, reduces the high nucleotide affinity. As an additional consequence of the remodeled switch II, the methyl group of Ala62 is moved into a position that overlaps with the magnesium binding site, and thus is incompatible with binding of the metal ion (Fig. 4d). This explains why the ROP4–GDP–PRONE8 complex lacks magnesium. It has been shown previously that loss of the Mg^{2+} ion can decrease nucleotide affinity in G proteins by a factor of 500–1,000 (Hall and Self 1986; Klebe et al. 1995). All together, PRONE lowers nucleotide affinity of ROP by removing the positive charge of K19 from the nucleotide β-phosphate while simultaneously juxtaposing the negative charge of E65 close to it, and by expelling Mg^{2+} through a repositioned Ala62. In addition, PRONE rearranges switch I of ROP in a way that it opens up the nucleotide binding pocket in order to facilitate dissociation and subsequent rebinding of nucleotide (Fig. 4d). The structural features of the ROP4–GDP–PRONE8 complex and biophysical measurements support the notion that the structure represents a predissociation intermediate of the GEF reaction in which the affinity between G protein and GEF is relatively high, while the nucleotide affinity is already significantly lowered (Thomas et al. 2007).

Further insights into the mechanism of PRONE-catalyzed nucleotide exchange subsequently came from the structure of a nucleotide-free binary complex between ROP7 and PRONE8 (Thomas et al. 2009). Some exchange factors lower nucleotide affinity by indirectly disrupting the structural integrity of the P-loop (Dong et al. 2007) or by displacing it (Renault et al. 2001). In one extreme case, binding of the exchange factor leads to unfolding of the P-loop and remaining parts of the nucleotide-binding pocket (Itzen et al. 2006). Other binary G protein–GEF complex structures display a collapsed P-loop due to the absence of nucleotide (Boriack-Sjodin et al. 1998; Goldberg 1998). Yet, there are exceptions: the P-loop in structures of Rac1-Tiam1 (Worthylake et al. 2000), Cdc42-SopE (Buchwald et al. 2002), Ran-RCC1 (Renault et al. 2001), RhoA-LARG (Kristelly et al. 2004), and in one structure of Sec4p-Sec2p (Sato et al. 2007b) is intact, but this is presumably due to the presence of a polyanion which mimics the features of the bound nucleotide. In Cdc42-Dbs, the empty nucleotide-binding pocket, including the P-loop, instead is occupied with water molecules (Rossman et al. 2002). However, in the nucleotide-free binary complex of ROP7 with PRONE8, the P-loop is devoid of nucleotide or any stabilizing polyanion, yet it retains its native structure (Thomas et al. 2009). The ROP7–PRONE8 complex is the intermediate of the GEF reaction cycle which succeeds the ternary ROP4–GDP–PRONE8 complex after nucleotide release (Figs. 4a and e). The P-loop in this binary complex is stabilized by interactions with the two consecutive tryptophans of the WW-loop, which have been shown to be essential for nucleotide exchange (Thomas et al. 2007). This GEF-mediated stabilization of the P-loop apparently facilitates reassociation with a new nucleotide, and thereby further promotes the overall exchange reaction (Thomas et al. 2009). The remaining elements of the nucleotide-binding pocket, including the essential G4 and G5 regions (see the Section "Nucleotide Binding, GTPase Activity and Commonly Used Mutants"), surprisingly also resemble those in the ternary complex. This corroborates the notion that the ternary complex structure represents an intermediate of the GEF reaction just before nucleotide dissociation, in which the major PRONE-induced structural rearrangements in ROP have already been finalized.

The structural perturbations in ROP induced by PRONE correspond to those observed in complexes of other G protein–GEF systems. The ionic interaction between the invariant P-loop lysine and the glutamate residue of switch II has also been observed in structures of Ras-SOS (Boriack-Sjodin et al. 1998) and Rac1-Tiam1 (Worthylake et al. 2000). A similar mechanism is employed by ArfGEFs (Goldberg 1998) and the Rab exchange factor Rabex-5 (Delprato and Lambright 2007), where a negatively charged residue is supplied in *trans*. The fundamental mechanistic role of the switch II glutamate for GEF-catalyzed nucleotide exchange on most Ras-like G proteins has recently been demonstrated biochemically (Gasper et al. 2008). GEF-induced expulsion of the Mg^{2+} ion by residues of the switch regions has also been observed for other exchange factors, like SOS (Boriack-Sjodin et al. 1998) and Sec2 (Dong et al. 2007; Sato et al. 2007b). Finally, the displacement of switch I is also a characteristic feature of other GEF-catalyzed reactions, with the most dramatic structural remodeling observed in Ras-SOS (Boriack-Sjodin et al. 1998) and Sec4p-Sec2p (Dong et al. 2007).

In summary, RopGEFs share common catalytic principles with other GEFs, yet exhibit pronounced differences in details of the catalytic mechanism. Like other exchange factors, RopGEFs reduce nucleotide affinity primarily through structural changes in the switch regions of the G protein: switch I is pushed aside to open up the nucleotide-binding pocket, while switch II is moved toward the P loop to weaken phosphate and magnesium binding, inducing sterically or electrostatically unfavorable conditions for nucleotide binding. A hallmark of the RopGEF catalytic cycle is a P-loop that is stabilized in its native conformation once the nucleotide has been dislodged. The GEF-mediated stabilization by an essential sequence motif of two consecutive tryptophans finds no match in other guanine nucleotide exchange reactions and provides another example for the unique features of the plant ROP system.

3.4 Substrate Specificity of RopGEFs

Another interesting feature of the ROP system is the restricted specificity of the RopGEFs for their plant G protein substrates, while both plant and human Rho proteins are good in vitro substrates for Dbl-type GEFs such as Tiam1 (Berken et al. 2005). Despite the high homology between ROPs and their human homologues, especially Rac, the PRONE domain was shown to be exclusively active toward ROPs. In a mutational analysis, it was possible to narrow down the substrate specificity determinants to three ROP-typical regions (see the Section "Plant- and Isoform-Specific Structural Features"): N68 and the insert region appear to be crucial for substrate recognition by PRONE, while R76 seems to strongly enhance catalysis by itself (Fricke and Berken 2009). These findings can be rationalized in the light of the ROP–PRONE complex structures: N68 in most ROPs is a negatively charged aspartate in the animal Rho proteins, which might lead to a repulsion with the side chain of a highly conserved glutamate (E249) in the α10-helix of PRONE. The guanidino group of the long side chain of R76 in the switch II loop interacts with the main-chain carbonyl oxygen of a serine (Ser313) in subdomain 2 of PRONE8 and is presumably involved in the remodeling of switch II during GEF catalysis. This important function cannot be compensated by a proline at the respective position in Rac, Cdc42, or Rho. In this context, substrate recognition might be controlled by phosphorylation of the proximal serine (S74). This serine is also located in the ROP–PRONE interface, and phosphorylation could possibly inhibit RopGEF binding (Fricke and Berken 2009). The two ROP–PRONE structures further reveal that the insert region of ROP4/ROP7 is contacted by a stretch of the WW-loop of PRONE8 via a main-chain interaction. This part of the insert contains small amino acids in most ROPs (proline 133/glycine 134) while larger residues are often present in animal Rho proteins. The respective methionine 134 and the following lysine 135 in RhoA (Fig. 2), for example, could sterically clash with the WW-loop of PRONE8, and thus impede complex formation. The central role of the insert region as a specificity determinant in RopGEF catalysis is further confirmed by the observation that the more unusual insert region of ROP9 (see the Section "Plant- and Isoform-Specific Structural Features"; Fig. 2) leads to a significantly reduced PRONE-catalyzed nucleotide exchange (Fricke and Berken 2009). The involved WW-loop of different RopGEFs also differs considerably in sequence and length. In RopGEFs from *Arabidopsis*, the length varies between 54 (PRONE4) and 40 (PRONE11 and 12) amino acids. This feature of RopGEFs most likely also contributes to specific substrate recognition of different PRONE domains.

Beyond the factors that govern PRONE substrate specificity at the molecular level, there is probably an additional layer of subcellular regulation. The expression levels of RopGEFs are known to vary in different plant tissues (Berken et al. 2005; Kaothien et al. 2005; Gu et al. 2006; Zhang and McCormick 2007). The idea that the spatial expression pattern of RopGEFs, in addition to substrate specificity, contributes to a differential activation of ROP-dependent signaling pathways is therefore appealing. Different temporal expression patterns are also likely to additionally influence the specific ROP activation (Berken et al. 2005).

3.5 RopGEFs in the Physiological Context

While the mechanism of ROP activation is now well established based on 3D structures, the regulation of this process in cells is still unknown. PRONE is constitutively active in vitro and does not need any activating factors for catalyzing nucleotide exchange on ROP proteins (Berken et al. 2005). According to an autoinhibitory model which was first proposed by Gu et al. (2006), the variable noncatalytic *C*-terminus of RopGEFs is able to block GEF activity through an intramolecular interaction with PRONE. In addition, the variable *N*-terminus is thought to have some regulatory function, too. This model gained support by a recent study (Zhang and McCormick 2007), which showed that overexpression of a *C*-terminally truncated version, but not a full-length construct, of RopGEF12 from *Arabidopsis* affects the polarity of pollen tube growth. Furthermore, by using a phosphate-mimicking mutation of an invariant serine in the same study, it was proposed that a phosphorylation event in the *C*-terminus of RopGEF12 and other pollen-specific isoforms could abolish autoinhibition. In addition to GEF activity regulation, phosphorylation might as well modulate interactions with other signaling proteins (Berken 2006). In this context, it is most interesting to note that the *C*-terminal portion of pollen RopGEFs is able to interact with the cytoplasmic domain of pollen-specific receptor-like kinases (RLKs) known as pollen receptor kinases (PRKs) (Kaothien et al. 2005; Zhang and McCormick 2007). RLKs are the primary transmembrane receptors for extracellular signals in plants and are involved in such fundamental processes as embryogenesis, pathogen resistance, hormone perception, and self-incompatibility (Johnson and Ingram 2005; Shiu and Bleecker 2001). They consist of an extracellular domain of leucine-rich repeats, a transmembrane region, and a cytoplasmic domain with Ser/Thr-specific kinase activity. Despite their importance, the cytoplasmic signaling pathways initiated by RLKs are poorly understood. ROP proteins are known to be part of RLK-dependent signaling complexes, including, for example CLAVATA1, which regulates shoot meristem fate in *Arabidopsis* (Trotochaud et al. 1999), or the PRKs that control polar pollen tube growth (Wengier et al. 2003).

The structures of the ROP–PRONE complexes suggest another possible interface for receptor interaction, namely the part of subdomain 2 that is not involved in ROP-binding (Fig. 3a). In addition to conserved residues visible in the X-ray crystallographic model, it contains a highly variable region (residues 269–287) that could not be built due to ill-defined electron density (Thomas et al. 2007). Apart from the WW-loop, it is the only variable region of PRONE. Subdomain 2 of both protomers in the PRONE dimer has the same orientation as the *C*-termini of the bound ROP molecules (Fig. 3a) which are known to be lipid-modified and membrane-anchored (Lavy et al. 2002). Assuming an interaction of PRONE with

RLKs via its second subdomain, the exchange factor would have the proper orientation to bind to the membrane-immobilized ROP proteins.

In light of the described findings, a model of receptor-mediated ROP activation can be drawn according to which RopGEFs are recruited from the cytosol to the membrane by activated RLKs. Autoinhibition of the exchange factor is then released by RLK-mediated phosphorylation, enabling the GEF to catalyze nucleotide exchange on membrane-anchored ROPs. GTP-loaded ROPs subsequently trigger intracellular signaling cascades by interacting with downstream effectors. RopGEFs thus appear to be the missing and primary link between extracellular signals perceived by RLKs and ROP-mediated signaling pathways inside the plant cell.

Interestingly, RopGEF homologs in Rhodophyta have been predicted to contain a microtubule interacting and trafficking (MIT) domain *N*-terminal of the PRONE domain (Elias 2008). MIT domains function as protein interaction modules and are able to bind endosomal ESCRT-III proteins (Scott et al. 2005; Tsang et al. 2006), suggesting that in rhodophytes, GEF-catalyzed ROP activation and vesicular transport are linked (Elias 2008).

4 Conclusions

In recent years, ROP proteins have emerged as key players in plant signal transduction. Although sharing basic common structural and functional principles with their counterparts in animals and fungi, ROPs exhibit unique features distinctive enough to be regarded as a separate class of Rho proteins. The ROP signal transduction protein repertoire includes unconventional regulatory and effector proteins. Among these are the RopGEFs with their characteristic catalytic PRONE domain, which folds into an unprecedented 3D structure, yet employs the universal mechanism of GEFs to catalyze nucleotide exchange on its substrates. Future structural investigations of the ROP molecular switch and its interaction partners, in conjunction with biochemical and in vivo data, will help to fully comprehend a very important part of plant signal transduction at the molecular level.

References

Basu D, Le J, Zakharova T, Mallery EL, Szymanski DB (2008) A SPIKE1 signaling complex controls actin-dependent cell morphogenesis through the heteromeric WAVE and ARP2/3 complexes. Proc Natl Acad Sci USA 105:4044–4049

Berken A (2006) ROPs in the spotlight of plant signal transduction. Cell Mol Life Sci 63: 2446–2459

Berken A, Thomas C, Wittinghofer A (2005) A new family of RhoGEFs activates the Rop molecular switch in plants. Nature 436:1176–1180

Berken A, Wittinghofer A (2008) Structure and function of Rho-type molecular switches in plants. Plant Physiol Biochem 46:380–393

Boriack-Sjodin PA, Margarit SM, Bar-Sagi D, Kuriyan J (1998) The structural basis of the activation of Ras by Sos. Nature 394:337–343

Bos JL, Rehmann H, Wittinghofer A (2007) GEFs and GAPs: critical elements in the control of small G proteins. Cell 129:865–877

Bourne HR, Sanders DA, McCormick F (1991) The GTPase superfamily: conserved structure and molecular mechanism. Nature 349:117–127

Brembu T, Winge P, Bones AM, Yang Z (2006) A RHOse by any other name: a comparative analysis of animal and plant Rho GTPases. Cell Res 16:435–445

Buchwald G, Friebel A, Galán JE, Hardt WD, Wittinghofer A, Scheffzek K (2002) Structural basis for the reversible activation of a Rho protein by the bacterial toxin SopE. EMBO J 21:3286–3295

Cherfils J, Chardin P (1999) GEFs: structural basis for their activation of small GTP-binding proteins. Trends Biochem Sci 24:306–311

Cool RH, Schmidt G, Lenzen CU, Prinz H, Vogt D, Wittinghofer A (1999) The Ras mutant D119N is both dominant negative and activated. Mol Cell Biol 19:6297–6305

Côté JF, Vuori K (2007) GEF what? Dock180 and related proteins help Rac to polarize cells in new ways. Trends Cell Biol 17:383–393

Delprato A, Lambright DG (2007) Structural basis for Rab GTPase activation by VPS9 domain exchange factors. Nat Struct Mol Biol 14:406–412

DerMardirossian C, Bokoch GM (2005) GDIs: central regulatory molecules in Rho GTPase activation. Trends Cell Biol 15:356–363

Dong G, Medkova M, Novick P, Reinisch KM (2007) A catalytic coiled coil: structural insights into the activation of the Rab GTPase Sec4p by Sec2p. Mol Cell 25:455–462

Dvorsky R, Ahmadian MR (2004) Always look on the bright site of Rho: structural implications for a conserved intermolecular interface. EMBO Rep 5:1130–1136

Elias M (2008) The guanine nucleotide exchange factors Sec2 and PRONE: candidate synapomorphies for the Opisthokonta and the Archaeplastida. Mol Biol Evol 25:1526–1529

Feig LA (1999) Tools of the trade: use of dominant-inhibitory mutants of Ras-family GTPases. Nat Cell Biol 1:25–27

Freeman JL, Abo A, Lambeth JD (1996) Rac "insert region" is a novel effector region that is implicated in the activation of NADPH oxidase, but not PAK65. J Biol Chem 271:19794–19801

Fricke I, Berken A (2009) Molecular basis for the substrate specificity of plant guanine nucleotide exchange factors for ROP. FEBS Lett 583:75–80

Gasper R, Thomas C, Ahmadian MR, Wittinghofer A (2008) The role of the conserved switch II glutamate in guanine nucleotide exchange factor-mediated nucleotide exchange of GTP-binding proteins. J Mol Biol 379:51–63

Goldberg J (1998) Structural basis for activation of ARF GTPase: mechanisms of guanine nucleotide exchange and GTP-myristoyl switching. Cell 95:237–248

Gu Y, Li S, Lord EM, Yang Z (2006) Members of a novel class of *Arabidopsis* Rho guanine nucleotide exchange factors control Rho GTPase-dependent polar growth. Plant Cell 18: 366–381

Hall A, Self AJ (1986) The effect of Mg^{2+} on the guanine nucleotide exchange rate of p21N-ras. J Biol Chem 261:10963–10965

Itzen A, Pylypenko O, Goody RS, Alexandrov K, Rak A (2006) Nucleotide exchange via local protein unfolding-structure of Rab8 in complex with MSS4. EMBO J 25:1445–1455

Ivanchenko M, Vejlupkova Z, Quatrano RS, Fowler JE (2000) Maize ROP7 GTPase contains a unique, CaaX box-independent plasma membrane targeting signal. Plant J 24:79–90

John J, Frech M, Wittinghofer A (1988) Biochemical properties of Ha-ras encoded p21 mutants and mechanism of the autophosphorylation reaction. J Biol Chem 263:11792–11799

Johnson KL, Ingram GC (2005) Sending the right signals: regulating receptor kinase activity. Curr Opin Plant Biol 8:648–656

Kaothien P, Ok SH, Shuai B, Wengier D, Cotter R, Kelley D, Kiriakopolos S, Muschietti J, McCormick S (2005) Kinase partner protein interacts with the LePRK1 and LePRK2 receptor kinases and plays a role in polarized pollen tube growth. Plant J 42:492–503

Karnoub AE, Der CJ, Campbell SL (2001) The insert region of Rac1 is essential for membrane ruffling but not cellular transformation. Mol Cell Biol 21:2847–2857

Kawasaki T, Koita H, Nakatsubo T, Hasegawa K, Wakabayashi K, Takahashi H, Umemura K, Umezawa T, Shimamoto K (2006) Cinnamoyl-CoA reductase, a key enzyme in lignin biosynthesis, is an effector of small GTPase Rac in defense signaling in rice. Plant Cell 103:230–235

Klebe C, Bischoff FR, Ponstingl H, Wittinghofer A (1995) Interaction of the nuclear GTP-binding protein Ran with its regulatory proteins RCC1 and RanGAP1. Biochemistry 34:639–647

Kristelly R, Gao G, Tesmer JJ (2004) Structural determinants of RhoA binding and nucleotide exchange in leukemia-associated Rho guanine-nucleotide exchange factor. J Biol Chem 279:47352–47362

Kwon T, Kwon DY, Chun J, Kim JH, Kang SS (2000) Akt protein kinase inhibits Rac1-GTP binding through phosphorylation at serine 71 of Rac1. J Biol Chem 275:423–428

Lavy M, Yalovsky S (2006) Association of *Arabidopsis* type-II ROPs with the plasma membrane requires a conserved C-terminal sequence motif and a proximal polybasic domain. Plant J 46:934–947

Lavy M, Bracha-Drori K, Sternberg H, Yalovsky S (2002) A cell-specific, prenylation-independent mechanism regulates targeting of type II RACs. Plant Cell 14:2431–2450

Lo Conte L, Chothia C, Janin J (1999) The atomic structure of protein–protein recognition sites. J Mol Biol 285:2177–2198

Meller N, Merlot S, Guda C (2005) CZH proteins: a new family of Rho-GEFs. J Cell Sci 118:4937–4946

Milburn MV, Tong L, deVos AM, Brünger A, Yamaizumi Z, Nishimura S, Kim SH (1990) Molecular switch for signal transduction: structural differences between active and inactive forms of protooncogenic ras proteins. Science 247:939–945

Nibau C, Wu HM, Cheung AY (2006) RAC/ROP GTPases: 'hubs' for signal integration and diversification in plants. Trends Plant Sci 11:309–315

Paduch M, Jelen F, Otlewski J (2001) Structure of small G proteins and their regulators. Acta Biochim Pol 48:829–850

Qiu JL, Jilk R, Marks MD, Szymanski DB (2002) The *Arabidopsis* SPIKE1 gene is required for normal cell shape control and tissue development. Plant Cell 14:101–118

Renault L, Kuhlmann J, Henkel A, Wittinghofer A (2001) Structural basis for guanine nucleotide exchange on Ran by the regulator of chromosome condensation (RCC1). Cell 105:245–255

Rossman KL, Worthylake DK, Snyder JT, Siderovski DP, Campbell SL, Sondek J (2002) A crystallographic view of interactions between Dbs and Cdc42: PH domain-assisted guanine nucleotide exchange. EMBO J 21:1315–1326

Rossman KL, Der CJ, Sondek J (2005) GEF means go: turning on RHO GTPases with guanine nucleotide-exchange factors. Nat Rev Mol Cell Biol 6:167–180

Sato Y, Shirakawa R, Horiuchi H, Dohmae N, Fukai S, Nureki O (2007a) Asymmetric coiled-coil structure with guanine nucleotide exchange activity. Structure 15:245–252

Sato Y, Fukai S, Ishitani R, Nureki O (2007b) Crystal structure of the Sec4p-Sec2p complex in the nucleotide exchanging intermediate state. Proc Natl Acad Sci USA 104:8305–8310

Scheffzek K, Ahmadian MR (2005) GTPase activating proteins: structural and functional insights 18 years after discovery. Cell Mol Life Sci 62:3014–3038

Scheffzek K, Ahmadian MR, Kabsch W, Wiesmuller L, Lautwein A, Schmitz F, Wittinghofer A (1997) The Ras-RasGAP complex: structural basis for GTPase activation and its loss in oncogenic Ras mutants. Science 277:333–338

Schmidt A, Hall A (2002) Guanine nucleotide exchange factors for Rho GTPases: turning on the switch. Genes Dev 16:1587–1609

Schmidt G, Lenzen C, Simon I, Deuter R, Cool RH, Goody RS, Wittinghofer A (1996) Biochemical and biological consequences of changing the specificity of p21ras from guanosine to xanthosine nucleotides. Oncogene 12:87–96

Scott A, Gaspar J, Stuchell-Brereton MD, Alam SL, Skalicky JJ, Sundquist WI (2005) Structure and ESCRT-III protein interactions of the MIT domain of human VPS4A. Proc Natl Acad Sci USA 102:13813–13818

Seabra MC, Wasmeier C (2004) Controlling the location and activation of Rab GTPases. Curr Opin Cell Biol 16:451–457

Shiu SH, Bleecker AB (2001) Plant receptor-like kinase gene family: diversity, function, and signaling. Sci STKE 2001:RE22

Sigal IS, Gibbs JB, D'Alonzo JS, Temeles GL, Wolanski BS, Socher SH, Scolnick EM (1986) Mutant ras-encoded proteins with altered nucleotide binding exert dominant biological effects. Proc Natl Acad Sci USA 83:952–956

Sorek N, Poraty L, Sternberg H, Bar E, Lewinsohn E, Yalovsky S (2007) Activation status-coupled transient S acylation determines membrane partitioning of a plant Rho-related GTPase. Mol Cell Biol 27:2144–2154

Sørmo CG, Leiros I, Brembu T, Winge P, Os V, Bones AM (2006) The crystal structure of *Arabidopsis thaliana* RAC7/ROP9: the first RAS superfamily GTPase from the plant kingdom. Phytochemistry 67:2332–2340

Takai Y, Sasaki T, Matozaki T (2001) Small GTP-binding proteins. Physiol Rev 81:153–208

Thapar R, Karnoub AE, Campbell SL (2002) Structural and biophysical insights into the role of the insert region in Rac1 function. Biochemistry 41:3875–3883

Thomas C, Fricke I, Scrima A, Berken A, Wittinghofer A (2007) Structural evidence for a common intermediate in small G protein-GEF reactions. Mol Cell 25:141–149

Thomas C, Fricke I, Weyand M, Berken A (2009) 3D structure of a binary ROP-PRONE complex: the final intermediate for a complete set of molecular snapshots of the RopGEF reaction. Biol Chem 390(5–6):427–435

Trotochaud AE, Hao T, Wu G, Yang Z, Clark SE (1999) The CLAVATA1 receptor-like kinase requires CLAVATA3 for its assembly into a signaling complex that includes KAPP and a Rho-related protein. Plant Cell 11:393–406

Tsang HT, Connell JW, Brown SE, Thompson A, Reid E, Sanderson CM (2006) A systematic analysis of human CHMP protein interactions: additional MIT domain-containing proteins bind to multiple components of the human ESCRT III complex. Genomics 88:333–346

Vetter IR, Wittinghofer A (2001) The guanine nucleotide-binding switch in three dimensions. Science 294:1299–1304

Walker SJ, Brown HA (2002) Specificity of Rho insert-mediated activation of phospholipase D1. J Biol Chem 277:26260–26267

Wei Y, Zhang Y, Derewenda U, Liu X, Minor W, Nakamoto RK, Somlyo AV, Somlyo AP, Derewenda ZS (1997) Crystal structure of RhoA-GDP and its functional implications. Nat Struct Biol 4:699–703

Wengier D, Valsecchi I, Cabanas ML, Tang WH, McCormick S, Muschietti J (2003) The receptor kinases LePRK1 and LePRK2 associate in pollen and when expressed in yeast, but dissociate in the presence of style extract. Proc Natl Acad Sci USA 100:6860–6865

Wennerberg K, Rossman KL, Der CJ (2005) The Ras superfamily at a glance. J Cell Sci 118:843–846

Winge P, Brembu T, Kristensen R, Bones AM (2000) Genetic structure and evolution of RAC-GTPases in *Arabidopsis thaliana*. Genetics 156:1959–1971

Wittinghofer A (2000) The functioning of molecular switches in three dimensions. In: Hall A (ed) GTPases. Oxford University Press, New York, pp 244–310

Worthylake DK, Rossman KL, Sondek J (2000) Crystal structure of Rac1 in complex with the guanine nucleotide exchange region of Tiam1. Nature 408:682–688

Wu G, Li H, Yang Z (2000) *Arabidopsis* RopGAPs are a novel family of Rho GTPase-activating proteins that require the Cdc42/Rac-interactive binding for Rop-specific GTPase stimulation. Plant Physiol 124:1625–1635

Yalovsky S, Bloch D, Sorek N, Kost B (2008) Regulation of membrane trafficking, cytoskeleton dynamics, and cell polarity by ROP/RAC GTPases. Plant Physiol 147:1527–1543

Yang Z (2002) Small GTPases: versatile signaling switches in plants. Plant Cell 14:S375–S388

Yang Z, Fu Y (2007) ROP/RAC GTPase signaling. Curr Opin Plant Biol 10:1–5

Zhang Y, McCormick S (2007) A distinct mechanism regulating a pollen-specific guanine nucleotide exchange factor for the small GTPase Rop in *Arabidopsis thaliana*. Proc Natl Acad Sci USA 104:18830–18835

Zheng Y (2001) Dbl family guanine nucleotide exchange factors. Trends Biochem Sci 26:724–732

Protein–Lipid Modifications and Targeting of ROP/RAC and Heterotrimeric G Proteins

Nadav Sorek and Shaul Yalovsky

Abstract ROP/RAC GTPases and heterotrimeric G protein are soluble proteins that function at cellular membranes, primarily the plasma membranes. Attachment to the membrane takes place by virtue of the posttranslational lipid modifications: prenylation, *S*-acylation, and *N*-myristoylation, as well as by lysine and arginine-rich positively charged domain, referred to as polybasic region. The lipid modifications and the polybasic regions have important regulatory roles in G protein signaling. In this chapter, we first describe the characteristic of each of the three lipid modifications. We then discuss their regulatory roles and how they synergistically modulate signaling by ROP/RAC GTPases and heterotrimeric G proteins.

1 Introduction

Signaling by ROP/RAC GTPases (For the sake of clarity, the ROP nomenclature is used in subsequent discussion in this chapter) and heterotrimeric G proteins requires their attachment to cellular membranes, primarily the plasma membrane. Membrane attachment takes place by virtue of the posttranslational lipid modifications: prenylation, *S*-acylation, and *N*-myristoylation (Fig. 1) and in ROPs also by lysine and arginine rich-domain, referred to as polybasic region (PBR) found at the *C*-terminal end, proximal to the lipid modified residues. Prenylation is an irreversible modification that is required for recruiting proteins to membranes. *S*-acylation is, in principle, a reversible modification that is often associated with either prenylation or *N*-myristoylation or two or more proximal residues that become *S*-acylated. The reversibility of *S*-acylation serves as a regulatory mechanism to

N. Sorek and S. Yalovsky (✉)
Department of Plant Sciences, Tel Aviv University, 603 Britannia Building, Ramat Aviv, 69978 Tel Aviv, Israel
e-mail: shauly@tauex.tau.ac.il

S. Yalovsky et al. (eds.), *Integrated G Proteins Signaling in Plants*, Signaling and Communication in Plants,
DOI 10.1007/978-3-642-03524-1_4,

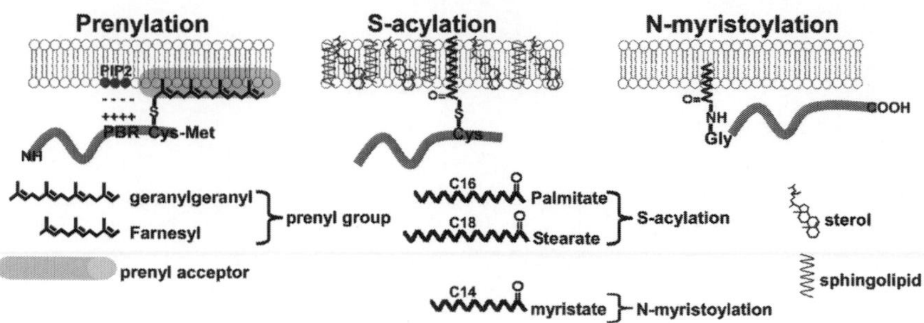

Fig. 1 *Three different lipid modifications. Prenylation* involves attachment of the 15-carbon farnesyl dihosphate or the 20-carbon geranylgeranyl diphosphate by nonreversible thioether linkages to conserved cysteine residues in a conserved C-terminal CaaX box motif. Following prenylation, the last three amino acids are cleaved and the free carboxyl group of the isoprenyl cysteine is, in turn, methylated (Met). The prenyl groups are probably not imbedded in the lipid bilayer but rather associated with membrane prenyl acceptrors. In ROPs, polybasic regions (PBR) enhance prenylation by PGGT-I and association with the plasma membrane via PtdIns 4,5-P2 (PIP2). *S-acylation* involves attachment of saturated acyl lipids such as the 16-carbon palmitic and the 18-carbon stearic acids to cysteine residues. Unlike prenylation, there is no canonical sequence motif for *S*-acylation. *S*-acylation often promotes partitioning of proteins into sterol and sphingolipid-rich membrane microdomains. *N-myristoylation* involves cotranslational attachment of the 14-carbon acyl group myristate to an N-terminal glycine. Stable membrane attachment of N-myristoylated proteins requires an additional lipid modification, usually *S*-acylation

modulate the stability of protein interaction with membranes and partitioning into discrete membrane domains. *N*-myristoylation is cotranslational modification of *N*-terminal glycine residues and is required for recruitment to the membrane but is not sufficient for stable association with membranes. All three lipid modifications also affect subcellular targeting pathways as well as protein–protein interactions, and thus provide additional levels of regulation to G protein signaling. The polybasic domains interact with certain lipid moieties in a cooperative manner, and thus serve as sensitive modules to their level and distribution in membranes. Prenylation and *S*-acylation are receiving much attention in biomedical research due to their important role in several human illnesses (Perez-Sala 2007). The enzymes that catalyze the lipid modifications are conserved between plants and animals (Caldelari et al. 2001; Hemsley et al. 2005; Yalovsky et al. 1997) and Arabidopsis mutants in the genes encoding these enzymes are viable (Cutler et al. 1996; Hemsley et al. 2005; Johnson et al. 2005; Pierre et al. 2007; Running et al. 2004). These mutants provide researchers with invaluable tools for studying the role of the lipid modifications in targeting and function of their protein substrates (Zeng et al. 2007). Since ROPs and heterotrimeric G proteins are conserved with their animal counterparts (see chapters "3 Structure and Function of ROPs and Their GEFs" and 14 "ROP Evolution and ROPs in Grasses"), study of their lipid medications has implication beyond the plant field. In this chapter, we focus on lipid modifications of ROPs and heterotrimeric G proteins and their implications to the function of these proteins in regulation of cell polarity and signaling (Table 1).

Table 1 Lipid modifications and protein domains affecting subcellular localization of plant signaling G proteins. PM – corresponds to Plasma membrane

Proteins	Lipid modifications/ domain	Role of the lipid modification/domain
Type I ROPs	Prenylation and *S*-acylation	Prenylation targets to the PM, transient *S*-acylation induces partitioning into DRMs and possibly conformational changes in the G-domain
Type II ROPs	*S*-acylation	*S*-acylation of *C* terminal GCCG box cysteines is crucial for PM localization
Heterotrimeric Gα	*S*-acylation and *N*-myristoylation	*N*-myristoylation-recruitment to the PM; *S*-acylation – stabilization of PM association
Heterotrimeric Gγ	Prenylation and S-acylation	Prenylation-recruitment to the PM; *S*-acylation of AGG2 stabilizes its association with the membrane and transport from endomembranes to PM
Type-I and type-II ROPs	Polybasic domain	Enhance prenylation by PGGT-I; promote association with thc PM via PtdIns-P2 and PdIns-P3 and PA

2 The Lipid Modifications

2.1 Prenylation and CaaX Processing

2.1.1 Prenylation

Protein prenylation is a posttranslational protein modification that involves the formation of covalent thioether bonds between the cysteines near the *C*-termini of target proteins and the 15-carbon isoprenoid farnesyldiphosphate (FPP) or 20-carbon isoprenoid geranylgeranyldiphosphate (GGPP) (Fig. 1). Prenylation is catalyzed by three protein prenyltransferases: protein farnesyltransferase (PFT) and protein geranylgeranyltransferase-I (PGGT-I), collectively termed the CaaX prenyltransferases, and by Rab-GGTase, whose substrates are limited to members of the Rab subfamily of G proteins. Both PFT and PGGT-I recognize a conserved carboxy-terminal amino acid sequence motif known as CaaX box in which "C" is a cysteine, "a" usually represents an aliphatic amino acid, and "X" is usually a serine, methionine, cysteine, alanine, glutamine, or leucine. If "X" is a leucine, the protein is geranylgeranylated by PGGT-I. If "X" is another amino acid, the protein is preferentially farnesylated by PFT (Reid et al. 2004). PFT and PGGT-I exist as heterodimers, sharing a common α subunit and homologous ß subunits (Lane and Beese 2006). Evolutionary-conserved plants homologs of both PFT and PGGT-I have been characterized (Caldelari et al. 2001; Yalovsky et al. 1997). Arabidopsis mutants in the PFT/PGGT I α subunit and PFT and PGGT I β subunits have been characterized and are not embryonic or cellular lethal unlike that in mammals and yeast (Cutler et al. 1996; Johnson et al. 2005; Running et al. 2004).

2.1.2 CaaX Processing

Following prenylation, proteins undergo two additional posttranslational modifications, collectively referred to as CaaX processing. The first of these modifications involves proteolytic removal of the last three amino acids by either of two CaaX proteases called Ste24 and Rce1. In turn, the free carboxyl group of the isoprenyl cysteine is methylated by isoprenyl carboxy methyltransferase (ICMT) (Young et al. 2000). The methyl group is attached by a reversible methyl-ester linkage and was shown to have regulatory functions in plants (Huizinga et al. 2008). Homologs of all CaaX processing enzymes have been identified and characterized in Arabidopsis (Bracha-Drori et al. 2008; Bracha et al. 2002; Cadinanos et al. 2003; Narasimha Chary et al. 2002; Rodriguez-Concepcion et al. 2000). Similar to their animal and yeast homologs, the Arabidopsis CaaX proteases and ICMTs are localized at the ER (Bracha-Drori et al. 2008; Bracha et al. 2002; Rodriguez-Concepcion et al. 2000). This suggests that following prenylation in the cytoplasm, prenylated proteins are targeted to the ER. It is unknown whether transport from the ER to the plasma membrane occurs along the secretory pathway or by a different route.

2.1.3 Protein Prenylation – A Crossroad Between Signaling and Metabolism

Both prenyl group donors FPP and GGPP are early intermediates in metabolic pathways that produce myriad of compounds, including the plant hormones abscisic acid (ABA), gibberellins (GA), cytokinins, and brassinosteroids (BR), as well as important metabolites such as the phytil side-chain of chlorophylls-carotenoids, and all membrane sterols. Thus, prenylation of signaling G proteins may link between the central metabolic pathways and diverse signaling cascades. Plants possess two distinct isoprenoid biosynthetic pathways: the cytosolic mevalonate (MVA) pathway and the plastidial 2-*C*-methyl-D-erythritol 4-phosphate (MEP) pathway (Eisenreich et al. 2001; Kuzuyama and Seto 2003; Lichtenthaler 2000; Rodriguez-Concepcion and Boronat 2002; Rohmer 1999). Recent studies demonstrated that in BY2 cells, the plastidial MEP pathway provides the isoprenyl moiety for protein geranylgeranylation (Gerber et al. 2009). Thus, a delicate balance may exist between the MVA and MEP pathways and G protein signaling in plants. Future studies on the MVA and MEP pathways will be required to elucidate such mode of regulation.

2.1.4 Prenylation of G Proteins

Two methods of ROP classification are described in the literature (Christensen et al. 2003; Winge et al. 1997), see also chapter 14 "ROP Evolution and ROPs in Grasses"). One classification method divided ROPs into two subgroups, designated type-I and type-II, according to the structure of the *C*-terminal hypervariable domain (Winge et al. 1997). For the sake of clarity, this method will be used subsequently in this chapter. All type-I ROPs and the two Arabidopsis Gγ subunit homologs AGG1 and AGG2 terminate with a CaaL box motif and are preferentially geranylgeranylated PGGT-I (Sorek et al. 2007; Zeng et al. 2007). Type-II ROPs are not prenylated and attach to the membrane by virtue of *S*-acylation (Lavy et al. 2002). *S*-acylation is required for stable membrane attachment of AGG2 (Zeng et al. 2007) and for partitioning of type-I ROPS into discrete membrane microdomains upon their activation (Sorek et al. 2007) (Fig. 2). A short overview of *S*-acylation is given in the next section.

2.2 *S*-Acylation

S-acylation, more commonly referred to as palmitoylation, involves the attachment of palmitate (C16:0 – referring to the number of carbon residues: number of double bonds) or other saturated lipids to cysteine residues through a reversible thioester linkage (Linder and Deschenes 2007; Smotrys and Linder 2004) (Fig. 1). Unlike prenylation, *S*-acylation has no single sequence requirement outside of the presence of a cysteine residue. *S*-acylation is catalyzed by protein *S*-acyl transferases (PATs), using acyl-CoA as the acyl donor. The active site of PATs contains a DHHC-motif cysteine-rich domain. D represents aspartate, H histidine and C the cysteine residues (Lobo et al. 2002; Roth et al. 2002). Three types of PATs have been characterized: ankyrin-repeat-containing (Hemsley et al. 2005; Huang et al. 2004), heterodimeric (Lobo et al. 2002; Swarthout et al. 2005), and monomeric PATs (Hemsley and Grierson 2008; Keller et al. 2004). Most PATs contain only the DHHC domain and belong to either the Erf2p-like heterodimeric PATs or the GODZ-like monomeric PATs (Hemsley and Grierson 2008; Keller et al. 2004; Lobo et al. 2002). Arabidopsis contains 23 putative PATs but only one ankyrin-repeat-containing PAT (TIP1) (Hemsley et al. 2005). *tip1* mutants have reduced cell size, and reduced root hair growth and cell polarity defects, suggesting that *S*-acyltransferases play important roles within the plant (Hemsley et al. 2005). It is unclear whether plants contain both monomeric GODZ-like and heterodimeric Erf2p-like PATs. Expressed sequence tags (ESTs) from other plant species indicate that DHHC proteins are found in both lower and higher plants. The subcellular localization of plant PATs is currently unknown (Hemsley and Grierson 2008; Hemsley et al. 2005). *S*-acylation frequently requires prior membrane association of the protein since PATs are integral membrane proteins (Politis et al. 2005). The most commonly described function of *S*-acylation is to increase the affinity of a soluble protein for membranes, which can thereby affect the protein localization and function. It has been shown that in some cases prenylation alone is not sufficient for stable membrane localization but along with *S*-acylation can stabilize the plasma membrane localization (Rocks et al. 2005; Shahinian and Silvius 1995). *S*-acylation also modulates protein–protein interactions and enzyme activity. *S*-acylation is the only known reversible lipid modification of proteins. The reversal

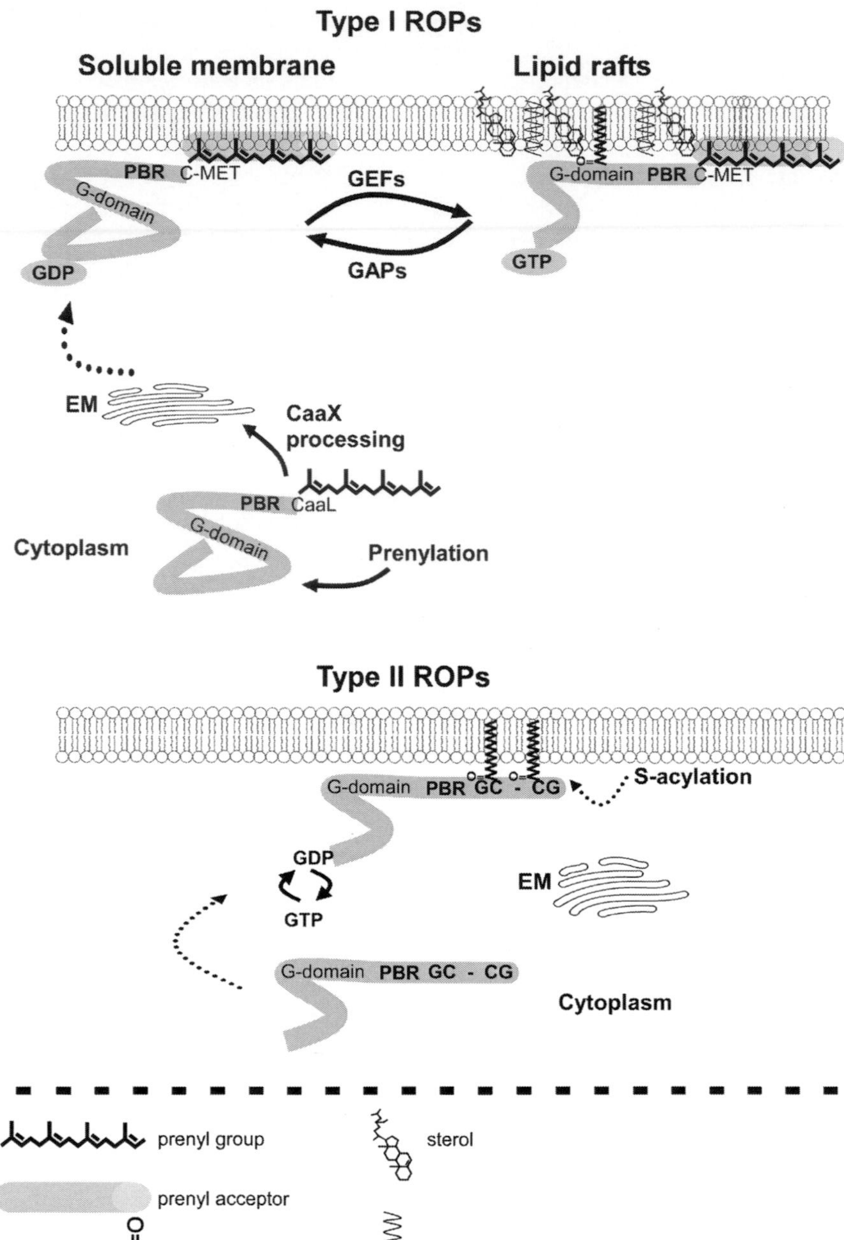

Fig. 2 *Differences in membrane attachment between type-I and type-II ROPs.* Type-I ROPs are prenylated in the cytoplasm, and in turn undergo CaaX processing in the ER. The prenylated and carboxy-methylated type-I ROPs may be transported from the ER to the plasma membrane along the secretory pathway, or by another unknown mechanism. Upon activation by guanine nucleotide exchange factors (GEFs), G-domain cysteine residue(s) of at least some ROPs are reversibly

of *S*-acylation is catalyzed by thioesterases that cleave the thioester bond between the *S*-acylated protein and the acyl group (Camp and Hofmann 1993). The reversibility of *S*-acylation makes it an attractive mechanism for regulating protein activity and subcellular localization and a number of *S*-acylated proteins have been observed cycling on and off membranes or in and out of lipid microdomains (Prior et al. 2001; Rotblat et al. 2004; Sorek et al. 2007).

S-acylation occurs on both types of ROPs as well as on the Arabidopsis Gα subunit homolog GPA1 and the Gγ subunit homolog AGG2.

2.3 *N*-Myristoylation

N-myristoylation is an acylation process absolutely specific to the *N*-terminal glycine residues (Khandwala and Kasper 1971; Resh 2004). Myristic acid (C14:0) represents less than 1% of all fatty acids in cells, but its specific length provides the possibility for reversible interactions with other proteins or membranes. The rare C14:0 saturated fatty acid is linked most often cotranslationally via amide bond specifically to the *N*-terminal glycine residue (Maurer-Stroh et al. 2002b) (Fig. 1). *N*-myristoylation is required, but not sufficient, for membrane anchoring (Maurer-Stroh et al. 2002b) and often occurs together with *S*-acylation of proximal cysteines residue (Martinez et al. 2008). The most abundant form of *N*-myristoylation is catalyzed by the *N*-myristoyltransferase (NMT) that is absolutely dependent on the *N*-terminal glycine residue (Boutin 1997; Maurer-Stroh et al. 2002a). Conserved NMTs have also been identified in plants and have been shown to affect different aspects of plant development (Pierre et al. 2007). Similar to Gα proteins from other eukaryotes, mutational analysis showed that the Arabidopsis G hetrotrimeric α subunit homolog GPA1 is likely *N*-myristoylated and *S*-acylated (Adjobo-Hermans et al. 2006).

3 Lipid Modifications and Subcellular Targeting of ROPs

3.1 Subcellular Distribution and Function of ROPs

ROPs function as molecular switches in a multitude of signaling cascades involved in the regulation of the actin filament and microtubule cytoskeleton, vesicle

Fig. 2 (continued) *S* acylated resulting in transient partitioning of these proteins into sterol and sphingolipid-rich membrane domains known as lipid rafts. GTP hydrolysis, which is enhanced by GTPase activating proteins (GAPs), causes deacylation and partitioning of proteins into detergent soluble membranes. By contrast, type-II ROPs are attached to the plasma membrane by virtue of stable *S*-acylation of two or more cysteine residues in the C-terminal GCCG box. Stable *S*-acylation of this domain does not depend on the activation status of ROPs. PBR corresponds to polybasic region

trafficking, and of plant responses to hormones, stresses, and light (Berken 2006; Nibau et al. 2006; Yang 2002; Yang and Fu 2007), see also chapters 2 "Regulatory and Cellular Functions of Plant RhoGAPs and RhoGDIs, 3-Structure and Function of ROPs and Their GEFs," 5 "ROP GTPases and the Cytoskeleton," 6 "RAC/ROP GTPases in the Regulation of Polarity and Polar Cell Growth" and 11 "The Role of Seven Transmembrane Domain MLO Proteins, Heterotrimeric G Proteins and Monomeric RAC/ROPs in Plant Defense"). The ability of ROPs to interact with membranes allows these proteins to regulate actin polymerization and vesicle trafficking at discrete sites of the plasma membrane and internal membranes, which is essential for their role in the control of cell polarity (Ridley 2006). Subcellular localization ROPs has been investigated by indirect immunofluorescence, GFP tagging, and cell fractionation/immuno-blotting. Imaging experiments have shown that these GTPases are associated with the plasma membrane in variety of cell types, and display enhanced membrane association at growth sites in pollen tubes, root hairs, and leaf epidermal cells (Bloch et al. 2005; Fu et al. 2002; Ivanchenko et al. 2000; Jones et al. 2002; Lavy et al. 2002; Lavy and Yalovsky 2006; Molendijk et al. 2001; Sorek et al. 2007). By immunoblotting, ROPs were exclusively detected in the membrane fraction of extracts of vegetative cells (Sorek et al. 2007), whereas they were found in both the membrane and the cytoplasmic fractions of pollen tube extracts (Kost et al. 1999).

3.2 *Prenylation of Type-I ROPs*

3.2.1 Identification of the Modifying Prenyl Group

Gas chromatography-coupled mass spectrometry (GC/MS) of purified recombinant type I ROP AtROP6 that was overexpressed in Arabidopsis showed that it is preferentially geranylgeranylated (Fig. 2). Low levels of farneylated AtROP6 were also detected (Sorek et al. 2007). Plant PFT can prenylate proteins with a CaaL box motif but cannot use GGPP as a prenyl group donor (Yalovsky et al. 1997). PGGT-I, however, can use FPP as a prenyl group donor, although at lower efficiency (Caldelari et al. 2001). Thus, the farnesylation of AtROP6 could be due to activities of either PFT or PGGT-I.

3.2.2 The Prenyl Receptor Hypothesis

Accumulating data indicate that contrary to previous views, the prenyl groups are associated with membrane receptors rather than being imbedded in the lipid bilayer (Ashery et al. 2006; Belanis et al. 2008). This view is also supported by existence of several proteins with prenyl-binding pockets, such as RhoGDI (Hoffman et al. 2000; Scheffzek et al. 2000), the CaaX proteases RCE1 and STE24, and the prenyl-dependent methyltransferase ICMT (Young et al. 2000). Plasma membrane attachment of the activated form of AtROP6 was unstable in *pggt-Ib* mutant plants. A GC/MS analysis showed that only a farnesylated and *S*-acylated form of the protein existed in both membrane and soluble fractions (our unpublished data). Because AtROP6 was both farnesylated and *S*-acylated in the *pggt-Ib* mutant plants, it is likely that the differences in membrane stability did not simply result from lower hydrophobicity of the 15C farnesyl versus the 20C geranylgeranyl. It rather suggests that an additional factor such as a geranylgeranyl-specific membrane receptor stabilizes interaction of modified proteins with the membrane (Fig. 1).

3.2.3 Function of ROPs in PFT and PGGT-I Mutant Plants

Although the partial redundancy between PFT and PGGT-I activities has been known for a long time, the specific function of each of these modifications was not understood, in part since the *pggt-Ib* mutant is lethal also in yeast (Trueblood et al. 1993). Thus, the differences in the membrane stability of farnesylated versus geranylgeranylated AtROP6 also answer an old question in the protein prenylation field. The relatively weak phenotype of *pggt-Ib* mutants in *Arabidopsis* possibly results from partial function of farnesylated type-I ROPs and the fact that type-II ROPs are only *S*-acylated but not farnesylated. The very strong phentotype of *pluripetala* (*plp*) mutants that lack the common PFT and PGGT-I α-subunit and that do not possess any CaaX prenyltransferase activity (Running et al. 2004) indicates that indeed farnesylation of type-I ROP can likely partially substitute geranylgeranylation. The *plp* mutants are viable, develop lobed epidermal pavement cells in their leaves, elongated root hairs, and have limited self-fertilization capability, indicating that they can form pollen tubes. Since attachment of type-I ROPs to the membrane by prenylation is required for their activity (Sorek et al. 2007), they cannot be functional in *plp* background. Furthermore, type-I ROPs (our unpublished data) as well all the hetrotrimeric G protein γ subunits AGG1 and AGG2 become unstable in *plp* (Zeng et al. 2007). Type II ROPs, on the other hand, are attached to the membrane in *plp* (our unpublished data) and some membrane-associated AGG2 were also detected (Zeng et al. 2007). Taken together, it appears that function of type-II ROPs in pavement cells, root hair, and pollen tubes can facilitate growth of these cells synergistically with type-I ROPs. The viability of *plp* plants can likely be accounted to the activity of type-II ROPs and the partial function of AGG2.

3.3 Transient S-Acylation of Type-I ROPs

Membrane-associated type-I ROPs partition between nonionic detergent soluble (TSM – Triton X-100 Soluble Membrane) and insoluble (DRM – Detergent Resistant Membrane) fractions (Fig. 2). Constitutive active AtROP6^{CA} was exclusively localized in DRMs and GDP/GTP exchanges induced dynamic partitioning of endogenous ROPs between DRMs and TSMs (Sorek et al. 2007). Analysis by GC/MS demonstrated that recombinant AtROP6 purified from TSMs was only geranylgeranylated while AtROP6 or AtROP6^{CA} purified from DRMs were geranylgeranylated and S-acylated by palmitic (C16) and stearic (C18) acids (Sorek et al. 2007). Transient *S*-acylation takes place on highly conserved cysteines residues within the G-domain (Fig. 2, see also Sorek et al. 2007) and chapter 3 "Structure and Function of ROPs and Their GEFs"), and may induce additional conformational changes in the activated, GTP-bound protein. Immunoblotting of WT *Arabidopsis* protein extracts using polyclonal antibodies that recognize both Type I and Type II ROPs showed that endogenous ROPs partition between TSM and DRM. Infiltration of GTPγS into WT plants induced accumulation of endogenous ROPs in the DRMs, whereas GDP infiltration had the opposite effect, inducing accumulation of ROPs in the TSM (Sorek et al. 2007). These results suggest that type II ROPs may also undergo activation-dependent transient *S*-acylation. Although the G-domain cysteines are conserved in the Rho superfamily, *S*-acylation of non-plant Rho proteins has not been reported to date.

3.4 Stable S-Acylation of Type-II ROPs

3.4.1 Identification of the GCCG Box

Type-II ROPs were likely formed by insertion of an intron in the last exon of an ancestral *ROP* gene ((Winge et al. 1997), see also chapter 14 "ROP Evolution and ROPs in Grasses"). The hypervariable region (HVR) of type-II ROPs consists of a unique sequence motif designated the GCCG box, and a proximal polybasic domain that are conserved in different plant species (Lavy and Yalovsky 2006). The GCCG box is comprised of two cysteines, which undergo *S*-acylation (Fig. 2) that are separated by five to six mostly aliphatic amino acids (Lavy et al. 2002; Lavy and Yalovsky 2006). The cysteines are flanked by glycines. In contrast to the transient *S*-acylation of the G-domain described above, GCCG box S-acylation is stable (Lavy et al. 2002; Lavy and Yalovsky 2006). In addition to the lipid-modified GCCG box cysteines, the aliphatic residues between them, the glycines flanking them, and the polybasic domain are required for membrane binding of type-II ROPs. Deletion of the GCCG box intervening nonpolar residues or their substitution with polar amino acids or introduction of a helix-breaking proline residue strongly compromised association with the plasma membrane (Lavy et al. 2002; Lavy and Yalovsky 2006), indicating that the GCCC box forms a short non-polar helix structure, which is required for *S*-acylation and membrane attachment. Similar to the *Arabidopsis*, the maize type-II ROP ZmROP7 is also not prenylated (Ivanchenko et al. 2000). Using a GFP–ZmROP7 fusion protein, it has been shown that association of ZmROP7 with the plasma membrane depended on the HVR and *C*-terminal cysteines but not on the CaaX box cysteine (Ivanchenko et al. 2000).

3.4.2 Functional Differences Between Type-I and Type-II ROPs

The functional differences between the prenylated type-I ROPs and the *S*-acylated type-II ROPs are not well understood. Although *Arabidopsis* type II ROPs AtROP9 and AtROP10, terminate with a CxxX box, they are not prenylated but *S*-acylated (Lavy et al. 2002). The a_1 glycine and, especially, the a_2 lysine residues in the CGKN sequence motif of AtROP10 completely abolish prenylation (Trueblood et al. 1997) and it is thus not a functional prenylation CaaX box. The CTAA motif of AtROP9 is a functional prenylation CaaX box and this protein was prenylated in yeast but not in plants (Lavy et al. 2002). This suggests that the *S*-acylation of the GCCG box cysteines prevents prenylation by PFT, possibly resulting in different membrane targeting pathway. Interestingly, an *S*-acylated, nonprenylated splice isoform of Cdc42 has recently been identified in neurons and was shown to have distinct subcellular localization and activity (Kang et al. 2008). It would be interesting if such differences will also be found in plants.

3.5 Role of the Polybasic Domain for Plasma Membrane Targeting

3.5.1 Function of Polybasic Domains in Prenylation and Membrane Attachment

Many small GTPases in the Ras and Rho families have a *C*-terminal PBR comprised of multiple lysines and/or arginines. The PBR-dependent membrane association of small GTPases may depend on the net charge of the PBR, or on the specific order of charged amino acids within the PBR (Heo et al. 2006). A polybasic domain near a prenylation site appears to act as a strong dominant targeting signal for the plasma membrane (Fig. 1). The polybasic domain has two essential functions: (1) it enhances PGGT-I mediated prenylation by about an order of magnitude (Caldelari et al. 2001; James et al. 1995), and (2) it facilitates membrane interaction (Del Pozo et al. 2002; Lavy and Yalovsky 2006). In plants, in the absence of lipid modifications, the PBR functions as a nonspecific nuclear targeting sequence (Lavy et al. 2002; Lavy and Yalovsky 2006; Rodriguez-Concepcion et al. 1999), indicating that attachment to the membrane is first achieved by the lipid modifications.

3.5.2 Function of Polybasic Domain in the Interaction with Phosphatidyl Phospho Inositides

It is now well established that PBRs in proteins function as interaction modules with phosphatidylinositol 3,4,5 triphosphate (PtdIns 3,4,5-P3), which has not been detected in plant cells to date, and phosphatidylinositol 4,5 bisphosphate (PtdIns 4,5-P2) (Fig. 1) (Heo et al. 2006; Kaadige and Ayer 2006; Orlando et al. 2008; Papayannopoulos et al. 2005; Sun et al. 2007). Importantly, it has been specifically demonstrated that the PBRs of Rho proteins and other small GTPases interact with both PtdIns 4,5-P2 and PtdIns 3,4,5-P3 (Heo et al. 2006). A PBR in N-WASP interacts with PtdIns 4,5-P2 in a multivalent cooperative manner. This facilitates a highly sensitive switch-like mechanism that induces membrane recruitment of N-WASP above a specific PtdIns 4,5-P2 threshold level (Papayannopoulos et al. 2005). In yeast, the Cdc42 effector GIC2 interacts with PtdIns 4,5-P2 in the membrane via a PBR and with Cdc42 through a CRIB domain. The interaction with PtdIns 4,5-P2 is required for polar localization of GIC2 and for its function in polar cell growth (Orlando et al. 2008).

Deleting the PBR of AtROP10 abrogated its interaction with the membrane (Lavy and Yalovsky 2006). Pollen tube ROPs were shown to physically interact with a phosphatidylinositol monophosphate kinase (PtdIns P-K) activity in extracts of tobacco pollen tubes, and PtdIns 4,5-P2, the product of PtdIns P-K activity, colocalizes with ROPs at the apical plasma membrane of these cells (Kost et al. 1999). Based on these observations, it appears possible that ROPs and PtdIns 4,5-P2 maintain a positive feedback loop (Fig. 1). Since PBRs function in a cooperative multivalent manner, PtdIns 4,5-P2 may serve as sensitive switch that above a threshold level triggers accumulation of ROPs in the membrane. In turn, accumulation of ROPs would lead to the production of more PtdIns 4,5-P2, promoting the recruitment of additional ROP molecules. This positive feedback loop may be tightly controlled by the ROP GTPase switch, and is potentially further enhanced by the ability of PtdIns 4,5-P2 to destabilize the interactions between ROPs and RhoGDIs. Work in animal cells has shown that through this mechanism, PtdIns 4,5-P2 can promote Rho membrane association and subsequent activation (Fauré et al. 1999).

3.5.3 Function of Polybasic Domains in Interaction with Phosphatidic Acid

A recent finding showed that chemotaxis of neutrophils in the immune system requires accumulation of the RacGEF DOCK2 at discrete domains in the plasma membrane. Translocation of DOCK2 to the plasma membrane is PdIns 3,4,5-P3 dependent. DOCK2 accumulation at the leading edge, however, is mediated by interaction of a PBR in this protein with phosphatidic acid (PA). Production of PA requires hydrolysis of phosphatidylcholine by phosholipase D (PLD). The PdIns 3,4,5-P3- and PA-dependent accumulation of DOCK2 is required for localized activation of RAC at the leading edge and consequent actin polymerization and formation of pseudopodia (Nishikimi et al. 2009). PA was shown to affect polar growth of pollen tubes (Monteiro et al. 2005, see also chapter 5 “Regulatory and cellular functions of plant RhoGAPs and RhoGDIs”), perhaps by directing ROP association through their PBRs.

4 Plasma Membrane Microdomains

4.1 The Lipid Raft Hypothesis

Biological membranes are composed of many lipid types, including phospholipids, sphingolipids, and sterols. The lipid raft hypothesis suggests that the various types of lipids are not uniformly distributed in eukaryotic plasma membrane but spatially organized in lateral patches of distinct molecular form (Grennan 2007; Parton and Hancock 2004; Tarahovsky et al. 2008). The lipid raft hypothesis is still bitterly debated (Munro 2003). Yet, it is becoming accepted that DRMs are inherent property of biological membranes (Grennan 2007; Hancock 2006). A unifying model for animal cells has been proposed that attempts to resolve the existing controversies about lipid rafts (Hancock 2006). The model predicts that lipid rafts are short-lived entities that are stabilized by their protein constituents, i.e., *S*-acylated proteins attract sterols and sphingolipids, which in turn attract more proteins to form nanoclusters. According to this view, lipid rafts are short-lived microdomains that form and disintegrate. In line with this hypothesis, in yeast, the levels and composition of sterols and sphingolipids are tightly linked (Pichler and Riezman 2004), suggesting that increase in one component attracts the other.

4.2 Accumulation of ROPs in Membrane Microdomains

Transient *S*-acylation of ROPs may be responsible for temporally attracting certain proteins and molecules to discrete membrane domains. Predictions based on modeling of the Ras-activated MAP kinase pathway in mammalian cells suggest that nanoclustering of Ras facilitates a mechanism, which converts graded ligand inputs into fixed outputs. The signal transmission is predicted to be fully dependent on Ras nanoclustering (Tian et al. 2007). Could a similar mechanism function in ROP signaling in plants? An analogous situation may be the auxin gradient-induced accumulation of ROPs in trichoblasts at the future position of root hair formation that was detected using indirect immunofluorescence and GFP tagging (Carol et al. 2005; Fischer et al. 2006). In fact, auxin has been shown to activate ROPs ((Tao et al. 2002), see chapter 6 "RAC/ROP GTPases in the Regulation of Polarity and Polar Cell Growth"). It would be of interest to determine whether auxin-induced nanoclustering into DRMs is involved in the stimulation of Rho signaling by this hormone. Accumulation of endogenous ROPs into DRM upon GTPγS treatment was detected within 16 minutes (Sorek et al. 2007). Thus, in plants too dynamic formation and breakdown of membrane microdomains maybe taking place.

5 Lipid Modifications and RhoGDI

Under physiological conditions, prenylation facilitates the interaction between Rho proteins and RhoGDI (DerMardirossian and Bokoch 2005; Di-Poi et al. 2001). In cocrystal structures, the geranylgeranyl moiety of Cdc42, Rac1, and Rac2 was shown to insert into a hydrophobic pocket formed by the immunoglobulin-like β sandwich of the RhoGDI (Grizot et al. 2001; Hoffman et al. 2000; Scheffzek et al. 2000). Given their structural conservation, plant RhoGDIs are predicted to function similar to their homologues in other organisms (Berken and Wittinghofer 2008), see chapters 2 "Regulatory and cellular functions of plant RhoGAPs and RhoGDIs" and 3 "Structure and function of ROPs and their GEFs"). Because type-II ROPs are not prenylated in plants (Lavy et al. 2002), they might be regulated by a RhoGDI-independent mechanism. In mammalian cells, *S*-acylation of RhoA in the hypervariable domain inhibited its interaction with RhoGDI (Michaelson et al., 2001). It could be that transient G-domain *S*-acylation of activated type-I ROPs inhibits their accessibility for interaction with RhoGDI. *S*-acylation may, thus, destabilize Rho interactions with RhoGDIs, similar to RhoGDI displacement factors such as PtdIns 4,5-P2 (Fauré et al. 1999) and different proteins, including integrins (Del Pozo et al. 2002).

6 Lipid Modifications and Targeting of Heterotrimeric G Proteins

6.1 *Modification of the Gα and Function of Hetertrimeric G Protein in Plants*

Critical to their signaling function is the localization of heterotrimeric G proteins to the cytoplasmic face of plasma membrane in animal cells (Casey 1995). This subcellular localization is often facilitated by *N*-myristoylation and *S*-acylation of the Gα subunit (GPA1) (Adjobo-Hermans et al. 2006) and prenylation and *S*-acylation of the Gγ subunit homolog AGG1 and AGG2 (Adjobo-Hermans et al. 2006; Hemsley et al. 2008; Zeng et al. 2007), see also Table 1 and Figs. 1 and 3). Gβ and Gγ are tightly associated as a functional unit and therefore plasma membrane localization of the Gβ relay on the Gγ (Adjobo-Hermans et al. 2006; Zeng et al. 2007). Mutational analysis of GPA1–GFP fusion protein suggested that *N*-myristoylation and *S*-acylation of GPA1 is necessary for its localization at the plasma membrane (Adjobo-Hermans et al. 2006). The reversibility of *S*-acylation might provide a mechanism that regulates cycling of Gα between the membrane and cytosol. Interestingly, the same study showed that GPA1–AGB–AGG1/AGG2 complexes may not dissociate upon GTP binding by the Gα subunit, providing a deviation from the dogma on heterotrimeric G protein signaling in yeast and mammals.

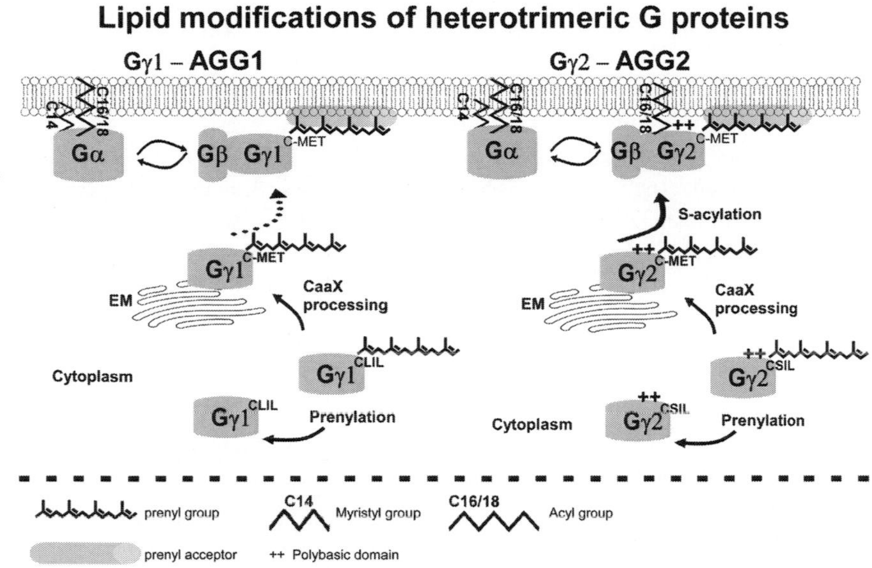

Fig. 3 *Lipid modifications and membrane targeting of hetrotrimeric G proteins*. Gα subunit GPA1 associates with the plasma membrane through *N*-myristoylation and *S*-acylation. The Gγ subunit AGG1 associates with the plasma membrane through prenylation only, whereas prenylation promotes association of the Gγ subunit AGG2 with ER and Golgi membranes and *S*-acylation is required for promoting its association with the plasma membrane. The Gβ subunit AGB1 is not lipid modified and relays on Gγ for plasma membrane association. The association between Gα and Gβγ complex depends on the GDP/GTP state of Gα. A possible deviation from this mode of action has been demonstrated for plant heterotrimeric G protein (Adjobo-Hermans et al. 2006)

6.2 Prenylation and S-Acylation of Gγ Subunits

Both *Arabidopsis* Gγ proteins AGG1 and AGG2 are prenylated, and prenylation is essential but not sufficient for their proper plasma membrane association. *S*-acylation functions as an important second signal to efficiently target AGG2 to the plasma membrane (Zeng et al. 2007). When *S*-acylation was inhibited, AGG2 accumulated in the ER and Golgi. (Zeng et al. 2007). Reversible *S*-acylation may, thus, provide a means to shuttle AGG2 between the plasma and internal membranes and through this regulates heterotrimeric G protein signaling. AGG1 was found both in the plasma and internal membranes. Both AGG1 and AGG2 were found in the plasma membrane when expressed in GPA1 mutant background, indicating that their recruitment was independent of the α-subunit. Membrane association of both AGG1 and AGG2 was reduced in the *pggt-Ib* mutants (Zeng et al. 2007), indicating that both Gγ subunits were preferentially geranylgeranylated, and that in the absence of PGGT-I, the proteins were likely farnesylated. Interestingly, some AGG2 was still found in the plasma membrane in *plp* PFT/PGGT-I α-subunit mutant, possibly maintaining the basal heterotrimeric G protein signaling in these plants (Zeng et al. 2007).

7 Conclusions

Signaling G proteins can be viewed as switches with protein interaction and subcellular targeting modules. This feature allows orchestration of signaling cascades at discrete subcellular domains, response to external stimuli, and coordination between different signaling and metabolic pathways. Localization at discrete domains is, especially, important for the function of ROPs in cell polarity. The synergistic effects of geranylgeranylation, *S*-acylation of the polybasic domain, and the activation-status enable ROP to accumulate at specific subcellular domains in response to changes in membrane composition. Together, the protein interaction and subcellular targeting module may serve as basis for the formation of positive feedback loops that enhance and stabilize polarity; for example, the activation of PtdIns P–K by ROPs and the interactions between ROPs and PtdIns 4,5-P2, the product of PtdIns P–K activity. Existence of positive feedback loops has been suggested by the observation that activated ROPs can change the properties of the plasma membrane and inhibit endocytic vesicle trafficking (Bloch et al. 2005), thus maintaining more ROPs at the membrane. The modularity of lipid modifications and membrane association domains also enables additional levels of regulation as in the case of transient S-acylation of type-I ROPs, AGG2, and possibly also GPA1. It would be exciting to learn how these mechanisms are associated with external stimuli such as directional flow of auxin, direction of light, and gravitational force.

References

Adjobo-Hermans MJ, Goedhart J, Gadella TW Jr (2006) Plant G protein heterotrimers require dual lipidation motifs of Galpha and Ggamma and do not dissociate upon activation. J Cell Sci 119:5087–5097

Ashery U, Yizhar O, Rotblat B, Elad-Sfadia G, Barkan B, Haklai R, Kloog Y (2006) Spatiotemporal organization of Ras signaling: rasosomes and the galectin switch. Cell Mol Neurobiol 26:471–495

Belanis L, Plowman SJ, Rotblat B, Hancock JF, Kloog Y (2008) Galectin-1 is a novel structural component and a major regulator of h-ras nanoclusters. Mol Biol Cell 19:1404–1414

Berken A (2006) ROPs in the spotlight of plant signal transduction. Cell Mol Life Sci 63: 2446–2459

Berken A, Wittinghofer A (2008) Structure and function of Rho-type molecular switches in plants. Plant Physiol Biochem 46:380–393

Bloch D, Lavy M, Efrat Y, Efroni I, Bracha-Drori K, Abu-Abied M, Sadot E, Yalovsky S (2005) Ectopic expression of an activated RAC in Arabidopsis disrupts membrane cycling. Mol Biol Cell 16:1913–1927

Boutin JA (1997) Myristoylation. Cell Signal 9:15–35

Bracha-Drori K, Shichrur K, Lubetzky TC, Yalovsky S (2008) Functional analysis of Arabidopsis postprenylation CaaX processing enzymes and their function in subcellular protein targeting. Plant Physiol 148:119–131

Bracha K, Lavy M, Yalovsky S (2002) The Arabidopsis AtSTE24 is a CaaX protease with broad substrate specificity. J Biol Chem 277:29856–29864

Cadinanos J, Varela I, Mandel DA, Schmidt WK, Diaz-Perales A, Lopez-Otin C, Freije JM (2003) AtFACE-2, a functional prenylated protein protease from *Arabidopsis thaliana* related to mammalian Ras-converting enzymes. J Biol Chem 278:42091–42097

Caldelari D, Sternberg H, Rodriguez-Concepcion M, Gruissem W, Yalovsky S (2001) Efficient prenylation by a plant geranylgeranyltransferase-I requires a functional CaaL box motif and a proximal polybasic domain. Plant Physiol 126:1416–1429

Camp LA, Hofmann SL (1993) Purification and properties of a palmitoyl-protein thioesterase that cleaves palmitate from H-Ras. J Biol Chem 268:22566–22574

Carol RJ, Takeda S, Linstead P, Durrant MC, Kakesova H, Derbyshire P, Drea S, Zarsky V, Dolan L (2005) A RhoGDP dissociation inhibitor spatially regulates growth in root hair cells. Nature 438:1013–1016

Casey PJ (1995) Protein lipidation in cell signaling. Science 268:221–225

Christensen TM, Vejlupkova Z, Sharma YK, Arthur KM, Spatafora JW, Albright CA, Meeley RB, Duvick JP, Quatrano RS, Fowler JE (2003) Conserved subgroups and developmental regulation in the monocot rop gene family. Plant Physiol 133:1791–1808

Cutler S, Ghassemian M, Bonetta D, Cooney S, McCourt P (1996) A protein farnesyl transferase involved in abscisic acid signal transduction in Arabidopsis. Science 273:1239–1241

Del Pozo MA, Kiosses WB, Alderson NB, Meller N, Hahn KM, Schwartz MA (2002) Integrins regulate GTP-Rac localized effector interactions through dissociation of Rho-GDI. Nat Cell Biol 4:232–239

DerMardirossian C, Bokoch GM (2005) GDIs: central regulatory molecules in Rho GTPase activation. Trends Cell Biol 15:356–363

Di-Poi N, Faure J, Grizot S, Molnar G, Pick E, Dagher MC (2001) Mechanism of NADPH oxidase activation by the Rac/Rho-GDI complex. Biochemistry 40:10014–10022

Eisenreich W, Rohdich F, Bacher A (2001) Deoxyxylulose phosphate pathway to terpenoids. Trends Plant Sci 6:78–84

Fauré J, Vignais PV, Dagher MC (1999) Phosphoinositide-dependent activation of Rho A involves partial opening of the RhoA/Rho-GDI complex. Eur J Biochem 262:879–889

Fischer U, Ikeda Y, Ljung K, Serralbo O, Singh M, Heidstra R, Palme K, Scheres B, Grebe M (2006) Vectorial information for Arabidopsis planar polarity is mediated by combined AUX1, EIN2, and GNOM activity. Curr Biol 16:2143–2149

Fu Y, Li H, Yang Z (2002) The ROP2 GTPase controls the formation of cortical fine F-actin and the early phase of directional cell expansion during Arabidopsis organogenesis. Plant Cell 14:777–794

Gerber E, Hemmerlin A, Hartmann M, Heintz D, Hartmann MA, Mutterer J, Rodriguez-Concepcion M, Boronat A, Van Dorsselaer A, Rohmer M et al (2009) The plastidial 2-C-methyl-D-erythritol 4-phosphate pathway provides the isoprenyl moiety for protein geranylgeranylation in tobacco BY-2 cells. Plant Cell 21:285–300

Grennan AK (2007) Lipid rafts in plants. Plant Physiol 143:1083–1085

Grizot S, Faure J, Fieschi F, Vignais PV, Dagher MC, Pebay-Peyroula E (2001) Crystal structure of the Rac1-RhoGDI complex involved in nadph oxidase activation. Biochemistry 40:10007–10013

Hancock JF (2006) Lipid rafts: contentious only from simplistic standpoints. Nat Rev Mol Cell Biol 7:456–462

Hemsley PA, Grierson CS (2008) Multiple roles for protein palmitoylation in plants. Trends Plant Sci 13:295–302

Hemsley PA, Kemp AC, Grierson CS (2005) The TIP GROWTH DEFECTIVE1 S-acyl transferase regulates plant cell growth in Arabidopsis. Plant Cell 17:2554–2563

Hemsley PA, Taylor L, Grierson CS (2008) Assaying protein palmitoylation in plants. Plant Methods 4:2

Heo WD, Inoue T, Park WS, Kim ML, Park BO, Wandless TJ, Meyer T (2006) PI(3, 4, 5)P3 and PI (4, 5)P2 lipids target proteins with polybasic clusters to the plasma membrane. Science 314:1458–1461

Hoffman GR, Nassar N, Cerione RA (2000) Structure of the Rho family GTP-binding protein Cdc42 in complex with the multifunctional regulator RhoGDI. Cell 100:345–356

Huang K, Yanai A, Kang R, Arstikaitis P, Singaraja RR, Metzler M, Mullard A, Haigh B, Gauthier-Campbell C, Gutekunst CA et al (2004) Huntingtin-interacting protein HIP14 is a palmitoyl transferase involved in palmitoylation and trafficking of multiple neuronal proteins. Neuron 44:977–986

Huizinga DH, Omosegbon O, Omery B, Crowell DN (2008) Isoprenylcysteine methylation and demethylation regulate abscisic acid signaling in Arabidopsis. Plant Cell 20:2714–2728

Ivanchenko M, Vejlupkova Z, Quatrano RS, Fowler JE (2000) Maize ROP7 GTPase contains a unique, CaaX box-independent plasma membrane targeting signal. Plant J 24:79–90

James GL, Goldstein JE, Brown SK (1995) Polylysine and CVIM sequences of K-RasB dictate specificity of prenylation and confer resistence to benzodiazepine peptidomimetic in vitro. J Biol Chem 270:6221–6226

Johnson CD, Chary SN, Chernoff EA, Zeng Q, Running MP, Crowell DN (2005) Protein geranylgeranyltransferase I is involved in specific aspects of abscisic acid and auxin signaling in Arabidopsis. Plant Physiol 139:722–733

Jones MA, Shen J-J, Fu Y, Li H, Yang Z, Grierson CS (2002) The Arabidopsis Rop2 GTPase is a positive regulator of both root hair initiation and tip growth. Plant Cell 14:763–776

Kaadige MR, Ayer DE (2006) The polybasic region that follows the plant homeodomain zinc finger 1 of Pf1 is necessary and sufficient for specific phosphoinositide binding. J Biol Chem 281:28831–28836

Kang R, Wan J, Arstikaitis P, Takahashi H, Huang K, Bailey AO, Thompson JX, Roth AF, Drisdel RC, Mastro R et al (2008) Neural palmitoyl-proteomics reveals dynamic synaptic palmitoylation. Nature 456:904–909

Keller CA, Yuan X, Panzanelli P, Martin ML, Alldred M, Sassoe-Pognetto M, Luscher B (2004) The gamma2 subunit of GABA(A) receptors is a substrate for palmitoylation by GODZ. J Neurosci 24:5881–5891

Khandwala AS, Kasper CB (1971) The fatty acid composition of individual phospholipids from rat liver nuclear membrane and nuclei. J Biol Chem 246:6242–6246

Kost B, Lemichez E, Spielhofer P, Hong Y, Tolias K, Carpenter C, Chua NH (1999) Rac homologues and compartmentalized phosphatidylinositol 4, 5-bisphosphate act in a common pathway to regulate polar pollen tube growth. J Cell Biol 145:317–330

Kuzuyama T, Seto H (2003) Diversity of the biosynthesis of the isoprene units. Nat Prod Rep 20:171–183

Lane KT, Beese LS (2006) Thematic review series: lipid posttranslational modifications. Structural biology of protein farnesyltransferase and geranylgeranyltransferase type I. J Lipid Res 47:681–699

Lavy M, Bracha-Drori K, Sternberg H, Yalovsky S (2002) A cell-specific, prenylation-independent mechanism regulates targeting of type II RACs. Plant Cell 14:2431–2450

Lavy M, Yalovsky S (2006) Association of Arabidopsis type-II ROPs with the plasma membrane requires a conserved C-terminal sequence motif and a proximal polybasic domain. Plant J 46:934–947

Lichtenthaler HK (2000) Non-mevalonate isoprenoid biosynthesis: enzymes, genes and inhibitors. Biochem Soc Trans 28:785–789

Linder ME, Deschenes RJ (2007) Palmitoylation: policing protein stability and traffic. Nat Rev Mol Cell Biol 8:74–84

Lobo S, Greentree WK, Linder ME, Deschenes RJ (2002) Identification of a Ras palmitoyltransferase in *Saccharomyces cerevisiae*. J Biol Chem 277:41268–41273

Martinez A, Traverso JA, Valot B, Ferro M, Espagne C, Ephritikhine G, Zivy M, Giglione C, Meinnel T (2008) Extent of N-terminal modifications in cytosolic proteins from eukaryotes. Proteomics 8:2809–2831

Maurer-Stroh S, Eisenhaber B, Eisenhaber F (2002a) N-terminal N-myristoylation of proteins: prediction of substrate proteins from amino acid sequence. J Mol Biol 317:541–557

Maurer-Stroh S, Eisenhaber B, Eisenhaber F (2002b) N-terminal N-myristoylation of proteins: refinement of the sequence motif and its taxon-specific differences. J Mol Biol 317: 523–540

Michaelson D, Silletti J, Murphy G, D'Eustachio P, Rush M, Philips MR (2001) Differential localization of Rho GTPases in live cells: regulation by hypervariable regions and RhoGDI binding. J Cell Biol 152:111–126

Molendijk AJ, Bischoff F, Rajendrakumar CS, Friml J, Braun M, Gilroy S, Palme K (2001) *Arabidopsis thaliana* Rop GTPases are localized to tips of root hairs and control polar growth. EMBO J 20:2779–2788

Monteiro D, Liu Q, Lisboa S, Scherer GE, Quader H, Malhó R (2005) Phosphoinositides and phosphatidic acid regulate pollen tube growth and reorientation through modulation of $[Ca^{2+}]c$ and membrane secretion. J Exp Bot 56:1665–1674

Munro S (2003) Lipid rafts: elusive or illusive? Cell 115:377–388

Narasimha Chary S, Bultema RL, Packard CE, Crowell DN (2002) Prenylcysteine alpha-carboxyl methyltransferase expression and function in *Arabidopsis thaliana*. Plant J 32:735–747

Nibau C, Wu HM, Cheung AY (2006) RAC/ROP GTPases: 'hubs' for signal integration and diversification in plants. Trends Plant Sci 11:309–315

Nishikimi A, Fukuhara H, Su W, Hongu T, Takasuga S, Mihara H, Cao Q, Sanematsu F, Kanai M, Hasegawa H et al (2009) Sequential regulation of DOCK2 dynamics by two phospholipids during neutrophil chemotaxis. Science 324:384–387

Orlando K, Zhang J, Zhang X, Yue P, Chiang T, Bi E, Guo W (2008) Regulation of GIC2 localization and function by pi(4, 5)p2 during the establishment of cell polarity in budding yeast. J Biol Chem 283:14205–14212

Papayannopoulos V, Co C, Prehoda KE, Snapper S, Taunton J, Lim WA (2005) A polybasic motif allows N-WASP to act as a sensor of PIP(2) density. Mol Cell 17:181–191

Parton RG, Hancock JF (2004) Lipid rafts and plasma membrane microorganization: insights from Ras. Trends Cell Biol 14:141–147

Perez-Sala D (2007) Protein isoprenylation in biology and disease: general overview and perspectives from studies with genetically engineered animals. Front Biosci 12:4456–4472

Pichler H, Riezman H (2004) Where sterols are required for endocytosis. Biochim Biophys Acta 1666:51–61

Pierre M, Traverso JA, Boisson B, Domenichini S, Bouchez D, Giglione C, Meinnel T (2007) N-myristoylation regulates the SnRK1 pathway in Arabidopsis. Plant Cell 19:2804–2821

Politis EG, Roth AF, Davis NG (2005) Transmembrane topology of the protein palmitoyl transferase Akr1. J Biol Chem 280:10156–10163

Prior IA, Harding A, Yan J, Sluimer J, Parton RG, Hancock JF (2001) GTP-dependent segregation of H-ras from lipid rafts is required for biological activity. Nat Cell Biol 3:368–375

Reid TS, Terry KL, Casey PJ, Beese LS (2004) Crystallographic analysis of CaaX prenyltransferases complexed with substrates defines rules of protein substrate selectivity. J Mol Biol 343:417–433

Resh MD (2004) Membrane targeting of lipid modified signal transduction proteins. Subcell Biochem 37:217–232

Ridley AJ (2006) Rho GTPases and actin dynamics in membrane protrusions and vesicle trafficking. Trends Cell Biol 16:522–529

Rocks O, Peyker A, Kahms M, Verveer PJ, Koerner C, Lumbierres M, Kuhlmann J, Waldmann H, Wittinghofer A, Bastiaens PI (2005) An acylation cycle regulates localization and activity of palmitoylated Ras isoforms. Science 307:1746–1752

Rodriguez-Concepcion M, Boronat A (2002) Elucidation of the methylerythritol phosphate pathway for isoprenoid biosynthesis in bacteria and plastids. A metabolic milestone achieved through genomics. Plant Physiol 130:1079–1089

Rodriguez-Concepcion M, Toledo-Ortiz G, Yalovsky S, Caldelari D, Gruissem W (2000) Carboxyl-methylation of prenylated calmodulin CaM53 is required for efficient plasma membrane targeting of the protein. Plant J 24:775–784

Rodriguez-Concepcion M, Yalovsky S, Zik M, Fromm H, Gruissem W (1999) The prenylation status of a novel plant calmodulin directs plasma membrane or nuclear localization of the protein. EMBO J 18:1996–2007

Rohmer M (1999) The discovery of a mevalonate-independent pathway for isoprenoid biosynthesis in bacteria, algae and higher plants. Nat Prod Rep 16:565–574

Rotblat B, Prior IA, Muncke C, Parton RG, Kloog Y, Henis YI, Hancock JF (2004) Three separable domains regulate GTP-dependent association of H-ras with the plasma membrane. Mol Cell Biol 24:6799–6810

Roth AF, Feng Y, Chen L, Davis NG (2002) The yeast DHHC cysteine-rich domain protein Akr1p is a palmitoyl transferase. J Cell Biol 159:23–28

Running MP, Lavy M, Sternberg H, Galichet A, Gruissem W, Hake S, Ori N, Yalovsky S (2004) Enlarged meristems and delayed growth in plp mutants result from lack of CaaX prenyltransferases. Proc Natl Acad Sci USA 101:7815–7820

Scheffzek K, Stephan I, Jensen ON, Illenberger D, Gierschik P (2000) The Rac-RhoGDI complex and the structural basis for the regulation of Rho proteins by RhoGDI. Nat Struct Biol 7:122–126

Shahinian S, Silvius JR (1995) Doubly-lipid-modified protein sequence motifs exhibit long-lived anchorage to lipid bilayer membranes. Biochemistry 34:3813–3822

Smotrys JE, Linder ME (2004) Palmitoylation of intracellular signaling proteins: regulation and function. Annu Rev Biochem 73:559–587

Sorek N, Poraty L, Sternberg H, Bar E, Lewinsohn E, Yalovsky S (2007) Activation status-coupled transient S acylation determines membrane partitioning of a plant Rho-related GTPase. Mol Cell Biol 27:2144–2154

Sun JP, Luo Y, Yu X, Wang WQ, Zhou B, Liang F, Zhang ZY (2007) Phosphatase activity, trimerization, and the C-terminal polybasic region are all required for PRL1-mediated cell growth and migration. J Biol Chem 282:29043–29051

Swarthout JT, Lobo S, Farh L, Croke MR, Greentree WK, Deschenes RJ, Linder ME (2005) DHHC9 and GCP16 constitute a human protein fatty acyltransferase with specificity for H- and N-Ras. J Biol Chem 280:31141–31148

Tao LZ, Cheung AY, Wu HM (2002) Plant Rac-like GTPases are activated by auxin and mediate auxin-responsive gene expression. Plant Cell 14:2745–2760

Tarahovsky YS, Muzafarov EN, Kim YA (2008) Rafts making and rafts braking: how plant flavonoids may control membrane heterogeneity. Mol Cell Biochem 314:65–71

Tian T, Harding A, Inder K, Plowman S, Parton RG, Hancock JF (2007) Plasma membrane nanoswitches generate high-fidelity Ras signal transduction. Nat Cell Biol 9:905–914

Trueblood CE, Boyartchuk VL, Rine J (1997) Substrate specificity determinants in the farnesyltransferase beta-subunit. Proc Natl Acad Sci USA 94:10774–10779

Trueblood CE, Ohya Y, Rine J (1993) Genetic evidence for in vivo cross-specificity of the CaaX box protein prenyltransferases farnesyltransferase and geranylgeranyltransferase-I in *Saccharomyces cerevisiae*. Mol Cell Biol 13:4260–4275

Winge P, Brembu T, Bones AM (1997) Cloning and characterization of rac-like cDNAs from *Arabidopsis thaliana*. Plant Mol Biol 35:483–495

Yalovsky S, Trueblood CE, Callan KL, Narita JO, Jenkins SM, Rine J, Gruissem W (1997) Plant farnesyltransferase can restore yeast Ras signaling and mating. Mol Cell Biol 17:1986–1994

Yang Z (2002) Small GTPases: versatile signaling switches in plants. Plant Cell 14:S375–S388

Yang Z, Fu Y (2007) ROP/RAC GTPase signaling. Curr Opin Plant Biol 10:490–494

Young SG, Ambroziak P, Kim E, Clarke S (2000) Postprenylation protein processing: CXXX (CaaX) endoproteases and isoprenylcyteine carboxyl methyltransferase. In: Tamanoi F, Sigman DS (eds) Protein lipidation. San Diego, Academic Press, pp 156–213

Zeng Q, Wang X, Running MP (2007) Dual lipid modification of Arabidopsis Ggamma-subunits is required for efficient plasma membrane targeting. Plant Physiol 143:1119–1131

ROP GTPases and the Cytoskeleton

Ying Fu

Abstract Plant Rho-family GTPases, named ROPs (Rho-like GTPases from plants), like their counterparts in fungi and animals, have the conserved function in the regulation of the cytoskeleton, a fundamental dynamic cellular structure important for the regulation of plant growth, development, and responses to the environment. In the regulation of the cytoskeleton, ROP signaling pathways involve specific regulators and effectors that are unique to plants, in addition to conserved components that are shared with their counterparts in other organisms. This chapter summarizes the current knowledge of ROP signaling networks that regulate actin filaments and microtubules, with an emphasis on recent reports of conserved and/or plant-specific signaling pathways.

1 Introduction

The cytoskeleton is a three-dimensional filamentous protein network that provides a framework for cellular organization. In plants, the cytoskeleton is composed of at least two major systems: actin filaments (AFs or F-actin) and microtubules (MTs). AFs and MTs are highly dynamic structures and are both regulated in time and space to fulfill a vast variety of cellular functions that are linked to cell division, cell growth, cell shape formation, and cellular responses to symbiotic association, pathogen invasion, or abiotic stresses. Therefore, the cytoskeleton undergoes dynamic changes and reorganization in responses to various developmental and environmental cues. The dynamics and reorganization of the cytoskeleton are known to depend on a battery of cytoskeleton-associated proteins, which are believed to serve as important effectors of signaling cascades. How these cytoskeleton-associated proteins are

Y. Fu
State Key Laboratory of Plant Physiology and Biochemistry, College of Biological Sciences, China Agricultural University, Beijing, 100193, People's Republic of China
e-mail: yingfu@cau.edu.cn

S. Yalovsky et al. (eds.), *Integrated G Proteins Signaling in Plants*, Signaling and Communication in Plants,
DOI 10.1007/978-3-642-03524-1_5,

regulated by the developmental and environmental cues is poorly understood, but recent evidence has demonstrated an important role for the conserved Rho-family GTPases and MAP kinase cascades in the regulation of cytoskeleton signaling (reviewed in Fu et al. 2008; Hussey and Hashimoto 2008). This chapter focuses on cytoskeleton signaling mediated by the monomeric Rho-family G proteins.

Monomeric G proteins (also known as small GTPases) in the Ras superfamily cycle between the GDP-bound "OFF" and the GTP-bound "ON" states. Among five families belonging to the Ras superfamily (Ras, Rho, Rab, Arf, and Ran), Rho and Ras families are considered the key signaling molecules that are directly involved in relaying extracellular signals and regulating many important cellular processes (Hall 1998; Brembu et al. 2006; Berken and Wittinghofer 2008). Signaling small GTPases in fungi and animal systems often work coordinately to control a specific cellular process. For example, polarity development in the budding yeast can be divided into three steps: the polar site selection, the polarity establishment, and the polarity maintenance; each step is regulated by Ras, Cdc42, and Rho GTPase, respectively (Fu and Yang 2001). Similarly, studies in mammalian cells reveal a general model for the regulation of cell migration: Cdc42 orients cell upon signals, Rac induces the formation of the leading edge at the front, and Rho controls retraction in the rear (Ridley et al. 2003; Raftopoulou and Hall 2004). In plants, however, neither Ras GTPase homologs nor CDC42, RAC, and RHO subfamilies of the conserved Rho-family small GTPases have been found, but a sole subfamily of signaling small GTPases, ROP (Rho-related GTPase from plants, also termed as RAC) is present throughout the plant kingdom. Phylogenetic analysis suggests that ROP is distinct from the other three subfamilies of Rho GTPases and is specific to plants (Zheng and Yang 2000; Vernoud et al. 2003). Although sharing conserved structural features, regulators, effectors with their counterparts in other systems, ROP GTPases contain unique structural domains and have specific regulators and functional partners (Zheng and Yang 2000; Vernoud et al. 2003; Berken 2006; Brembu et al. 2006; Yang and Fu 2007; Berken and Wittinghofer 2008; Fu et al. 2008). The ROP GTPase subfamily has been recognized to be the key signaling components that orchestrate a wide range of signaling pathways to control plant development and plant innate immunity. Several aspects of ROP signaling with regards to the structure, function, and regulation of the "ON/OFF" status of ROP GTPases are covered in recent reviews (Berken 2006; Brembu et al. 2006; Yang and Fu 2007; Berken and Wittinghofer 2008; Fu et al. 2008; Kost 2008; Yalovsky et al. 2008) (also see Chapter "Regulatory and Cellular Functions of Plant RhoGAPs and RhoGDIs" and "Structure and Function of ROPs and Their GEFs"). This chapter focuses on recent insights into ROP regulation of cytoskeleton reorganization in plants.

2 Regulation of AFs

AFs are made of an abundant intracellular protein, actin, which is highly conserved in eukaryotic cells. AFs can be reversely assembled into many different types of structures with distinct functions. For example, in fibroblasts, actin can be assembled into stress fibers in the focal adhesion regions and a meshwork at the cell periphery in the leading edge. Spike-like protrusions called filopodia are rich in AFs as well (Hall 1998). The shape formation of fibroblasts relies on the actin meshwork to produce membrane protrusions (lamellipodia and filopodia) in the front and contractile stress fibers to pull cell body up in the rear. The organization and remodeling of the actin meshwork is mediated by Rho-family GTPase signaling. In higher plants, actin structures include nuclear baskets, subcortical bundles, cortical networks (either random or more organized) (McCurdy and Staiger 2000). Recently, fine AFs are reported to associate with polar growth sites in tip-growing cells (pollen tubes and root hairs) (Fu et al. 2001; Carol and Dolan 2002; Jones et al. 2002; Gu et al. 2005; Lee et al. 2008), as well as in cells undergoing diffuse growth, like trichome and pavement cells (Frank and Smith 2002; Fu et al. 2002, 2005; Frank et al. 2003). These fine AFs associated with the polar cortical site are also dependent upon the plant Rho-family GTPases, ROPs. Although cell shape formation in plants is intimately linked to cell wall patterns due to the presence of turgor pressure, this process is governed by a unifying mechanism: Rho GTPase signaling regulates cell polarity at least in part through polarized cortical AFs (Raftopoulou and Hall 2004; Bannigan and Baskin 2005; Szymanski 2005; Berken 2006; Brembu et al. 2006; La Carbona et al. 2006; Moseley and Goode 2006; Park and Bi 2007; Yang and Fu 2007; Fu et al. 2008; Iden and Collard 2008; Li and Gundersen 2008).

2.1 *Conserved Rho GTPase Downstream Pathways in the Regulation of AFs*

Actin nucleation machinery/factor is a target of Rho GTPase signaling in the regulation of the actin cytokeleton in fungi and animals (Li and Yang 2000; Ridley et al. 2003; Lodish et al. 2007) (Fig. 1). Two proteins for actin nucleation are conserved in plants: the ARP2/3 complex and formins. ARP2/3 is an actin-nucleation complex that associates with the side of existing AFs to generate a new filament branching out at an angle of 70°(relative to the old AFs). In the front of animal cells, the active form of CDC42 activates the ARP2/3 complex through the regulatory protein WASP to produce filopodia formation, whereas active RAC (which is colocalized with CDC42) activates the ARP2/3 complex through WAVE complex (Eden et al. 2002; Pollard and Borisy 2003; Disanza et al. 2005; Lodish et al. 2007). In plants, conserved homologs of subunits of ARP2/3 and WAVE complexes have been identified (Brembu et al. 2005; Deeks et al. 2005). Phenotypical analysis on mutants of *Arabidopsis* ARP2/3 and WAVE complex subunits reveals highly resembled defects in cell morphology of different epidermal cell types. *Arabidopsis* trichomes are branched single cells protruding from leaf epidermis. In the above mutants, striking distorted trichome phenotypes are displayed

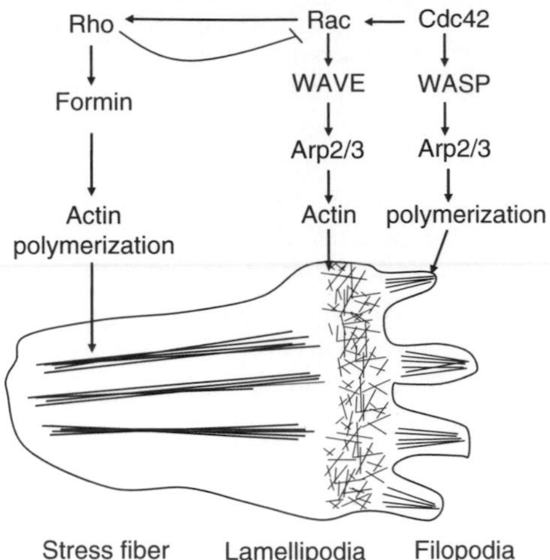

Fig. 1 Coordination of RHO, RAC, and CDC42 in regulating actin cytoskeleton in a fibroblast. In the leading edge of fibroblast, activation of CDC42 induces formation of AFs in filopodia through WASP-regulated ARP2/3 complex. RAC can be activated by CDC42 and induces AFs in lamelipodia through WAVE-regulated ARP2/3 complex. RAC activates RHO to promote the stress fibers formation in the rear, through formin-based actin polymerization. RHO, in turn, inhibits RAC activity

including underdeveloped or twisted branches, swollen stalk, as well as abnormal actin organization (reviewed by Szymanski 2005).

Mutations in ARP2/3 and WAVE complexes also alter the morphology of interlocking leaf pavement cells, although discrepancies exist in the reports of different groups, regarding the severity of pavement cell defects (Le et al. 2003; Li et al. 2003; Mathur et al. 2003a, b; Brembu et al. 2004). Interestingly, the pavement cell phenotype in these mutants exhibits some similarities to that in ROP2 and ROP4 loss-of-function mutants (Fu et al. 2002, 2005), hinting a possible role for ROP2 and ROP4 in the regulation of the WAVE complex leading to the activation of the ARP2/3 complex. Evidence for direct interaction between ROP2 and WAVE complex subunits has been controversial. Szymanski's group detected interaction between ROP2 and PIR121/SRA1/KLK in their yeast two-hybrid assay (Basu et al. 2004). In contrast, when Uhrig and colleagues performed yeast two-hybrid assay as well as Bimolecular Fluorescence Complementation (BiFC) to analyze protein–protein interaction between components of WAVE and ARP2/3 complexes and ROP GTPases, none of the tested five ROPs (ROP2, ROP5, ROP7, ROP8, and ROP11) interacted with PIR121/SRA1/KLK (Uhrig et al. 2007). The controversy may be due to different experimental setups (such as vectors, yeast strains, and assay conditions) in two separate studies.

Other studies suggest that ROP signaling may regulate the ARP2/3 and WAVE complexes but apparently by a different mechanism from the RAC activation of the WAVE complex in animals. SPK1 is the only DOCK family guanine nucleotide exchange factor (GEF) found in plants (Schmidt and Hall 2002; Yang and Fu 2007; Fu et al. 2008). It associates with the nucleotide-depleted or the GDP-bound forms of ROP2 and itself is sufficient for GEF activity, which is different from other DOCKs, such as Dock180 and Zizimin (Brugnera et al. 2002; Meller et al. 2002; Cote and Vuori 2007; Basu et al. 2008). Loss of function in *spk1* mutants exhibits several phenotypes similar to *wave* and *arp2/3* mutants, including defective cell morphology of trichomes, pavement cells, and hypocotyls as well as cell–cell adhesion in epidermis. Genetic analysis indicated that SPK1, ROP, and SRA1 function in a common pathway to regulate actin-dependent growth (Qiu et al. 2002; Szymanski 2005; Basu et al. 2008). In addition to ROP2, other ROPs may be needed to activate WAVE-ARP2/3-AFs pathway, since *CA-rop2, spk1* displayed additive phenotypes compared with single *spk1* or *CA-rop2* mutants. This hypothesis is supported by the fact that SPK1 binds to ROP3, ROP4, ROP5, and ROP10 in the nucleotide-depleted or the GDP-bound forms and that SPK1 can activate ROP3, ROP4, and ROP6 (Basu et al. 2008). Furthermore, Hulskamp's group found that SCAR (another WAVE complex component) interacted with ROP5, ROP7, ROP8, and ROP11, despite no interactions between SCARs and other Rho-family GTPases have been reported (Uhrig et al. 2007). Evidence also indicates that SPK1 can interact with NAP1 (a component of WAVE complex) as well. It has been suggested that GEF binding to Rho GTPases downstream effectors may provide a general scaffolding strategy for the regulation of signaling specificity (Schmidt and Hall 2002; Basu et al. 2008).

2.2 *Plant-Specific Players in the ROP-Dependent Regulation of AFs*

Subtle phenotypes caused by mutations in subunits of the ARP2/3 and WAVE complexes in plants are surprising, because similar mutations result in severe, sometimes lethal, phenotypes in other organisms like yeast, *Drosophila* and *C. elegans* (Morrell et al. 1999; Winter et al. 1999; Hudson and Cooley 2002; Sawa et al. 2003). This indicates that the WAVE-ARP2/3 pathway is important for cell morphogenesis in selected plant cell types such as trichomes and pavement cells, but is essential neither for tip-growing pollen tubes and root hairs nor for overall plant growth and development (reviewed by Bannigan and Baskin 2005; Mathur 2005; Brembu et al. 2006). Since ROP-dependent fine AFs are required for tip growth, it is likely that plant-specific ROP downstream pathways are required for the regulation of AFs in plants.

RIC4 belongs to a plant-specific family of ROP effectors, termed RICs (ROP-interactive CRIB motif containing proteins) (Wu et al. 2001). RIC4 was found to directly interact with ROP1 and ROP2. Overexpression of RIC4 has been shown to promote the formation of fine AFs in both pollen tubes and pavement cells (Fu et al. 2005; Gu et al. 2005). Knocking down *RIC4* mRNA levels depleted tip AFs and inhibited tip growth in pollen tubes, as did ROP1 inactivation (Fu et al. 2001; Gu et al. 2005; Hwang et al. 2005). Similarly, shallower lobes as well as fewer fine AFs were detected in pavement cells from both *ric4* and *rop2* loss of function mutants (Fu et al. 2001, 2005). These indicate that RIC4 is an effector of ROP1 and ROP2 that activates the assembly of cortical AFs. RIC4-dependent fine AFs are required for tip growth in pollen tubes as well as lobe outgrowth in pavement cells. However, the precise mechanism by which RIC4 regulates AFs is still unclear. The involvement of formins, another actin nucleator, has been proposed. In animals and yeast, diaphanous-related formins are Rho effectors containing a GTPase binding domain (GBD). Upon binding to Rho, formins stimulate assembly of unbranched AFs (Deeks et al. 2002; Wallar and Alberts 2003; Zigmond 2004). Homologs of formins have been found in plants. Overexpression of the *Arabidopsis* formin AFH1 in pollen tube triggered abnormal actin cables and swollen tubes similar to those induced by overexpression of wild type or active form of ROP (Cheung and Wu 2004), which suggests that ROP and formins may act in the same pathway. However, plant formins do not contain a GBD, and there is no evidence for a direct link between formin and ROP (Deeks et al. 2002; Cheung and Wu 2004; Berken 2006). RIC4 (like other RICs) contains a conserved CRIB motif (Wu et al. 2001), which is a GBD found in CDC42 and RAC effectors (Burbelo et al. 1995). ROP regulation of formins may indirectly through the plant-specific RIC4 to assembly fine AFs associated with tip growth in pollen tubes and lobe outgrowth in pavement cells. Nevertheless, it can not be excluded that formins are subject to ROP regulation via another unknown novel effector.

The RIC4-mediated signaling pathway is not the sole player in the ROP control of polarized cell growth in plants. RIC4 coordinates with RIC1-mediated MT regulation in the pavement cell system (see *below*), as well as RIC3-mediated Ca^{2+} gradient regulation in the pollen tube system (Fu et al. 2005; Gu et al. 2005). A recent report from Yang's lab demonstrates that the RIC4-promoted actin assembly was responsible for vesicle accumulation at the pollen tube tip, whereas RIC3-promoted AFs disassembly (through tip-focused Ca^{2+} gradient) was responsible for exocytosis to the tip (Hwang et al. 2005; Lee et al. 2008). These two pathways counteract with each other to coordinate targeted vesicle accumulation and exocytosis. The spatiotemporal coordination is achieved by the oscillation of the apical ROP1 activity (Hwang et al. 2005; Lee et al. 2008). In addition to the ROP-dependent actin dynamics, another ROP-dependent pathway that regulates exocytosis has been reported by Lavy et al. (2007). This pathway is mediated by ICR1 (Interactor of Constitutive Active ROPs 1), a novel ROP effector that binds to the active form of ROP GTPases. The interaction between ICR1 and SEC3 (an exocyst vesicle tethering complex subunit) has been demonstrated using yeast two-hybrid assay and BiFC system. Deformation of pavement cells and root hairs can be seen in both ICR1 loss-of-function mutants and gain-of-function mutants. Together these results support a coordinative regulation of exocytosis for polarized growth in pavement cells and root hairs: ROP-dependent vesicle targeting (mediated by RIC4-regulated actin dynamics) is coordinated with the ROP-dependent vesicle tethering (mediated by ICR1 and SEC3) (Fu et al. 2005; Lavy et al. 2007).

Besides RIC4, ROP was found to associate with phosphatidylinositol monophosphate kinase (PIPK) activity (Kost et al. 1999; Helling et al. 2006). The PIPK product, phosphatidyl inositol 4,5-bisphosphate (PI 4,5-P_2), was proposed to be a ROP effector and fulfill multiple important functions in tip-growing pollen tubes and root hairs. PI 4,5-P_2 was believed to regulate actin organization through modulating the activity of actin-binding proteins and subsequently the exocytosis. PI 4,5-P_2 may also mediate ROP-controlled vesicle fusion (exocytosis) through recruiting exocyst as well. Furthermore, PI 4,5-P_2 serves as a substrate of phospholipase C (PLC), and its product inositol 1,4,5-triphosphate (IP3) affects Ca^{2+} gradient. Last, PI 4,5-P_2 was known to modulate the fission of endocytic vesicles at the apical plasma membrane in pollen tubes (reviewed by Kost 2008; Yalovsky 2008; Berken 2006). For further description of the roles of ROP on vesicle trafficking and lipid modifications, see Chap. "ROPs, Vesicle Trafficking and Lipid Modifications."

3 Regulation of Microtubules

Cortical MTs in plants are highly dynamic structures that are associated with the plasma membrane. Nucleation of MT occurs at disperse sites throughout the cortex in a MT-dependent way (Yuan et al. 1994; Shaw et al. 2003; Dixit and Cyr 2004; Murata et al. 2005; Dixit et al. 2006; Murata and Hasebe 2007). Organization of cortical MT array is regulated by self-organization and sustained treadmilling as well as by rotary movement (Shaw et al. 2003; Dixit and Cyr 2004; Dixit et al. 2006; Chan et al. 2007), but the signaling leading to MT ordering is not understood.

One of the RIC proteins, RIC1 was found to be a novel MT-associated protein (MAP). RIC1 promotes well-organized transverse MTs to restrict cell expansion (Fu et al. 2005). RIC1 has been demonstrated to directly interact with both ROP1 and ROP2 (Wu et al. 2001; Fu et al. 2005). It is interesting that RIC1 is colocalized with ROP1 in pollen tubes (Wu et al. 2001), but in pavement cells, RIC1 binds to cortical MTs. Overexpression of RIC1 promotes well-ordered MTs and restricts outgrowth of lobes, whereas knocking out of RIC1 leads to random-orientated MT cortical array and wider indented neck region (Fu et al. 2005). On the other hand, RIC1-mediated ordered MTs reduce fine AFs formation in pavement cells. But active ROP2 at the lobe tip inhibits RIC1 to associate with MTs, and subsequently excludes RIC1 activity from outgrowing lobe tip (Fu et al. 2005). To identify the activator of RIC1, available *Arabidopsis rop* knockout mutants were analyzed for pavement cell-shape phenotype. A *rop6* loss-of-function mutant exhibited a phenotype similar to that of the *ric1-1* knockout mutant, suggesting that ROP6 may activate RIC1 to promote MT ordering (Fu et al., unpublished data). New evidence also demonstrates that RIC1-MT may control the elongation of hypocotyl epidermal cells (cells that undergo simple diffuse growth) as well, since knocking out RIC1 increased lateral expansion of these cells (Fu et al., unpublished data). Further investigation of RIC1-mediated signaling network in the regulation of MT ordering may contribute to our knowledge of plant cell morphogenesis and development.

4 Crosstalk Between AFs, MTs, and ROPs

AFs and MTs are usually organized into distinct cortical arrays in plant cells. However, crosstalk between AFs and MTs has been observed in various cell types. In some cases, AFs and MTs are colocalized (Andersland et al. 1998) or localized close parallel to each other (Ridge 1988). Pharmacological disruption of one filament system often has dramatic effects on the other filament system (Seagull 1990; Fu et al. 2005). Mutants that have defects in MTs organization such as *dv*, *ms17*, *mor1*, *spk1*, and *ric1* display altered AFs as well (Staiger and Cande 1990, 1991; Whittington et al. 2001; Qiu et al. 2002; Fu et al. 2005). It has been reported that AFs and MTs are functionally and physically linked by proteins that interact with both AFs and MTs. These include a tobacco 190-kD polypeptide, a potato protein SB401, a cotton CH domain-containing Kinesin GhKCH1, and a ARM domain-containing Kinesin MRH2 (Igarashi et al. 2000; Preuss et al. 2004; Huang et al. 2007; Yang et al. 2007). Recent evidences suggest that ROP signaling may provide a mechanism for cross talk between AFs and MTs, just like other Rho family members do in yeast and animal cells (Etienne-Manneville 2004; Rosales-Nieves et al. 2006; Minc et al. 2009).

As discussed above, overexpression of RIC1 (a MAP) in pavement cells suppresses ROP2 activity and, subsequently, inhibits the accumulation of cortical fine AFs by its promoting well-ordered MTs. On the other hand, ROP2 activity inhibits RIC1 function by interrupting its association with MTs (Fu et al. 2005). It has been well known that ROP GTPase-mediated AFs dynamics is critical for pollen tube tip growth (Kost et al. 1999; Fu et al. 2001; Gu et al. 2005; Hwang et al. 2005; Lee et al. 2008), and apical fine AFs are suggested to participate in a positive feedback loop to activate ROP GTPase at the pollen tube tip (Hwang et al. 2005). Although treatment with MT disruption drug has little effect on pollen germination or elongation, increasing evidence implicates MTs in pollen tube tip growth. It is believed that subapical cortical MTs may suppress ROP1 by lateral inhibition (Yang 2008). In addition, it was recently reported that SB401, a pollen-specific protein in potato, can bind and bundle both MTs and AFs, which suggests a potential coordination of MTs and AFs in controlling pollen tube tip growth (Huang et al. 2007).

Tip-growing root hair is another model system to study cell polarity and cell morphogenesis. ROP-dependent cytoskeleton dynamics, Ca^{2+} gradient, and membrane cycling are essential for root hair polarized growth as well (Molendijk et al. 2001; Jones et al. 2002; Bloch et al. 2005; Xu and Scheres 2005). A recent study from Zheng's group reported that a plant-specific Armadillo (ARM) domain-containing gene *MRH2* encodes a putative kinesin MRH2 that linked MTs and AFs together. A mutation in *MRH2* enhanced swollen root hair phenotype in *CA-rop2* background, as well as induced significant MT fragmentation and random orientation in both wild type plants and *CA-rop2* mutants. Interestingly, in vitro cosedimentation assay suggests that a fragment of MRH2-containing ARM domain binds AFs, whereas the fragment containing the motor domain could bind to MTs. *mrh2-3* root hairs also showed enhanced sensitivity to LatB, an actin disrupting drug

(Yang et al. 2007). On the basis of these observations, the role of ROP signaling to both AFs-dependent polar growth and cortical MT-dependent polarity determination was proposed. The crosstalk between these two signaling pathways is a unifying mechanism that underscores polarized cell expansion in plants (Yang 2008).

5 Conclusion and Perspectives

Important progresses have been made in recent studies of ROP signaling to the cytoskeleton. ROP GTPase-mediated cytoskeleton dynamics and its interaction with vesicular trafficking are the cores of polarity establishment and maintenance during polar cell growth (Fig. 2). However, many questions about ROP-mediated signaling networks in the control of the cytoskeleton remain to be answered. For example, the ROP–RIC1–MTs pathway is the first well-characterized signaling pathway that regulates the ordering of cortical MTs, yet it is still a mystery how RIC1 regulates MT dynamics. RIC1 is localized to different subcellular compartment in different cell types, suggesting the existence of cell type-specific RIC1 partners and signaling pathways, which are yet to be elucidated. A major challenge is the identification of developmental signals that regulate ROP-dependent cytoskeleton signaling pathways. The mechanism underlying the crosstalk between

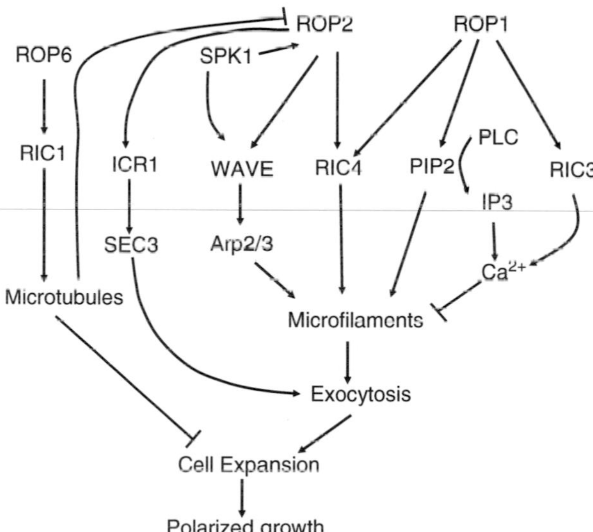

Fig. 2 ROPs signaling network controls polarized cell growth in *Arabidopsis*. ROP proteins coordinately regulate cytoskeleton dynamics, Ca^{2+} gradient, and membrane cycling, and eventually control polarized cell growth. Different ROP member may function redundantly like ROP2 and ROP4, or specifically acts in one particular cell type like ROP1 in pollen tube. It is very interesting that ROP6, which is closely related to ROP2 and ROP4, works antagonistically with ROP2 and ROP4

AFs, MTs, and ROP activity is another major future challenge. Investigation of new components in the ROP signaling networks will help to elucidate the relationship between ROP GTPases and the cytoskeleton during plant development and morphogenesis. In addition, cytoskeleton remodeling during plant immune responses is also mediated by ROP signaling (Schultheiss et al. 2005); however, the signaling pathway is still not clear. Finally, it would also be interesting to see whether ROP-cytoskeleton signaling networks mediate abiotic stresses.

References

Andersland JM, Dixon DC, Seagull RW, Triplett BA (1998) Isolation and characterization of cytoskeletons from cotton fiber cytoplasts. In Vitro Cell Dev Biol Plant 34:173–180

Bannigan A, Baskin TI (2005) Directional cell expansion–turning toward actin. Curr Opin Plant Biol 8:619–624

Basu D, El-Assal Sel D, Le J, Mallery EL, Szymanski DB (2004) Interchangeable functions of *Arabidopsis* PIROGI and the human WAVE complex subunit SRA1 during leaf epidermal development. Development 131:4345–4355

Basu D, Le J, Zakharova T, Mallery EL, Szymanski DB (2008) A SPIKE1 signaling complex controls actin-dependent cell morphogenesis through the heteromeric WAVE and ARP2/3 complexes. Proc Natl Acad Sci USA 105:4044–4049

Berken A (2006) ROPs in the spotlight of plant signal transduction. Cell Mol Life Sci 63: 2446–2459

Berken A, Wittinghofer A (2008) Structure and function of Rho-type molecular switches in plants. Plant Physiol Biochem 46:380–393

Bloch D, Lavy M, Efrat Y, Efroni I, Bracha-Drori K, Abu-Abied M, Sadot E, Yalovsky S (2005) Ectopic expression of an activated RAC in *Arabidopsis* disrupts membrane cycling. Mol Biol Cell 16:1913–1927

Brembu T, Winge P, Bones AM (2005) Catching the WAVEs of plant actin regulation. J Plant Growth Regul 24:55–66

Brembu T, Winge P, Bones AM, Yang Z (2006) A RHOse by any other name: a comparative analysis of animal and plant Rho GTPases. Cell Res 16:435–445

Brembu T, Winge P, Seem M, Bones AM (2004) NAPP and PIRP encode subunits of a putative wave regulatory protein complex involved in plant cell morphogenesis. Plant Cell 16:2335–2349

Brugnera E, Haney L, Grimsley C, Lu M, Walk SF, Tosello-Trampont AC, Macara IG, Madhani H, Fink GR, Ravichandran KS (2002) Unconventional Rac-GEF activity is mediated through the Dock180-ELMO complex. Nat Cell Biol 4:574–582

Burbelo PD, Drechsel D, Hall A (1995) A conserved binding motif defines numerous candidate target proteins for both Cdc42 and Rac GTPases. J Biol Chem 270:29071–29074

Carol RJ, Dolan L (2002) Building a hair: tip growth in *Arabidopsis thaliana* root hairs. Philos Trans R Soc Lond B Biol Sci 357:815–821

Chan J, Calder G, Fox S, Lloyd C (2007) Cortical microtubule arrays undergo rotary movements in *Arabidopsis* hypocotyl epidermal cells. Nat Cell Biol 9:171–175

Cheung AY, Wu HM (2004) Overexpression of an *Arabidopsis* formin stimulates supernumerary actin cable formation from pollen tube cell membrane. Plant Cell 16:257–269

Cote JF, Vuori K (2007) GEF what? Dock180 and related proteins help Rac to polarize cells in new ways. Trends Cell Biol 17:383–393

Deeks MJ, Cvrckova F, Machesky LM, Mikitova V, Ketelaar T, Zarsky V, Davies B, Hussey PJ (2005) *Arabidopsis* group Ie formins localize to specific cell membrane domains, interact with actin-binding proteins and cause defects in cell expansion upon aberrant expression. New Phytol 168:529–540

Deeks MJ, Hussey PJ, Davies B (2002) Formins: intermediates in signal-transduction cascades that affect cytoskeletal reorganization. Trends Plant Sci 7:492–498

Disanza A, Steffen A, Hertzog M, Frittoli E, Rottner K, Scita G (2005) Actin polymerization machinery: the finish line of signaling networks, the starting point of cellular movement. Cell Mol Life Sci 62:955–970

Dixit R, Chang E, Cyr R (2006) Establishment of polarity during organization of the acentrosomal plant cortical microtubule array. Mol Biol Cell 17:1298–1305

Dixit R, Cyr R (2004) Encounters between dynamic cortical microtubules promote ordering of the cortical array through angle-dependent modifications of microtubule behavior. Plant Cell 16:3274–3284

Eden S, Rohatgi R, Podtelejnikov AV, Mann M, Kirschner MW (2002) Mechanism of regulation of WAVE1-induced actin nucleation by Rac1 and Nck. Nature 418:790–793

Etienne-Manneville S (2004) Actin and microtubules in cell motility: which one is in control? Traffic 5:470–477

Frank MJ, Cartwright HN, Smith LG (2003) Three Brick genes have distinct functions in a common pathway promoting polarized cell division and cell morphogenesis in the maize leaf epidermis. Development 130:753–762

Frank MJ, Smith LG (2002) A small, novel protein highly conserved in plants and animals promotes the polarized growth and division of maize leaf epidermal cells. Curr Biol 12:849–853

Fu Y, Gu Y, Zheng Z, Wasteneys G, Yang Z (2005) *Arabidopsis* interdigitating cell growth requires two antagonistic pathways with opposing action on cell morphogenesis. Cell 120: 687–700

Fu Y, Kawasaki T, Shimamoto K, Yang Z (2008) ROP/RAC GTPases. In: Yang Z (ed) Intracellular signaling in plants, annual plant reviews, vol 33. Wiley-Blackwell, Hoboken, pp 64–99

Fu Y, Li H, Yang Z (2002) The ROP2 GTPase controls the formation of cortical fine F-actin and the early phase of directional cell expansion during *Arabidopsis* organogenesis. Plant Cell 14:777–794

Fu Y, Wu G, Yang Z (2001) Rop GTPase-dependent dynamics of tip-localized F-actin controls tip growth in pollen tubes. J Cell Biol 152:1019–1032

Fu Y, Yang Z (2001) Rop GTPase: a master switch of cell polarity development in plants. Trends Plant Sci 6:545–547

Gu Y, Fu Y, Dowd P, Li S, Vernoud V, Gilroy S, Yang Z (2005) A Rho family GTPase controls actin dynamics and tip growth via two counteracting downstream pathways in pollen tubes. J Cell Biol 169:127–138

Hall A (1998) Rho GTPases and the actin cytoskeleton. Science 279:509–514

Helling D, Possart A, Cottier S, Klahre U, Kost B (2006) Pollen tube tip growth depends on plasma membrane polarization mediated by tobacco PLC3 activity and endocytic membrane recycling. Plant Cell 18:3519–3534

Huang S, Jin L, Du J, Li H, Zhao Q, Ou G, Ao G, Yuan M (2007) SB401, a pollen-specific protein from *Solanum berthaultii*, binds to and bundles microtubules and F-actin. Plant J 51:406–418

Hudson AM, Cooley L (2002) A subset of dynamic actin rearrangements in Drosophila requires the Arp2/3 complex. J Cell Biol 156:677–687

Hussey P, Hashimoto T (2008) The cytoskeleton and signal transduction: role and regulation of plant actin- and microtubule-binding proteins. In: Yang Z (ed) Intracellular signaling in plants, annual plant reviews, vol 33. Wiley-Blackwell, Hoboken, pp 244–272

Hwang JU, Gu Y, Lee YJ, Yang Z (2005) Oscillatory ROP GTPase activation leads the oscillatory polarized growth of pollen tubes. Mol Biol Cell 16:5385–5399

Iden S, Collard JG (2008) Crosstalk between small GTPases and polarity proteins in cell polarization. Nat Rev Mol Cell Biol 9:846–859

Igarashi H, Orii H, Mori H, Shimmen T, Sonobe S (2000) Isolation of a novel 190 kDa protein from tobacco BY-2 cells: possible involvement in the interaction between actin filaments and microtubules. Plant Cell Physiol 41:920–931

Jones MA, Shen JJ, Fu Y, Li H, Yang Z, Grierson CS (2002) The *Arabidopsis* Rop2 GTPase is a positive regulator of both root hair initiation and tip growth. Plant Cell 14:763–776

Kost B (2008) Spatial control of Rho (Rac-Rop) signaling in tip-growing plant cells. Trends Cell Biol 18:119–127

Kost B, Lemichez E, Spielhofer P, Hong Y, Tolias K, Carpenter C, Chua NH (1999) Rac homologues and compartmentalized phosphatidylinositol 4, 5-bisphosphate act in a common pathway to regulate polar pollen tube growth. J Cell Biol 145:317–330

La Carbona S, Le Goff C, Le Goff X (2006) Fission yeast cytoskeletons and cell polarity factors: connecting at the cortex. Biol Cell 98:619–631

Lavy M, Bloch D, Hazak O, Gutman I, Poraty L, Sorek N, Sternberg H, Yalovsky S (2007) A novel ROP/RAC effector links cell polarity, root-meristem maintenance, and vesicle trafficking. Curr Biol 17:947–952

Le J, El-Assal Sel D, Basu D, Saad ME, Szymanski DB (2003) Requirements for *Arabidopsis* ATARP2 and ATARP3 during epidermal development. Curr Biol 13:1341–1347

Lee YJ, Szumlanski A, Nielsen E, Yang Z (2008) Rho-GTPase-dependent filamentous actin dynamics coordinate vesicle targeting and exocytosis during tip growth. J Cell Biol 181:1155–1168

Li H, Yang Z (2000) Rho GTPases and the actin cytoskeleton. In: Staiger CJ, Baluska F, Volkmann D, Barlow PW (eds) Actin: a dynamic framework for multiple plant cell functions, Developments in plant and soil sciences. Kluwer, Dordrecht/Boston/London, pp 301–321

Li R, Gundersen GG (2008) Beyond polymer polarity: how the cytoskeleton builds a polarized cell. Nat Rev Mol Cell Biol 9:860–873

Li S, Blanchoin L, Yang Z, Lord EM (2003) The putative *Arabidopsis* arp2/3 complex controls leaf cell morphogenesis. Plant Physiol 132:2034–2044

Lodish H, Berk A, Kaiser C, Krieger M, Scott M, Bretscher A, Ploegh H, Matsudaira P (2007) Molecular cell biology, 6th edn. W H Freeman and Company, New York

Mathur J (2005) The ARP2/3 complex: giving plant cells a leading edge. Bioessays 27:377–387

Mathur J, Mathur N, Kernebeck B, Hulskamp M (2003a) Mutations in actin-related proteins 2 and 3 affect cell shape development in *Arabidopsis*. Plant Cell 15:1632–1645

Mathur J, Mathur N, Kirik V, Kernebeck B, Srinivas BP, Hulskamp M (2003b) *Arabidopsis* CROOKED encodes for the smallest subunit of the ARP2/3 complex and controls cell shape by region specific fine F-actin formation. Development 130:3137–3146

McCurdy D, Staiger CJ (2000) Fimbrin. In: Staiger CJ, Baluska F, Volkmann D, Barlow PW (eds) Actin: a dynamic framework for multiple plant cell functions, Developments in plant and soil sciences. Kluwer, Dordrecht, pp 87–102

Meller N, Irani-Tehrani M, Kiosses WB, Del Pozo MA, Schwartz MA (2002) Zizimin1, a novel Cdc42 activator, reveals a new GEF domain for Rho proteins. Nat Cell Biol 4:639–647

Minc N, Bratman SV, Basu R, Chang F (2009) Establishing new sites of polarization by microtubules. Curr Biol 19:83–94

Molendijk AJ, Bischoff F, Rajendrakumar CS, Friml J, Braun M, Gilroy S, Palme K (2001) *Arabidopsis thaliana* Rop GTPases are localized to tips of root hairs and control polar growth. EMBO J 20:2779–2788

Morrell JL, Morphew M, Gould KL (1999) A mutant of Arp2p causes partial disassembly of the Arp2/3 complex and loss of cortical actin function in fission yeast. Mol Biol Cell 10: 4201–4215

Moseley JB, Goode BL (2006) The yeast actin cytoskeleton: from cellular function to biochemical mechanism. Microbiol Mol Biol Rev 70:605–645

Murata T, Hasebe M (2007) Microtubule-dependent microtubule nucleation in plant cells. J Plant Res 120:73–78

Murata T, Sonobe S, Baskin TI, Hyodo S, Hasezawa S, Nagata T, Horio T, Hasebe M (2005) Microtubule-dependent microtubule nucleation based on recruitment of gamma-tubulin in higher plants. Nat Cell Biol 7:961–968

Park HO, Bi E (2007) Central roles of small GTPases in the development of cell polarity in yeast and beyond. Microbiol Mol Biol Rev 71:48–96

Pollard TD, Borisy GG (2003) Cellular motility driven by assembly and disassembly of actin filaments. Cell 112:453–465

Preuss ML, Kovar DR, Lee YR, Staiger CJ, Delmer DP, Liu B (2004) A plant-specific kinesin binds to actin filaments and interacts with cortical microtubules in cotton fibers. Plant Physiol 136:3945–3955

Qiu JL, Jilk R, Marks MD, Szymanski DB (2002) The *Arabidopsis* SPIKE1 gene is required for normal cell shape control and tissue development. Plant Cell 14:101–118

Raftopoulou M, Hall A (2004) Cell migration: Rho GTPases lead the way. Dev Biol 265:23–32

Ridge RW (1988) Freeze-substitution improves the ultrastructural preservation of legume root hairs. Bot Mag Tokyo 101:427–441

Ridley AJ, Schwartz MA, Burridge K, Firtel RA, Ginsberg MH, Borisy G, Parsons JT, Horwitz AR (2003) Cell migration: integrating signals from front to back. Science 302:1704–1709

Rosales-Nieves AE, Johndrow JE, Keller LC, Magie CR, Pinto-Santini DM, Parkhurst SM (2006) Coordination of microtubule and microfilament dynamics by Drosophila Rho1, Spire and Cappuccino. Nat Cell Biol 8:367–376

Sawa M, Suetsugu S, Sugimoto A, Miki H, Yamamoto M, Takenawa T (2003) Essential role of the *C. elegans* Arp2/3 complex in cell migration during ventral enclosure. J Cell Sci 116:1505–1518

Schmidt A, Hall A (2002) Guanine nucleotide exchange factors for Rho GTPases: turning on the switch. Genes Dev 16:1587–1609

Schultheiss H, Hensel G, Imani J, Broeders S, Sonnewald U, Kogel KH, Kumlehn J, Huckelhoven R (2005) Ectopic expression of constitutively activated RACB in barley enhances susceptibility to powdery mildew and abiotic stress. Plant Physiol 139:353–362

Seagull RW (1990) The effects of microtubule and microfilament disrupting agents on cytoskeletal arrays and wall deposition in developing cotton fibers. Protoplasma 159:44–59

Shaw SL, Kamyar R, Ehrhardt DW (2003) Sustained microtubule treadmilling in *Arabidopsis* cortical arrays. Science 300:1715–1718

Staiger CJ, Cande WZ (1990) Microtubule distribution in *dv*, a maize meiotic mutant defective in the prophase to metaphase transition. Dev Biol 138:231–242

Staiger CJ, Cande WZ (1991) Microfilament distribution in maize meiotic mutants correlates with microtubule organization. Plant Cell 3:637–644

Szymanski DB (2005) Breaking the WAVE complex: the point of *Arabidopsis* trichomes. Curr Opin Plant Biol 8:103–112

Uhrig JF, Mutondo M, Zimmermann I, Deeks MJ, Machesky LM, Thomas P, Uhrig S, Rambke C, Hussey PJ, Hulskamp M (2007) The role of *Arabidopsis* SCAR genes in ARP2-ARP3-dependent cell morphogenesis. Development 134:967–977

Vernoud V, Horton AC, Yang Z, Nielsen E (2003) Analysis of the small GTPase gene superfamily of *Arabidopsis*. Plant Physiol 131:1191–1208

Wallar BJ, Alberts AS (2003) The formins: active scaffolds that remodel the cytoskeleton. Trends Cell Biol 13:435–446

Whittington AT, Vugrek O, Wei KJ, Hasenbein NG, Sugimoto K, Rashbrooke MC, Wasteneys GO (2001) MOR1 is essential for organizing cortical microtubules in plants. Nature 411: 610–613

Winter DC, Choe EY, Li R (1999) Genetic dissection of the budding yeast Arp2/3 complex: a comparison of the in vivo and structural roles of individual subunits. Proc Natl Acad Sci USA 96:7288–7293

Wu G, Gu Y, Li S, Yang Z (2001) A genome-wide analysis of *Arabidopsis* Rop-interactive CRIB motif-containing proteins that act as Rop GTPase targets. Plant Cell 13:2841–2856

Xu J, Scheres B (2005) Dissection of *Arabidopsis* ADP-RIBOSYLATION FACTOR 1 function in epidermal cell polarity. Plant Cell 17:525–536

Yalovsky S, Bloch D, Sorek N, Kost B (2008) Regulation of membrane trafficking, cytoskeleton dynamics, and cell polarity by ROP/RAC GTPases. Plant Physiol 147:1527–1543

Yang G, Gao P, Zhang H, Huang S, Zheng ZL (2007) A mutation in MRH2 kinesin enhances the root hair tip growth defect caused by constitutively activated ROP2 small GTPase in *Arabidopsis*. PLoS ONE 2:e1074

Yang Z (2008) Cell polarity signaling in *Arabidopsis*. Annu Rev Cell Dev Biol 24:551–575

Yang Z, Fu Y (2007) ROP/RAC GTPase signaling. Curr Opin Plant Biol 10:490–494

Yuan M, Shaw PJ, Warn RM, Lloyd CW (1994) Dynamic reorientation of cortical microtubules, from transverse to longitudinal, in living plant cells. Proc Natl Acad Sci USA 91:6050–6053

Zheng ZL, Yang Z (2000) The Rop GTPase: an emerging signaling switch in plants. Plant Mol Biol 44:1–9

Zigmond SH (2004) Formin-induced nucleation of actin filaments. Curr Opin Cell Biol 16:99–105

RAC/ROP GTPases in the Regulation of Polarity and Polar Cell Growth*

Hen-ming Wu, Christine Y. Chen, Qiaohong Duan, Daniel Kita, Candida Nibau, Li-zhen Tao, Yan-jiao Zou, and Alice Y. Cheung

Abstract Mounting evidence supports that RAC/ROP GTPases are central regulators for diverse signaling pathways for plant growth, development, and interactions with the environment. Their regulatory activities for key intracellular process, such as control of actin dynamics, membrane trafficking, and several hormone signal transduction pathways, suggest inevitable functional roles for RAC/ROPs in regulating cellular activities that underlie important growth and developmental events, especially those that involve cellular and morphological asymmetry. RAC/ROPs are well established as a regulator for the two most polarized cell growth processes in plants, pollen tube tip growth, and root hair elongation, and are known to be crucial for the differentiation of leaf epidermal cell patterning, which depends substantially on differential cell expansion around its periphery and asymmetric cell division. We focus here on discussing recent findings, especially those that relate to upstream regulators and downstream effectors of RAC/ROPs that illuminate how

*Electronic supplementary material The online version of this article (doi: 10.1007/978-3-642-03524-1_6) contains supplementary material, which is available to authorized users.

H.-m. Wu, C.Y. Chen, Q. Duan, D. Kita, C. Nibau, L.-z. Tao, Y.-j. Zou and A.Y. Cheung (✉)
Department of Biochemistry and Molecular Biology
e-mail: acheung@biochem.umass.edu

H.-m. Wu, C.Y. Chen, D. Kita and A.Y. Cheung
Molecular Cell Biology Program

C. Nibau and A.Y. Cheung
Plant Biology Graduate Program, University of Massachusetts, Amherst, MA 01003

C.Y. Chen
Millipore, 28820 Single Oak Drive, Temecula, CA 92590

C. Nibau
Institute of Biological, Environmental and Rural Sciences, Edward Llwyd Building, Aberyswyth University, Aberystwyth, SY23 3DA, UK

L.-z. Tao
College of Life Sciences, South China Agricultural University, Guangzhou, China

S. Yalovsky et al. (eds.), *Integrated G Proteins Signaling in Plants*, Signaling and Communication in Plants,
DOI 10.1007/978-3-642-03524-1_6,

these small GTPases and their interactors together contribute to polarity-dependent processes in plants.

1 Introduction

Rho GTPases are a large family of Ras-related monomeric small G proteins that serve important and diverse signaling functions in eukaryotic cells (Etienne-Manneville and Hall 2002). They act as molecular transducers by shuttling between the inactive GDP-bound to the activated and cell membrane-associated GTP-bound form to mediate multiple extracellular signals to diverse cellular responses (Brembu et al. 2006; Nibau et al. 2006; Yang and Fu 2007). Instead of speciation into distinct subfamilies such as the Rho, Rac, and Cdc42 in animal cells or Rho and Cdc42 in yeast cells, plants have evolved closely related Rho GTPases that collectively form a distinct family, referred to as RACs/ROPs (for their sequence relatedness to animal Racs/for Rhos of plants). Compared with the subfamilies of Rho GTPases in other species, RAC/ROPs are encoded by relatively large families in different plant species, e.g., there are 11 RAC/ROPs in *Arabidopsis*, 9 in maize, and 7 in rice (Christensen et al. 2003), see chapter "ROP Evolution and ROPs in Grasses"). As a group, RAC/ROPs are known to be important for myriad cellular processes ranging from cell growth and polarity establishment to hormone-, pathogen-, and abiotic stress-induced responses and are thus critical for plant growth, development, reproduction, and adaptation to environmental challenges. Multiple families of regulators control the shuttling between inactive and activated forms of RAC/ROPs (Shichrur and Yalovsky 2006; Kost 2008), see chapters "Regulatory and Cellular Functions of Plant RhoGAPs and RhoGDIs" and "Structure and Function of ROPs and Their GEFs") in mechanisms similar to those found in other organisms.

Guanine nucleotide exchange factors (GEFs) stimulate GDP to GTP exchange (Berken et al. 2005), activating RAC/ROPs; GTPase-activating proteins (GAPs) accelerate GTP hydrolysis (Wu et al. 2000) and guanine nucleotide dissociation inhibitors (GDIs) inhibit GDP release (Klahre et al. 2006), thus shifting their activity equilibrium toward the inactive state. NADPH oxidase and two other plant-specific families of proteins, ROP-interactive CRIB motif-containing proteins (RICs) and interactor of constitutive active (CA) ROPs (ICRs), have been identified as immediate effectors for RAC/ROPs (Wu et al. 2001; Lavy et al. 2007; Wong et al. 2007), see chapter "ROP GTPases and the Cytoskeleton"). Reviews on various aspects of RAC/ROP signaling can be found in recent literature and in chapters "Regulatory and Cellular Functions of Plant RhoGAPs and RhoGDIs," "Structure and Function of ROPs and Their GEFs," "ROPs, Vesicle Trafficking and Lipid Modifications," "ROP GTPases and the Cytoskeleton," and "The Role of Seven Transmembrane Domain MLO Proteins, Heterotrimeric G-Proteins, and Monomeric RAC/ROPs in Plant Defense." Here, we focus on their roles as regulators for polar cell growth and, on a more general level, as contributors to a polarized cellular organization that underlie a plant's architecture.

2 RAC/ROP, a Tip-Localized Regulator for the Polarized Pollen Tube Growth Process

Pollen tubes are undeniably among the most dramatic polarized cells in nature (Hepler et al. 2001; Cheung and Wu 2008). Whether in planta and upon landing on the stigmatic surface of the pistil, or in vitro upon hydration in chemically defined media, pollen grains germinate, each extruding a polarized outgrowth of its cytoplasm to form a pollen tube, which grows exclusively at the tip as it extends away from the grain. In the pistil, the polarized cell growth process continues until pollen tubes reach the female gametophytes located within the ovules in the most basal part of the pistil, which can be centimeters and sometimes even further, away from the stigma and may be achieved at very high rates, e.g., ~1 cm h^{-1} in the maize silk. Thus, the pollen tube growth process in planta has to maintain polarity from the start to the finish. In vitro, pollen tubes also maintain the strict polarized growth habit, indicating that the male gametophyte is self-sufficient in organizing the fundamental cellular activity needed for tip growth. Numerous studies, primarily in *Arabidopsis* and tobacco, have firmly established that RAC/ROP GTPases play a key role in maintaining pollen tube growth polarity (Fig. 1; see Cheung and Wu 2008; Kost 2008; Lee and Yang 2008 and chapter "Regulatory and Cellular Functions of Plant RhoGAPs and RhoGDIs" and "ROP GTPases and the Cytoskeleton").

When RAC/ROP GTPases are deregulated, focused growth at the pollen tube tip cannot be sustained and growth becomes depolarized when growth occurs over a broader surface area around the apical dome. Although still tubular, pollen tubes become wider and their growth rate is reduced. In the most extreme case, in particular when RAC/ROP GTPases are upregulated, growth becomes isotropic and the pollen tube apical region balloons out and growth is arrested. The normally highly organized actin cytoskeleton in tip-growing pollen tubes becomes severely disrupted in these depolarized tubes (Fig. 2). Many studies have shown that RAC/ROPs regulate actin dynamics, Ca^{2+} homeostasis, and vesicle trafficking, the three key cellular aspects that underlie the pollen tube growth process. As these have been reviewed extensively else where (see e.g., Cheung and Wu 2008; Kost 2008; Lee and Yang 2008; chapters "Regulatory and Cellular Functions of Plant Rho-GAPs and RhoGDIs" and "ROP GTPases and the Cytoskeleton"), only aspects essential to complete our discussion are included here.

Using a GFP- or YFP-labeled mutant form of RICs which, as immediate downstream targets for activated RAC/ROP GTPases have been adopted as reporters for activated RAC/ROPs, the Kost and Yang groups provided compelling evidence that pollen RAC/ROPs constitute an apically located signaling apparatus (Fig. 1; Hwang et al. 2005; Klahre and Kost 2006). The recruitment and maintenance of activated RAC/ROPs to the apical membrane is tightly regulated by their activity regulators GDI and GAP, respectively (Klahre and Kost 2006; Klahre et al. 2006; chapter "Regulatory and Cellular Functions of Plant RhoGAPs and RhoGDIs"). However, visualization of a spatially restricted location for active RAC/ROPs to a narrow zone of expanding apical membrane has been difficult,

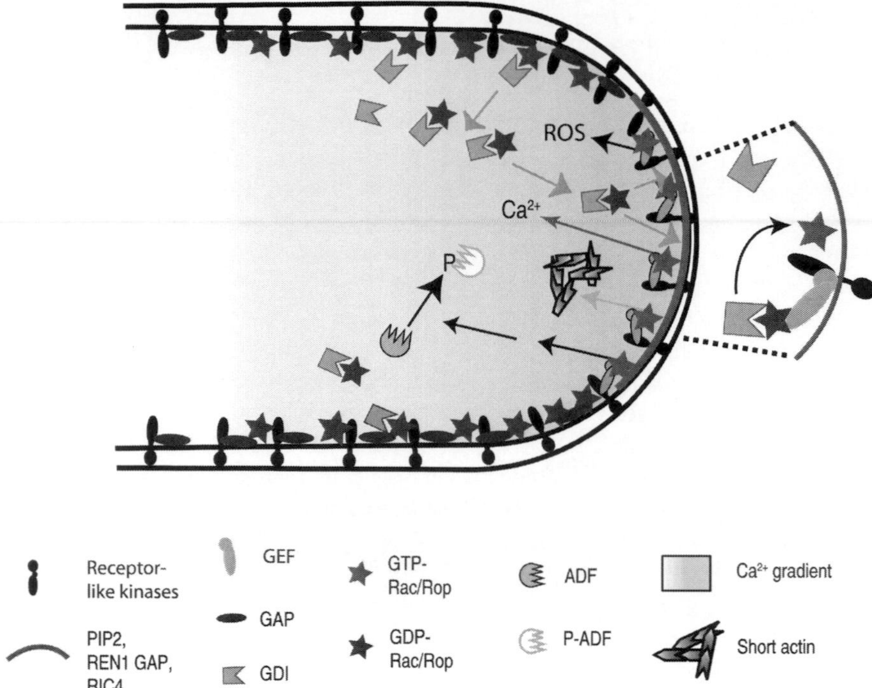

Fig. 1 A schematic model depicting the apical RAC/ROP signal transduction machinery at the pollen tube apex (modified from Cheung and Wu 2008). Transport vesicles and the actin cytoskeleton, cellular systems that critically underlie the polarized pollen tube growth process and ultimate targets for RAC/ROP signaling, are not included to avoid overcrowding the figure. Although details are not yet clear and may differ, it is most probable that the basic framework for RAC/ROP-signaled polarized root hair growth is similar. The linkage between RAC/ROP to ROS is extrapolated from information established in root hairs but remains to be established for pollen tube polar growth

since even subdepolarizing level of GFP-RAC/ROP expression already expands its localization to a significantly broader membrane area in the proximal region of morphologically normal pollen tubes (Kost et al. 1999). That activated RAC/ROPs indeed stimulate cell growth can be seen in a time lapse observation of a transformed pollen tube that was overexpressing a CA form of RAC/ROP, GFP-Ntrac1^{CA}, and undergoing transition from mild to severe but still subgrowth inhibiting level of depolarization (Fig. 3; Movie 1 (The online version of this article (doi: 10.1007/978-3-642-03524-1_6) contains supplementary material, which is available to authorized users)). GFP-Ntrac1^{CA} initially was associated with a broad membrane domain, consistent with growth occurring over a broad span of the apical membrane region and slightly expanded tip morphology. As growth at the tube apex become increasingly inhibited, an alternate growth front could be seen getting organized and this was coupled with a shift of GFP-Ntrac^{CA} to the new growth front.

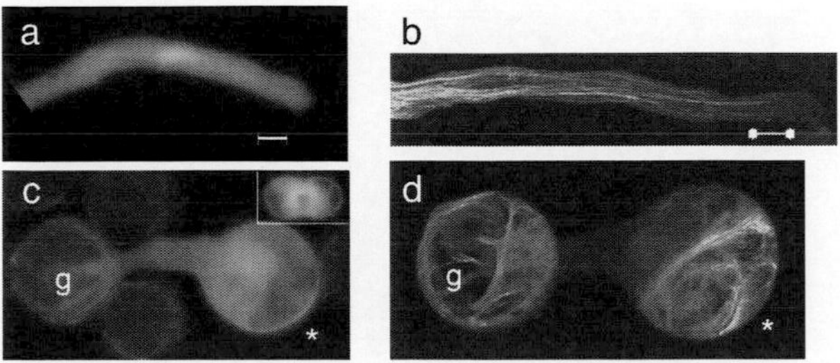

Fig. 2 Upregulating RAC/ROP GTPase activity induces depolarization of pollen tube tip growth. (**a**, **c**) show transformed tobacco pollen tubes expressing GFP (**a**) and GFP–AtROP–GEF1 (**c**). Inset in (**c**) shows a pollen tube that had ballooned upon emergence from the grain. This dumb-belled shaped pollen tube represents the most severe form of depolarized tip growth induced by upregulating RAC/ROP signaling. (**b**, **d**) showed transformed pollen tubes expressing the actin reporter NtpLIM2b-GFP (**b**) and co-expressing NtpLIM2b–GFP and AtROP–GEF1 (**d**) (Cheung et al. 2008). The "ballooning" of pollen tube is accompanied by severe bundling of the actin microfilaments and disorganization of the overall actin cytoskeleton, which is exemplified by long actin cables that run parallel to the long axis of the tube until the subapical region, where it is marked by a prominent subapical structure. The effect of overexpressing a GEF is comparable to directly increasing RAC/ROP activity by the expression of constitutively active RAC/ROPs (Kost et al. 1999; Fu et al. 2001; Chen et al. 2003; Cheung et al. 2003). *g* pollen grain; * pollen tube front

Since pollen tube growth is oscillatory with periods of increasing and declining growth rates (for a review, see Holdaway-Clarke and Hepler 2003), parameters that regulate and support the growth process are expected to likewise fluctuate. Using GFP-RIC4 as a reporter for activated RAC/ROPs, Yang and colleagues showed that activation of RAC/ROPs, as reflected by RIC4-GFP intensity at the apical cell membrane, indeed oscillates and peaks seconds ahead of peak growth rates (Hwang et al. 2005). It would seem the derivative that the inactivation of RAC/ROPs would follow suit. However, quite intriguingly, peak Rho·GAP activity at the apical membrane, as reflected by the peak apical intensity a GFP-REN1, a novel Rho-GAP encoded by At*Ren1* (*Rop1 enhancer 1*) (Hwang et al. 2008), actually appears to occur before peak growth rate ensues. Cycling of activation and inactivation of these small GTPases occurs in pollen tubes elongating with periods of growth rate oscillation in the range of 35–40 s. This suggests that relay from RAC/ROPS to cellular processes that underlie tip growth is extremely rapid and that rapid return of RAC/ROP signaling activity to the resting state is crucial for pollen tube growth. This is consistent with observations that the most severe RAC/ROP-induced phenotype, tube ballooning that often occurs even as the pollen tube just emerges, is most prevalently induced when RAC/ROP activity is at an over drive in pollen that constitutively maintains high levels of activated RAC/ROPs, such as by over-expressing CA RAC/ROPs or their upstream activator ROP-GEFs (Fig. 2; Gu et al. 2006; Cheung et al. 2008). It would be interesting to examine whether the apical membrane-associated REN1 and subapical membrane-localized, more classical "CRIB" domain-containing Rho·GAPs (Wu et al. 2000; Klahre and Kost 2006; see chapter "Regulatory and Cellular Functions of Plant RhoGAPs and RhoGDIs") act in concert to ensure spatial restriction and temporal modulation of RAC/ROP

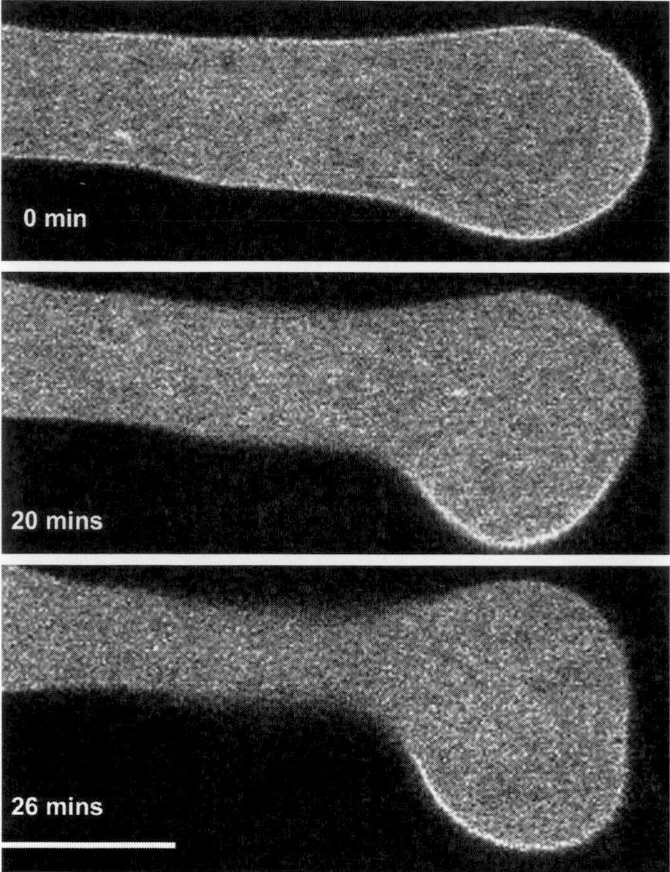

Fig. 3 Selected confocal images from a time series of a growing transformed tobacco pollen tube overexpressing GFP-Ntrac1^{CA}. The pollen tube was undergoing depolarization at the start of the time series. Overexpression of GFP-Ntrac1^{CA} showed a prominent cell membrane association that was beyond the restricted location of endogenous RAC/ROPs under normal tip-growth conditions. In lieu of total isotropic growth to form a ballooned tip, a new growth front was organized and growth continued along a different trajectory, reflecting a more modest GFP-Ntrac1^{CA} expression and RAC/ROP signaling defect. Scale bar = 20 μm. See Supplemental Movie 1 (The online version of this article (doi: 10.1007/978-3-642-03524-1_6) contains supplementary material, which is available to authorized users) for the entire time series. The movie was composed of images collected intermittently over a period of 30 mins. Six images, at 10 s intervals, were collected at the beginning of every 5 min intervals during the course of the 30 mins. It is most evident in the movie that, during the pollen tube growth, reorientation that GFP-Ntrac1^{CA} was recruited to the new growth front, as shown in the 20 min image, consistent with RAC/ROP signaling activity being spatially coupled to cell growth

activity for a tip-focused and oscillatory growth process. Interestingly, loss of function At*ren1* pollen is nonviable and when they managed to germinate in vitro, their tubes were highly defective, whereas loss of function mutations in other Rho·GAPs do not induce observable phenotype (see Hwang et al. 2008), thus suggesting that REN1 and other Rho·GAPs have nonoverlapping functions. It is plausible that local inactivation of RAC/ROPs at the apical membrane by REN1 is critical for polarity establishment during germination and their rapid inactivation during elongation before even stimulated apical growth has occurred and displaced them to the apical flank. On the other hand, the subapical membrane-located Rho·GAPs provide additional assurance that any remaining activated RAC/ROPs displaced to the apical flank membrane as the tube grows are inactivated, so lateral tube expansion is restricted.

Compared with tube growth, our understanding of how polarity is established during pollen germination is relatively poor. The overexpression of Rho·GAPs in tobacco pollen grains is highly inhibitory to germination (Chen 2002) suggests that establishment of a polarity axis for cytoplasmic protrusion from the pollen grain is dependent on a certain threshold level of activated RAC/ROPs. Recently, Li et al. (2008) overexpressed the *Arabidopsis* ROP1-interacting protein, RIP1 (ROP1 Interactive partner1)/ICR1 in *Arabidopsis* and tobacco pollen and observed that GFP–RIP1/ICR1 was initially located in the nucleus of pollen grains, but was translocated to the plasma membrane of a germination pore prior to pollen tube emergence, reflecting a role for RAC/ROP signaling in germination site determination. Overexpression of RIP1/ICR1 also induced pollen tube ballooning similar to RAC/ROP-induced phenotype. Under in vitro condition, ICR1/RIP1 interacted with wild-type and GTP-bound forms of AtROP6, AtROP10, and AtROP1 and to a substantially lower level that likely reflects background nonspecific binding with GDP-bound forms (Lavy et al. 2007; Li et al. 2008). Under in vivo condition, RIP1/ICR1 stabilized association of ROP1 with the membrane (Li et al. 2008), largely confirming previous findings on the interaction between ICR1 and different ROPs in leaf epidermal cells (Lavy et al. 2007). The stabilization of AtROP1 membrane association by ICR1/RIP1 led to the suggestion that ICR1/RIP1 potentially acts as an effector as well as a regulator for AtROP1 (Li et al. 2008). Since a pollen-related phenotype based on loss of function mutants in*Aticr1/rip1* (see below) has not yet been reported, its biological role in pollen remains to be determined.

3 RAC/ROPs as Regulators for Root Hair Tip Growth

Root hairs emerge from trichoblasts – the hair-forming cells on the root epidermis – and elongate by a tip growth process that is dependent on properly regulated RAC/ROP activity (Jones et al. 2002, 2007; Carol and Dolan 2006). In fact, according to microarray data, the level of RAC/ROP expression in *Arabidopsis* root hairs may only be second to that in pollen (Hruz et al. 2008). Unlike the effects of deregulated RAC/ROP signaling on pollen tube growth, several studies have shown that upregulation of RAC/ROP GTPases induces growth depolarization at the root hair tip (Molendijk et al. 2001; Tao et al. 2002; Carol et al. 2005; Jones et al. 2002, 2007). Overexpression of wild type or RAC/ROPs induces a continuum of weak to severe depolarization phenotypes that range from blunt-ended root hairs to root hair-ballooning upon or shortly after emergence (Fig. 4). On the other hand,

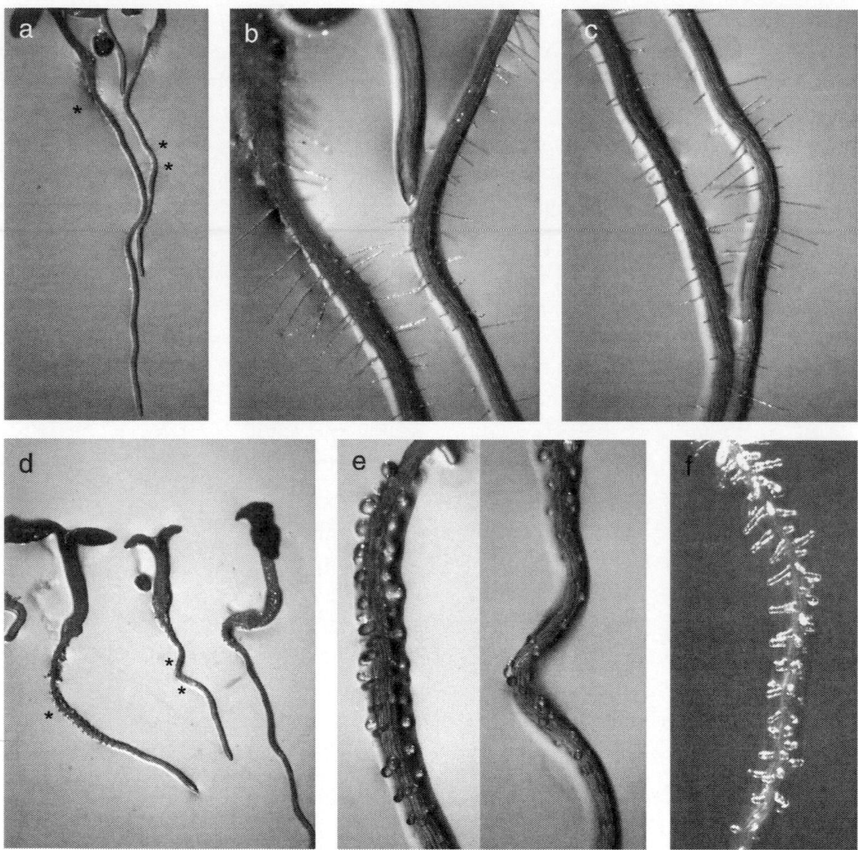

Fig. 4 Upregulating RAC/ROP GTPase activity induces depolarization of root hair growth. (**a–c**) are from wild-type 12 days old *Arabidopsis* seedlings; **b**, **c** show magnification of the * (**b**) and ** (**c**) regions shown in **a**, featuring normal tip-growth root hairs. (**d–f**) are from transgenic *Arabidopsis* expressing GFP-Ntrac1^{CA} (**d**, **e**) or GFP-ARAC3 (**f**). The panels in **e** show higher magnification of the * and ** regions shown in **d**, highlighting severely depolarized, completely "ballooned" root hairs. The effect of overexpressing wild-type forms of RAC/ROPs is mild to often nonobservable. The club-shaped root hairs in **f** are representative of the strongest phenotype observed among seedlings that overexpressed a wild-type form of RAC/ROPs

overexpressing DN RAC/ROPs result in more moderate defects mostly manifested as shorter root hairs. A most dramatic root hair phenotype, where multiple initials, instead of a single root hair, emerge from a trichoblast, is seen in the *Arabidopsis* mutant *supercentipede1* (*scn1*), which is defective in a Rho·GDI (Carol et al. 2005). With the proper regulation of RAC/ROPs severely compromised, the defective root hair phenotype in *scn1* indicates a major role for these small GTPases in both initiation and growth of root hairs. Moreover, hair cell membrane-associated RAC/ROPs are mislocalized and occupy a more extended cytoplasmic region surrounding the expanding growth front in *scn1* mutant root hairs, suggesting that, similar to pollen tubes, the apical membrane localization of these small GTPases is important for this polarized cell growth process. That root hair morphology is most severely affected in the *Arabidopsis tip1* mutant, which is defective in a *S*-acyl transferase (Hemsley et al. 2005), which is consistent with the notion that insertion of RAC/ROPs into cell membrane, which is dependent on lipid modification, is important for root hair growth (chapter "ROPs, Vesicle Trafficking and Lipid Modifications").

Recent studies have revealed important roles for reactive oxygen species (ROS) in cell growth and development (Carol and Dolan 2006; Knight 2007). The evidence supporting a critical role for ROS in polarized cell growth is particularly strong in root hairs, where NADPH oxidase-dependent ROS production is tightly linked to root hair initiation and elongation (Foreman et al. 2003; Monshausen et al. 2007). In analyzing the *Arabidopsis* NADPH oxidase mutant *rhd2* (root hair defective2)/*AtrbohC* (respiratory burst oxidase homolog C), which develops very short root hairs, Foreman et al. (2003) showed that in wild-type plants, ROS is detected on the tip of root hairs as they first emerge from the trichoblasts and it is maintained at the tip as they elongate. On the other hand, the level of detectable ROS in developing *rhd2* mutant root hairs is severely reduced. Definitive evidence that RAC/ROPs play a critical role in ROS-dependent root hair development is shown by the suppression of root hair phenotypes in *scn1* mutants by the NADPH oxidase deficient *rhd2* mutation, thus reversing the constitutively high levels of activated RAC/ROPs in these mutants (Carol et al. 2005).

A subsequent study also correlated overexpression of a CA- and DN Atrop2 with increased and decreased levels of ROS, respectively, in transgenic *Arabidopsis* root hairs (Jones et al. 2007). Moreover, Atrop2^{CA} overexpression fails to counteract the root hair growth defects in NADPH oxidase mutants, consistent with RAC/ROP acting upstream of ROS for polarized cell growth. However, overexpression of CA-At*Rop2* in the *rhd2* NADPH oxidase-deficient mutant background remains capable of inducing multiple root hair initials per trichoblast, a phenotype akin to that of the *scn1* mutant and consistent with its being the consequence of a constitutively high level of activated RAC/ROPs. This is somewhat surprising in light of the previous observation that both multiple root hair initiations per trichoblast and root hair growth defect in *scn1* are abolished by *rhd2* (Carol et al. 2005). This difference may simply have resulted from differences in growth conditions. Alternatively, it is possible that root hair emergence from the trichoblast and its polarized elongation have different degree of dependence on ROS and that any residual ROS in the *rhd2* background is adequate to mediate root hair differentiation from trichoblast when RAC/ROP signaling is highly magnified in the CA ROP2 overexpressing plants. Nevertheless, if root hair initiation and tip growth are indeed distinct in their dependence on ROS, it is tempting to suggest that these observations could imply organizing a polarized protrusion from an existing cell surface, whether it be from a trichoblast or from a pollen grain, has distinct downstream requirements than supporting restricted growth at the tip of a polarized cell. A role for RAC/ROP-regulated ROS production in pollen germination and tube growth has not been documented. However, ROS and oscillatory NAD(P)H have been detected in pollen tubes (Cardenas et al. 2006; Potocky et al. 2007); it is difficult to envision that these small GTPases not being the key players in regulating the oxido-reductive conditions in the pollen tube apex (Fig. 1).

4 RAC/ROPs as Regulators of Polarized Cellular Activity Associated with Differentiation, Development and Defense

Although far from the dramatic polarity seen in pollen tubes and root hairs, numerous morphological, developmental, and stress-related response processes rely on localized cellular activity to promote uneven expansion rate around the cell periphery. A role for RAC/ROP GTPases to promote localized cell growth is well documented in the differentiation of the jigsaw pattern of epidermal pavement cells (Fig. 5; Fu et al. 2005; Lavy et al. 2007). Apparently in two distinct downstream pathways mediated by RIC3 and RIC4, RAC/ROP GTPases regulate actin microfilaments to promote lobe outgrowth and microtubule organization to maintain cellular indentation, respectively (Fu et al. 2005).

Loss of ICR1/RIP1, or overexpression of a GFP-tagged counterpart, also abolish the epidermal jigsaw pattern (Lavy et al. 2007). Instead of interdigited by lobes and indentations, epidermal cells in *icr1* or in ICR1-overexpressing plants are almost rectangular, similar to the effect of overexpressing CA RAC/ROPs (Fig. 5). The induction of noninterdigited pavement cells by overexpression of ICR1 is dependent on its ability to interact with ROPs since expression of an ICR1 ROP noninteracting mutant failed to induce this phenotype (Lavy et al. 2007).

Interestingly, ICR1 interacts with SEC3, a constituent protein of the exocyst, to form a protein complex that is important for tethering secretory vesicles to the cell membrane for exocytosis (TerBush et al. 1996; Novick and Guo 2002; Hala et al. 2008). Even the overexpression of a non-SEC3-interacting form of ICR1 that

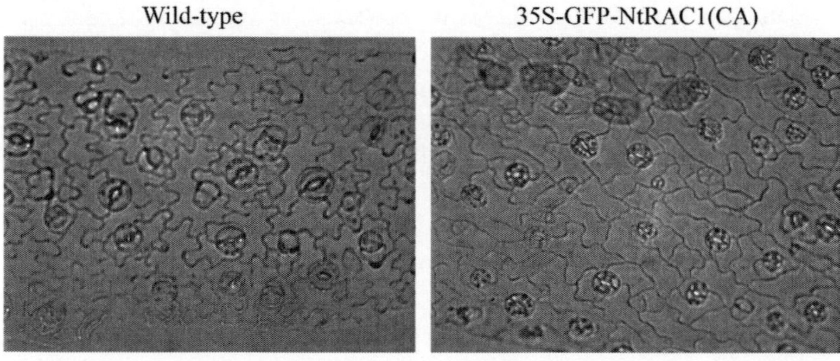

Fig. 5 Upregulating RAC/ROP GTPase activity reduces polarized growth in epidermal cells, disrupting the normal jigsaw puzzle pattern of the epidermis. Images are peels from the lower epidermis of the first true leaf from 3 weeks old wild type and transgenic seedlings that overexpressed GFP-Ntrac^{CA}. The pronounced lobe and indentation formed by differential outgrowth in neighboring membrane locations in wild-type leaf epidermis is substantially reduced, and in many cases abolished, by increasing RAC/ROP signaling

fails to induce an epidermal cell pattern phenotype provides strong implications that localized outgrowth regulated by RAC/ROP GTPases is closely linked to augmented membrane trafficking activity. An effect on membrane vesicle trafficking was also observed when Atrac10/rop11^{CA} was expressed in plants (Bloch et al. 2005).

While epidermal pattern differentiation during development may be a process that requires relatively subtle but, possibly, sustained RAC/ROP regulated activities, epidermal cells are, nonetheless, at the front line for defense, e.g., against a pathogen attack, and capable of launching rapid and drastic cellular responses at localized sites (see Chapter "The Role of Seven Transmembrane Domain MLO Proteins, Heterotrimeric G-Proteins and Monomeric RAC/ROPs in Plant Defense"). A role for RAC/ROP GTPases in defense response is well documented in the study of OsRac1 regulation of NADPH-dependent defense response pathway in rice (Kawasaki et al. 1999; Ono et al. 2001; Wong et al. 2007). A series of work in the defense response to the biotrophic pathogen powdery mildew fungus in barley also provided strong implications that modulated RAC/ROP activity is important in orchestrating polarized actin reorganization focused toward the infection site on the epidermis (Schultheiss et al. 2002, 2003, 2005; Opalski et al. 2005). Although the mechanism remains unknown, augmenting RAC/ROP signaling activity is associated with increased susceptibility to these pathogens in barley and the level of activated RAC/ROPs, as reflected by the activated RAC/ROP reporter RIC, is enhanced by attempted fungal penetration (Schultheiss et al. 2008).

That RAC/ROPs play the central roles in overall plant growth and development is evident from their ability to mediate several key hormone signaling pathways, such as those for auxin, absicsic acid (ABA), and brassinosteroid (Li et al. 2001; Lemichez et al. 2001; Zheng et al. 2001; Tao et al. 2002, 2005). Normal apical–basal determination during embryogenesis that underlies the shoot–root architecture of a plant is critically dependent on polar auxin transport, which is, in turn, dependent on the cellular polarization of the auxin efflux proteins, PINs, to specific membrane domains (Kleine-Vehn and Friml 2008). In addition to defects in root hairs, misregulating RAC/ROP signaling also results in numerous seedling phenotypes suggestive of auxin-related defects, such as fused cotyledons, defective shoot–root axis, and collapsed root meristems (Tao et al. 2002; Lavy et al. 2007). That these RAC/ROP-induced developmental consequences may, indeed, be mediated by defective auxin signaling is supported by the observations that auxin rapidly stimulates RAC/ROP activation, and these activated small GTPases, in turn, promote auxin-induced repressor proteolysis and subsequent gene derepression (Tao et al. 2002, 2005). How RAC/ROP signaling is connected to the overall auxin response may at least be partly based on the role of RAC/ROP GTPases on membrane trafficking (Bloch et al. 2005; Lee et al. 2008), which has a fundamental role in polarized auxin transport (Dhonukshe et al. 2008). In particular, regulated endocytosis and transcytosis of PINs play a crucial role for their localization to restricted membrane domains (Paciorek et al. 2005; Kleine-Vehn et al. 2008). Moreover, auxin itself inhibits endocytosis and PIN internalization, thus promotes its own efflux, so it acts as a modulator for its own homeostasis and the ultimate auxin response. Augmenting RAC/ROP signaling by overexpressing AtRAC10/ROP11 also inhibits endocytosis (Bloch et al. 2005) and, presumably, would also deregulate PIN localization and auxin signaling. Rather intriguingly, mutations in TIR1/AFB (auxin F-box) family of F-box proteins, which serve as auxin receptors in the nucleus and mediate auxin-induced gene derepression via 26 S proteasome-regulated proteolysis of their transcriptional repressors, also inhibit PIN internalization (Pan et al. 2009). RAC/ROP GTPases have also been shown to promote the assembly of the ubiquitin/26 S proteasome to form proteolytically active nuclear protein bodies capable of degrading the transcriptional repressors for auxin-induced gene expression (Tao et al. 2005). Taken together, although details are not known, it certainly appears that an important component for how RAC/ROPs impact overall plant growth and development may be based on how their regulation integrates into establishing and modulating polarized location of PINs, and thus intracellular auxin signaling capacity.

Activated RAC/ROPs are largely associated with growth-promoting activity and are also supported by observations that they serve as negative regulators for signaling ABA response, which in many aspects is antagonist to that of auxin (Lemichez et al. 2001; Zheng et al. 2001). ABA treatment has been shown to induce inactivation of RAC/ROP GTPases and the ability to downregulate RAC/ROP signaling correlates with the ABA-induced response of stomata closure and disruption of guard cell actin cytoskeleton (Lemichez et al. 2001). While knock-out mutations in At*ROP10* and its most closely related At*ROP11*/At*ROP10* enhance ABA responses in mutant seedlings (Zheng et al. 2001; Nibau 2005), revealing their role as negative regulators for ABA signaling, other RAC/ROPs, such as the slightly distant but still closely related AtROP9/ARAC7 actually acts as a positive regulator for ABA (Nibau, 2005). For a detailed understanding of the biological significance of RAC/ROP GTPases, these studies underscore the importance to dissect the individual function for each RAC/ROP isoform within a plant species.

5 Insights from Upstream RAC/ROP Regulators on Their Role in Polarized Cell Growth

As a family, RAC/ROP GTPases apparently serves as an integration point for multiple biotic, abiotic, and developmental signal inputs and diverse downstream signaling pathways and cellular targets (Nibau et al. 2006). The discovery of ROP–GEFs, the plant-specific guanine nucleotide exchange factors for RAC/ROPs (Berken et al. 2005; Berken 2006; chapter “Structure and function of ROPs and their GEFs”) provide links to the expectedly diverse upstream regulators for RAC/ROPs. Thus far, functional studies on these upstream regulators of RAC/ROPs, including several of the ROP–GEFs (Kaothien et al. 2005; Gu et al. 2006), the ROP–GEF-interacting tomato pollen receptor kinase 2, and its *Arabidopsis* counterpart (Zhang and McCormick 2007; Zhang et al. 2008), have essentially confirmed the role of RAC/ROP signaling in polarized pollen tube growth. Our own studies on a different family of ROP–GEF-interacting receptor kinases reveal that they mediate RAC/ROP signaling of ROS-dependent polarized root hair growth, pollen tube growth inside the ovules, and various stress-related responses (Duan et al. – manuscript in preparation; D. Kita and W-G. Zheng – unpublished observations). These studies further affirm that a major RAC/ROP signaling apparatus underlies polarity maintenance in two dramatic tip growth cell types and are important mediators for defense against biotic and abiotic stresses that may involve polarized cellular activities.

In the leaf epidermis, a unique GEF belonging to the DOCK family of proteins, SPIKE1, plays an important role in controlling the activity of the actin nucleating Arp2/3 complex (Basu et al. 2008). In plants, the biological role for the Arp2/3 complex appears to be rather subtle and mutants in Arp2/3 complex constituents or its regulators are noted mostly for their distorted trichome morphology. Thus, one of the most notable functions for the Arp2/3 complex is apparently to produce an actin cytoskeleton optimum for supporting polarized cellular activities that underlie differentiation of the epidermis, in particular, during the polarized elongation phase in trichome development (Szymanski 2005). Activation of the Arp2/3 complex is a Rho GTPase-dependent process (Welch and Mullins 2002; Basu et al. 2004; 2008). Genetic and biochemical evidence supports that SPIKE1 functions in a high molecular weight protein complex to activate RAC/ROPs, which in turn activate an immediate effector SRA1, a component of what is known as the WAVE complex and an upstream activator for Arp2/3 activity. In Arp2/3 complex-related mutants, phenotypes beyond the defective trichomes are very subtle. While cell size and elongation may be slightly reduced (Mathur et al. 2003), vigorous tip growing cells such as root hairs and pollen tubes are not visibly or functionally affected (Li et al. 2003; Szymanski 2005). However, hypocotyl and cotyledon epidermis develop clear gaps between adjacent cells, often near lobe tips of the epidermal pavement cells (Mathur et al. 2003; El-Assal et al. 2004; Le et al. 2003). Thus, it is possible that the SPIKE1-controlled and RAC/ROP-regulated Arp2/3 activity may underlie polarized cellular activities responsible for the secretion of adhesive materials that bond neighboring epidermal cells, another contribution of RAC/ROP GTPases to a polarity-dependent process.

6 Perspective

Tremendous amount of studies on Rho GTPases in plants and other eukaryotic systems has firmly established their importance in cellular processes that underlie polarity determination and maintenance in growth, development, and stress-related responses. For plant RAC/ROPs, where speciation into different functional subgroups, such as those for the mammalian and yeast Rho GTPases, has not occurred, the challenge has been and remains to be on specific functional assignment for each or a subset of the most closely related RAC/ROPs. The emerging number of the upstream regulators and downstream effectors suggests that RAC/ROP signaling pathways are diverse with potential for many feedback loops and numerous nodes for cross-talks between different pathways. Therefore, functional maps that link regulators, RAC/ROPs, and effectors that are coexpressed in specific cell or tissue types will need to be established in order to clearly resolve how and when particular RAC/ROPs are activated and in response to which signals. Even though plants have well-established forward and reverse genetic approaches for functional dissection of one, or even several closely related RAC/ROPs at a time, the ability to overcome difficulties due to functional redundancy among family members is not necessarily assured. Moreover, observations of often subtle phenotypes that might be revealed only under specific growth or challenged conditions are labor intensive. A productive approach to facilitate establishing specific functional linkages may be the combined use of transient cell systems, such as in protoplasts or pollen tubes, to express differentially tagged RAC/ROPs and their potential regulators or effectors, combined with biochemical pull-down assays and cellular imaging to identify authentic in vivo functional partners. Moreover, how these interactions are regulated temporally and spatially and how these regulations impact the RAC/ROP-regulated target responses are also more amenable for detail analysis in these transient cell systems. These, followed by studies in planta on the effect of down-regulating and upregulating specific RAC/ROP and its regulators should provide more definitive functional assignments for the various components of the RAC/ROP-signaling pathways. Establishing specific functional relationship for individual RAC/ROP with its upstream and downstream signaling molecules should ultimately lead to unveiling how as a family, RAC/ROPs integrate multiple upstream inputs and mediate diverse outputs carried out by cellular, biochemical, and molecular pathways that may be unique for individual stimulus or shared by multiple stimuli to effect myriad responses in growth and development.

Acknowledgments Works cited from our laboratory were supported by grants from the Cooperative State Research, Education and Extension Service, 2003-02913, #2004-03419, 2005-02396, and from the National Science Foundation IOB-0544222. CN was supported by the Fundacao para a Ciencia e Tecnologia, Portugal, Ph.D. grant #SFRH/BD/840/2000.

References

Basu D, El-Assai S, Le J, Mallery EL, Szymanski DB (2004) Interchangeable functions of *Arabidopsis* PIROGI and the human WAVE complex subunit SRA1 during leaf epidermal development. Development 131:4345–4355

Basu D, Le J, Zakharova T, Mallery EL, Szymanski DB (2008) A SPIKE1 signaling complex controls actin-dependent cell morphogenesis through the heteromeric WAVE and ARP2/3 complexes. Proc Natl Acad Sci USA 105:4044–4049

Berken A (2006) ROPs in the spotlight of plant signal transduction. Cell Mol Life Sci 63:2446–2459

Berken A, Thomas C, Wittinghofer A (2005) A new family of RhoGEFs activates the Rop molecular switch in plants. Nature 436:1176–1180

Bloch D, Lavy M, Efrat Y, Efroni I, Bracha-Drori K, Abu-Abied M, Sadot E, Yalovsky S (2005) Ectopic expression of an activated RAC in *Arabidopsis* disrupts membrane cycling. Mol Biol Cell 16:1913–1927

Brembu T, Winge P, Bones AM, Yang Z (2006) A RHOse by any other name: a comparative analysis of animal and plant Rho GTPases. Cell Res 16:435–445

Cardenas L, McKenna ST, Kunkel JG, Hepler PK (2006) NAD(P)H oscillates in pollen tubes and is correlated with tip growth. Plant Physiol 142:1460–1468

Carol RJ, Dolan L (2006) The role of reactive oxygen species in cell growth: lessons from root hairs. J Expt Bot 57:1829–1834

Carol RJ, Takeda S, Linstead P, Currant MC, Kakesova H, Derbyshire P, Drea S, Zarsky V, Dolan L (2005) A RhoGDP dissociation inhibitor spatially regulates growth in root hair cells. Nature 438:1013–1016

Chen C (2002) Functional analysis of Actin Depolymerizing Factor (ADF) in Rac-mediated pollen tube growth. Ph.D. Dissertation, University of Massachusetts, Amherst, MA.

Chen CY, Cheung AY, Wu H (2003) Actin-depolymerizing factor mediates Rac/Rop GTPase-regulated pollen tube growth. Plant Cell 15:237–249

Cheung AY, Chen C, Tao L-Z, Andreyeva T, Twell D, Wu H-M (2003) Regulation of pollen tube growth by Rac-like GTPases. J Exp Bot 54:73–81

Cheung AY, Duan Q-H, de Costa S, de Graaf B, di Stilio V, Wu H-M (2008) The dynamic pollen tube cytoskeleton: live cell studies using actin-binding and microtubule-binding reporter proteins. Mol Plant 1:686–702

Cheung AY, Wu H-M (2008) Structural and signaling networks for the polar cell growth machinery in pollen tubes. Annu Rev Plant Biol 59:547–572

Christensen TM, Vejlupkova Z, Sharma YK, Arthur KM, Spatafora JW, Albright CA, Meeley RB, Duvick JP, Quatrano RS, Fowler JE (2003) Conserved subgroups and developmental regulation in the monocot rop gene family. Plant Physiol 133:1791–1808

Dhonukshe P, Grigoriev I, Fischer R, Tominaga M, Robinson DG, Hasek J, Paciorek T, Petrasek J, Seifertova D, Tejos R, Meisel LA, Zazimalova E, Gadella TW Jr, Stierhof YD, Ueda T, Oiwa K, Akhmanova A, Brock R, Spang A, Friml J (2008) Auxin transport inhibitors impair vesicle motility and actin cytoskeleton dynamics indiverse eukaryotes. Proc Natl Acad Sci USA 105:4489–4494

El-Assal SE, Le J, Basu D, Mallery EL, Szymanski DB (2004) *DISTORTED2* encodes an ARPC2 subunit of the putative *Arabidopsis* ARP2/3 complex. Plant J 38:526–538

Etienne-Manneville S, Hall A (2002) Rho GTPases in cell biology. Nature 420:629–635

Foreman J, Demidchik V, Bothwell JHF, Mylona P, Miedema H, Torres MA, Linstead P, Costa S, Brownlee C, Jones JDG, Davies JM, Dolan L (2003) Reactive oxygen species produced by NADPH oxidase regulate plant cell growth. Nature 422:442–446

Fu Y, Gu Y, Zheng Z, Wasteneys G, Yang Z (2005) *Arabidopsis* interdigitating cell growth requires two antagonistic pathways with opposing action on cell morphogenesis. Cell 120:687–700

Fu Y, Wu G, Yang Z (2001) Rop GTPase-dependent dynamics of tip-localized F-actin controls tip growth in pollen tubes. J Cell Bio 1152:1019–1032

Gu Y, Li S, Lord EM, Yang ZB (2006) Members of a novel class of *Arabidopsis* Rho guanine nucleotide exchange factors control Rho GTPase-dependent polar growth. Plant Cell 18:366–381

Hala M, Cole R, Synek L, Drdova E, Pecenkova T, Nordheim A, Lamkemeyer T, Madlung J, Hochholdinger F, Fowler JE, Zarsky V (2008) An exocyst complex functions in plant cell growth in *Arabidopsis* and tobacco. Plant Cell 20:1330–1345

Hemsley PA, Kemp AC, Grierson CS (2005) The TIP GROWTH DEFECTIVE1 *S*-acyl transferase regulates plant cell growth in *Arabidopsis*. Plant Cell 17:2554–2563

Hepler PK, Vidali L, Cheung AY (2001) Polarized cell growth in higher plants. Ann Rev Cell Dev Biol 17:159–187

Holdaway-Clarke T, Hepler PK (2003) Control of pollen tube growth: role of ion gradients and fluxes. New Phytol 159:539–563

Hruz T, Laule O, Szabo G, Wessendorp F, Bleuler S, Oertle L, Widmayer P, Gruissem W, Zimmermann W (2008) Genevestigator V3: a reference expression database for the meta-analysis of transcriptomes. Adv Bioinform 2008:420–447

Hwang J, Gu Y, Lee Y, Yang J (2005) Oscillatory ROP GTPase activation leads the oscillatory polarized growth of pollen tubes. Mol Biol Cell 16:5385–5399

Hwang J-U, Vernoud V, Szumlanski A, Nielsen E, Yang N (2008) A tip-localized RhoGAP controls cell polarity by globally inhibiting Rho GTPase at the cell apex. Curr Biol 18:1907–1916

Jones MA, Shen J-J, Fu Y, Yang Z, Grierson CS (2002) The Arabidopsis Rop2 GTPase is a positive regulator of both root hair initiation and tip growth. Plant Cell 14:763–776

Jones MA, Raymond MJ, Yang Z, Smirnoff N (2007) NADPK oxidase-dependent reactive oxygen species formation required for root hair growth depends on ROP GTPase. J Expt Bot 58:1261–1270

Kaothien P, SH OK, Shuai B, Wengier D, Cotter R, Kelley D, Kiriakopolos S, Muschietti J, McCormick S (2005) Kinase partner protein interacts with the LePRK1 and LePRK2 receptor kinases and plays a role in polarized pollen tube growth. Plant J 42:492–503

Kawasaki T, Henmi K, Ono E, Hatakeyama S, Iwano M, Satoh H, Shimamoto K (1999) The small GTP-binding protein Rac is a regulator of cell death in plants. Proc Natl Acad Sci USA 96:10922–10929

Klahre U, Becker C, Schmitt AC, Kost B (2006) Nt-RhoGDI2 regulates Rac/Rop signaling and polar cell growth in tobacco pollen tubes. Plant J 46:1018–1031

Klahre U, Kost B (2006) Tobacco RhoGTPase ACTIVATING PROTEIN1 spatially restricts signaling RAC/ROP to the Apex of pollen tubes. Plant Cell 18:3033–3046

Kleine-Vehn J, Dhonukshe P, Sauer M, Brewer PB, Wisniewska J, Paciorek T, Benkova E, Friml J (2008) ARF GEF-dependent transcytosis and polar deliveryof PIN auxin carriers in *Arabidopsis*. Curr Biol 18:526–531

Kleine-Vehn J, Friml J (2008) Polar targeting and endocytic recycling inauxin-dependent plant development. Annu Rev Cell Dev Biol 24:447–473

Knight MR (2007) New ideas on root hair growth appear from the flanks. Proc Natl Acad Sci USA 104:20649–20650

Kost B (2008) Spatial control of Rho (Rac-Rop) signaling in tip-growing plant cells. Trends Cell Biol 18:119–127

Kost B, Lemichez E, Spielhofer P, Hong Y, Tolias K, Carpenter C, Chua NH (1999) Rac homologues and compartmentalized phosphatidylinsitol 4, 5-bisphosphate act in a common pathway to regulate polar pollen tube growth. J Cell Biol 145:317–330

Lavy M, Bloch D, Hazak O, Gutman I, Poraty L, Sorek N, Sternberg H, Yalovsky S (2007) A novel RAC/ROP effector links cell polarity, root meristem maintenance, and vesicle trafficking. Curr Biol 17:947–952

Le J, El-Assal SE, Basu D, Saad ME, Szymanski DB (2003) Requirements for *Arabidopsis* ATARP1 and ATARP3 during epidermal development. Curr Biol 13:1341–1347

Lee JL, Szumlanski A, Nielsen E, Yang Z (2008) Rho GTPase-dependent filamentous actin dynamics coordinate vesicle targeting and exocytosis during tip growth. J Cell Biol 181:1155–1168

Lee YJ, Yang Z (2008) Tip growth: signaling in the apical dome. Curr Opin Plant Biol 11:662–671

Lemichez E, Wu Y, Sanchez JP, Mettouchi A, Mathur J, Chua NH (2001) Inactivation of AtRac1 by abscisic acid is essential for stomatal closure. Genes Dev 15:1808–1816

Li S, Blanchoin L, Yang Z, Lord EM (2003) The putative *Arabidopsis* Arp2/3 complex controls leaf cell morphogenesis. Plant Physiol 132:2034–2044

Li S, Gu Y, Yan A, Lord E, Yang Z (2008) RIP1 (ROP-interactive Partner 1)/ICR1 marks pollen germination sites and may act in the ROP1 pathway in the control of polarized pollen growth. Mol Plant 1:1021–1035

Li H, Shen JJ, Zheng ZL, Lin Y, Yang Z (2001) The Rop GTPase switch controls multiple developmental processes in *Arabidopsis*. Plant Physiol 126:670–684

Mathur J, Mathur N, Kirik V, Kernebeck B, Srinivas BP, Hülskamp M (2003) *Arabidopsis CROOKED* encodes for the smallest subunit of the ARP2/3 complex and controls cell shape by region specific fine F-actin formation. Development 130:3137–3146

Molendijk AJ, Bischoff F, Rajendrakumar CS, Friml J, Braun M, Bilroy S, Palme K (2001) *Arabidopsis* thaliana Rop GTPases are localized to tips of root hairs and control polar growth. EMBO J 20:2779–2788

Monshausen GB, Bibikova RN, Messerli MA, Gilroy S (2007) Oscillations in extracellular pH and reactive oxygen species modulate tip growth of *Arabidopsis* root hairs. Proc Natl Acad Sci USA 104:20996–21001

Nibau C (2005) The role of the *Arabidopsis* small GTPase Arac7 (ROP9) in hormone signaling. Ph.D. Dissertation, University of Massachusetts, Amherst, MA

Nibau C, Wu H, Cheung AY (2006) RAC/ROP GTPases: hubs for signal integration and diversification in plants. Trends Plant Sci 11:309–315

Novick P, Guo W (2002) Ras family therapy: Rab, Rho and Ral talk to the exocyst. Trends Cell Biol 12:247–249

Ono E, Wong HL, Kawasaki T, Jasegawa M, Kodama O, Shimamoto K (2001) Essential role of the small GTPase Rac in disease resistance of rice. Proc Natl Acad Sci USA 98:759–764

Opalski KS, Schultheiss H, Kogel KH, Hückelhoven R (2005) The receptor-like MLO protein and the RAC/ROP family G protein RACB modulate actin reorganization in barley attacked by the biotropic powdery mildew fungus *Blumeria graminis* f.sp. hordei. Plant J 41:291–303

Paciorek T, Zazimalova E, Ruthardt N, Petrasek J, Stierhof YD, Kleine-Vehn J, Morris DA, Emans N, Jurgens G, Geldner N, Friml J (2005) Auxin inhibits endocytosis and promotes its ownefflux from cells. Nature 435:1251–1256

Pan J, Fujioka S, Peng J, Chen J, Li G, Chen R (2009) The E3 ubiqutin ligase $SDF^{tir1/AFB}$ and membrane sterols play key roles in auxin regulation of endocytosis, recycling, and plasma membrane accumulation of the auxin efflux transporter PIN2 in *Arabidopsis thaliana*. Plant Cell 21(2):568–580

Potocky M, Jones MA, Bezvoda R, Smirnoff N, Zarsky V (2007) Reactive oxygen species produced by NADPH oxidase are involved in pollen tube growth. New Phytol 174:742–751

Shichrur K, Yalovsky S (2006) Turning ON the switch – RhoGEFs in plants. Trends Plant Sci 11:57–59

Schultheiss H, Dechert C, Kogel KH, Huckelhoven R (2002) A small GTP-binding host protein is required for entry of powdery mildew fungus into epidermal cells of barley. Plant Physiol 128:1447–1454

Schultheiss H, Dechert C, Kogel KH, Huckelhoven R (2003) Functional analysis of barley RAC/ROP G protein family members in susceptibility to the powdery mildew fungus. Plant J 36:589–601

Schultheiss H, Hensel G, Imani J, Broeders S, Sonnewald U, Kogel KH, Kumlehn J, Huckelhoven R (2005) Ectopic expression of constitutively activated RACB in barley enhances susceptibility to powdery mildew and abiotic stress. Plant Physiol 139:353–362

Schultheiss H, Preuss J, Pircher T, Eichmann R, Huckelhoven R (2008) BarleyRIC171 interacts with RACB in plants and supports entry of the powdery mildew fungus. Cell Microbiol 10:1815–1826

Szymanski DB (2005) Breaking the WAVE complex: the point of *Arabidopsis* trichomes. Curr Opin Plant Biol 8:103–112

Tao L-z, Cheung AY, Wu H-M (2002) Plant Rac-like GTPases are activated by auxin and mediate auxin responsive gene expression. Plant Cell 14:2745–2760

Tao L-z, Cheung AY, Nibau SC, Wu H-M (2005) RAC GTPases in Tobacco and *Arabidopsis* mediate auxin-induced formation of proteolytically active nuclear protein bodies that contain AUX/IAA proteins. Plant Cell 17:2369–2383

TerBush DR, Maurice T, Roth D, Novick P (1996) The exocyst is a multiprotein complex required for exocytosis in *Saccharomyces*. EMBO J 15:6483–6494

Welch MD, Mullins RD (2002) Cellular control of actin nucleation. Annu Rev Cell Dev Biol 18:247–288

Wong HL, Pinontoan R, Hayashi K, Tabata R, Yaeno T, Hasegawa K, Kojima C, Yoshioka H, Iba K, Kawasaki T, Shimamoto K (2007) Regulation of Rice NADPH oxidase by binding of Rac GTPase to its *N*-terminal extension. Plant Cell 19:4022–4034

Wu G, Gu Y, Li S, Yang Z (2001) A genome-wide analysis of *Arabidopsis* Rop-interactive CRIB motif-containing proteins that act as Rop GTPase targets. Plant Cell 13:2841–2856

Wu G, Li H, Yang Z (2000) *Arabidopsis* RopGAPs are a novel family of rho GTPase-activating proteins that require the Cde42/Rac-interactive binding motif for Rop-specific GTPase stimulation. Plant Physiol 124:1625–1636

Yang Z, Fu Y (2007) ROP/RAC GTPase signaling. Curr Opin Plant Biol 10:490–494

Zhang D, Wengier D, Shuai B, Gui C, Muschietti J, McCormick S, Tang W-H (2008) The pollen receptor kinase LePRK2 mediates growth promoting signals and positively regulates pollen germination and tube growth. Plant Physiol 148:1368–1379

Zhang Y, McCormick S (2007) A distinct mechanism regulating a pollen-specific guanine nucleotide exchange factor for the small GTPase Rop in *Arabidopsis thaliana*. Proc Natl Acad Sci USA 104:18830–18835

Zheng ZL, Nafisi M, Tam A, Li H, Crowell DN, Chary SN, Schroeder JI, Shen J, Yang Z (2001) Plasma membrane-associated ROP10 small GTPase is a specific negative regulator of abscisic acid responses in *Arabidopsis*. Plant Cell 14:2787–2797

Heterotrimeric G Proteins and Plant Hormone Signaling in Rice

Katsuyuki Oki, Yukiko Fujisawa, Hisaharu Kato, and Yukimoto Iwasaki

Abstract Deficient mutants of the heterotrimeric G protein α subunit (Gα) gene in rice (*d1*) showed several characteristic phenotypes, such as erect and shortened leaves, shortened internodes, and shortened seeds. The characteristics of ten alleles of *d1* are summarized in this chapter. Sensitivity of *d1* to seven plant hormones (auxin, abscisic acid, gibberellin, cytokinin, brassinosteroid, ethylene, and jasmonate) was tested. *d1* clearly exhibited decreased sensitivity to only 24-epibrassinolide (24-epiBL), a brassinosteroid (BR) derivative. Growth inhibition of the seminal roots, elongation of the coleoptile and the second leaf sheath, and increase of lamina inclination were partially impaired in the *d1-1* mutant upon 24-epiBL treatment. However, no apparent epistasis was observed between *d1-1* and a BR-deficient mutant, *d61-7*. The feedback regulation of BR-biosynthetic genes in response to 24-epiBL was normal in *d1-1*. The amounts of BR-intermediates in *d1-1* were not different from those in T65, a recurrent parent of *d1-1*. These results suggest that a mutation in the rice Gα gene affects the BR signaling pathway, but Gα may not be a signaling molecule in BRI1-mediated perception/transduction.

1 Introduction

Heterotrimeric G proteins play important roles in a wide range of physiological responses by transducing extracellular information into intracellular signaling components. They are composed of three subunits, namely the G protein α subunit (Gα), β subunit (Gβ), and γ subunit (Gγ). It is considered that these subunits form a complex. When a ligand is recognized by a G protein-coupled receptor (GPCR), the GDP bound to Gα is exchanged to GTP by the guanine exchange factor (GEF)

K. Oki, Y. Fujisawa, H. Kato, and Y. Iwasaki (✉)
Department of Bioscience, Fukui Prefectural University, 4-1-1 Matsuoka Kenjyojima, Eiheiji-cho, Yoshida-gun Fukui 910-1195, Japan
e-mail: s0693001@s.fpu.ac.jp; fujisawa@fpu.ac.jp; vtec@fpu.ac.jp; iwasaki@fpu.ac.jp

S. Yalovsky et al. (eds.), *Integrated G Proteins Signaling in Plants*, Signaling and Communication in Plants,
DOI 10.1007/978-3-642-03524-1_7, © Springer-Verlag Berlin Heidelberg 2010

activity of GPCR. This complex subsequently dissociates into the GTP-bound Gα and the Gβγ dimer with subsequent regulation of effector proteins. In plants, the canonical Gα, Gβ, and Gγ are encoded by a small number of genes (Jones and Assmann 2004) unlike in animals (Offermanns 2000). *Arabidopsis* harbors one gene each for Gα (Ma et al. 1990), Gβ (Weiss et al. 1994), and Gγ (Mason and Botella 2000). The rice plant also harbors one gene each for Gα (Ishikawa et al. 1995), Gβ (Ishikawa et al. 1996), and Gγ (Kato et al. 2004). Although there is a limited number of subunits in the heterotrimeric G protein in plants, it is involved in the responses of plant not only to hormones but also to other external signals such as light and elicitors (Assmann 2002; Iwasaki et al. 2008). In this chapter, we focus on the function of rice Gα in plant hormone responses.

2 Analysis of the Rice *d1* Mutant Deficient in the Heterotrimeric G Protein α Subunit (Gα) Gene

The *d1* mutant, which is deficient in the heterotrimeric G protein α subunit (Gα) gene, in rice was isolated in 1999 (Ashikari et al. 1999; Fujisawa et al. 1999). Until now, we have obtained ten *d1* alleles through the screening of shortened seed mutants (Oki et al. 2009b). As many cultivars with different genetic background are used in rice research, it is necessary to take into account the genetic background of rice plants to investigate plant hormone responses. The ten *d1* alleles can be grouped on the basis of their recurrent parent (Fig. 1). The ten *d1* alleles showed several of the characteristic phenotypes, such as erect and shortened leaves, shortened internodes, and seeds with reduced lengths.

When cDNAs of Gα were synthesized and amplified by RT-PCR in the different *d1* alleles using primers that cover the first Met and the stop codon, products of various sizes were obtained depending on the presence of an insertion or a deletion in Gα (Oki et al. 2009b). These results showed that mRNAs for the mutant Gα did accumulate in all *d1* alleles. The different mutations in the *d1* allele were identified by sequencing the cDNAs. The three in-frame mutants, *d1-3*, *d1-4*, and *d1-8*, characterized by a 16-amino acid deletion, one amino acid exchange, and one amino acid deletion, respectively, will be useful for further biochemical studies. However, the Gα protein which is normally localized in the plasma membrane fraction of wild-type (WT) was not detected in the plasma membrane fraction of all *d1* alleles, except in *d1-4*. In this mutant, the mutated Gα protein was present in plasma membrane fraction but in very low amounts.

These studies brought out some interesting points. First, the translation products in the *d1* mutants do not accumulate (or very little) in the plasma membrane fraction, unlike their WT counterparts (Oki et al. 2009b). This suggests that rice Gα may be strictly regulated by a quality-control system at the protein conformation level, but the mechanism of this regulation is not currently studied. Second, *d1-8*, characterized by a single amino acid deletion in the putative effector-binding

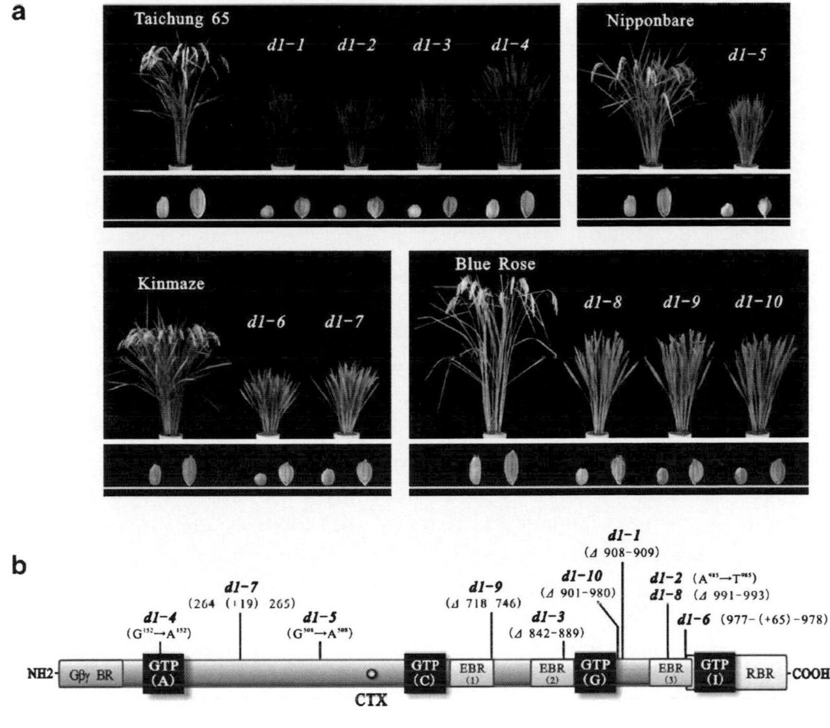

c

Allele	Mutation		Accumlation of Gα mRNA	Predicted size of Mutated Gα (kDa)	Accumlation of Mutated Gα Protein	Phenotype
d1-1	2 base deletion	preterm stop codon	Yes	35.3	No	severe
d1-2	1 base substitution	preterm stop codon	Yes	37.9	No	severe
d1-3	48 base deletion	16 amino acid deletion	Yes	42.2	No	severe
d1-4	1 base substitution	G''→E''	Yes	44.3	slightly accumulated	mild
d1-5	1 base substitution	preterm stop codon	Yes	19.1	No	severe
d1-6	65 base insertion	preterm stop codon	Yes	37.7	No	severe
d1-7	19 base insertion	preterm stop codon	Yes	10.7	No	severe
d1-8	3 base deletion	one amino acid deletion	Yes	44.1	No	severe
d1-9	29 base deletion	preterm stop codon	Yes	27.2	No	severe
d1-10	80 base deletion	preterm stop codon	Yes	35.2	No	severe

Fig. 1 *Overall morphology and schematic representation of the mutation positions of d1 alleles.* (**a**) *Top*; Four *d1* alleles, *d1-1*, *d1-2*, *d1-3*, and *d1-4*, the recurrent parent of which was Taichung 65. One *d1* allele, *d1-5*, the recurrent parent of which was Nipponbare. *Bottom*; two *d1* alleles, *d1-6* and *d1-7*, the recurrent parent of which was Kinmaze. Three *d1* alleles, *d1-8*, *d1-9*, and *d1-10*, the recurrent parent of which was Blue Rose. (**b**) Mutation positions of ten *d1* alleles. Gβγ BR, a putative Gβγ binding region; GTP(A), (C), (G) and (I), GTP-binding sites; EBR (1), (2) and (3), putative effector-binding regions; RBR, a putative receptor binding region; CTX, the modification site by cholera toxin. The A of the initiation codon ATG of rice Gα was designated as number 1. (**c**) List of the type of mutations, predicted translational products, and severity of the mutant phenotype

region 3 near its C terminus, has similar phenotypes to those of the null mutants of *d1* and does not accumulate Gα proteins in the plasma membrane fraction. This suggests that the putative effector binding region 3 is very important for the stability of rice Gα. Third, a mild allele of *d1*, *d1-4* was isolated. The internode length and seed size in *d1-4* were longer and bigger, respectively, than that in *d1-1*, *d1-2*, and *d1-3,* the recurrent parents of which are T65. Thus, *d1-4* showed mild phenotypes in plant height and seed size. The mutated Gα protein in *d1-4* may partially lose the guanine-nucleotide binding or hydrolysis activity, because a single amino acid mutation was generated in the GTP-binding box A, a highly conserved motif among Gα proteins throughout eukaryotes. As low levels of the mutated Gα protein can be detected in *d1-4*, the mild phenotypes of *d1-4* may be due to the accumulation of low levels of the mutated Gα protein. It was showed that rice Gα was present in a large complex localized in the plasma membrane fraction and that it could be dissociated into its monomeric components in the presence of GTPγS (Kato et al. 2004). The presence of the large complex was also shown in *Arabidopsis* (Wang et al. 2008). This complex may be a platform for the Gα signaling cascade in plants. In the future, it will be important to study the subunit composition of these complexes. It was shown that *Arabidopsis* Gα (AtGPA1) has a unique enzymatic characteristic, in that it is constitutively present in a GTP-bound form (Johnston et al. 2007). AtGPA1 may not require GPCR-mediated GEF activity to accomplish signal transduction as do other known Gα subunits. This suggests that *Arabidopsis* Gα in the large complex may be an active form. It will be necessary to analyze the enzymatic characteristics of the Gα localized in large complex.

3 Response of the Rice *d1* Mutants to Plant Hormones

Analyzes of the mutants deficient in *gpa1*, the heterotrimeric G protein α subunit (Gα) gene in *Arabidopsis*, have shown that GPA1 (the Gα subunit) is involved in many physiological responses, including those to abscisic acid (ABA) (Wang et al. 2001; Ullah et al. 2002; Lapik and Kaufman 2003; Mishra et al. 2006; Pandey et al. 2009), gibberellic acid (GA) (Ullah et al. 2002; Chen et al. 2004), brassinosteroid (BR) (Ullah et al. 2002; Chen et al. 2004), D-glucose (Huang et al. 2006), blue light (Warpeha et al. 2006; Warpeha et al. 2007), sphingosine-1-phosphate (S1P) (Coursol et al. 2003; Pandey and Assmann 2004), and ozone (Joo et al. 2005). GPA1 affects ABA signaling via ABA receptors. Recently, GPCR-type G proteins (GTG1 and GTG2) were identified as ABA receptors that interact with GPA1 (Pandey et al. 2009). GTGs participate in the multiple ABA responses, such as the inhibition of germination and primary root growth and the promotion of stomatal closure. GTGs possess an intrinsic GTPase activity which is inhibited by GPA1. GDP-bound state GTGs show a greater affinity to ABA and they are assumed to be able to transmit the ABA signal. Hypersensitivity of the *gpa1* mutant and hyposensitivity of the *gtg1/gtg2* double mutant to ABA are in good concordance with the functional relationship between GTGs and GPA1.

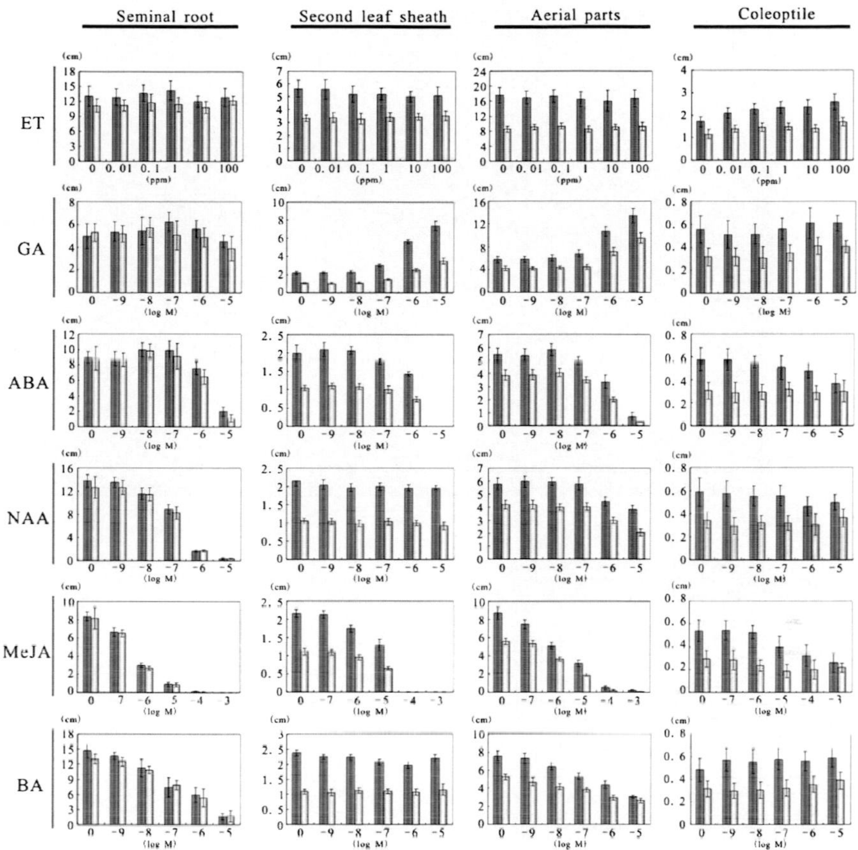

Fig. 2 *Plant hormones responses in d1-1.* Comparison of the responses to six plant hormones in T65 as WT and *d1-1*. Plants were grown for 1 week after germination on the agar medium with various concentrations of each plant hormone. The lengths of the seminal root, adventitious root, second leaf sheath, aerial parts, and coleoptile of these plants were measured. Gray bars and white bars correspond to values of WT and *d1-1* mutant respectively. Each value represents the means of 15 seedlings. (Error bars = SD)

The involvement of Gα in plant hormone responses was also investigated in rice, using the Gα deficient mutant, *d1* (Wang et al. 2006; Oki et al. 2008). In Fig. 2, we show the results of responses of *d1-1* to plant hormone treatments, namely GA, ABA, naphthaleneacetic acid (NAA, an auxin), 6-benzylaminopurine (BA, a cytokinin), ethylene (ET), and methyl jasmonate (MeJA). The sensitivity of *d1-1* (named T65d1 in a previous paper) to ET, GA, ABA, NAA, MeJA, and BA at young seedling stage was not significantly different from those of T65 (Taichung 65, the recurrent parent of *d1-1*). On the other hand, *d1-1* exhibited clearly decreased sensitivity to 24-epibrassinolide (24-epiBL), a brassinosteroid (BR) derivative (Fig. 3). Similar results have been described previously (Wang et al. 2006). Therefore, at the seedling stages of rice plants, G protein signaling appears to be related to the BR signaling pathway.

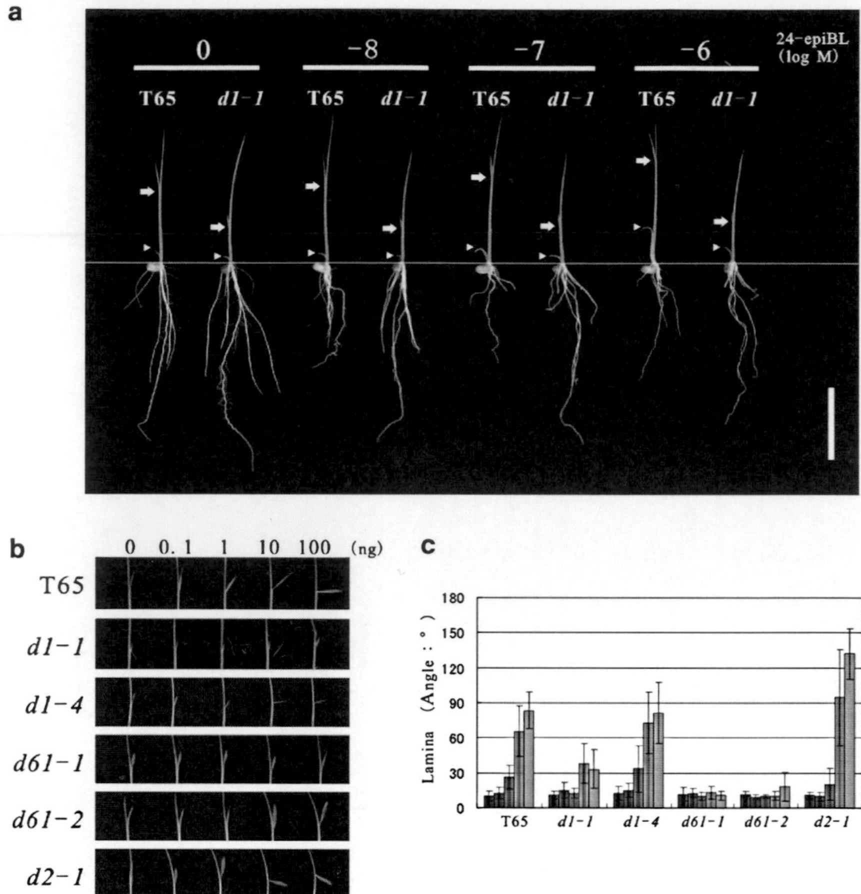

Fig. 3 *Sensitivity of d1-1 and BR-deficient mutants to exogenously applied 24-epiBL.* (**a**) Plant morphology of T65 and *d1-1* grown for 1 week after germination on the agar medium with the various concentrations of 24-epiBL. The second leaf sheath, the coleoptile and the adventitious roots are shown. *Arrows* represent the top part of the second leaf sheath. *Arrowheads* represent the top part of the coleoptile. (Bar, 3 cm). (**b**) Lamina joint inclination of *d1* and BR-deficient mutants upon 24-epiBL application. T65, *d1* mutants (*d1-1* and *d1-4*) and BR-deficient mutants (*d61-1*, *d61-2* and *d2-1*) were grown for 4 days. Subsequently, various concentrations of 24-epiBL were applied to the lamina region of second leaves and grown for 3 days. (**c**) Angles of the lamina region of (**b**). Dark to light gray bars represent the concentration of 0, 0.01, 0.1, 1, 10, and 100 ng of 24-epiBL, respectively. Each value is the means of 15 seedlings. (Error bar = SD)

The growth inhibition of seminal roots and the promotion of elongation of coleoptiles and the second leaf sheathes were partially impaired by 24-epiBL in *d1-1* (Fig. 3a). *d1-1* also showed decreased sensitivity to an enhancement of lamina joint inclination by 24-epiBL (Fig. 3b and c). These results indicate that rice Gα is involved in specific BR responses. The morphology of *d1-1* was compared with those of the BR-deficient mutants (BR receptor mutants *d61-1*, *d61-2*, and *d61-7*,

Fig. 4 *Phenotype of the d1-1/d61-7 double mutant.* Overall morphology of T65, *d1-1*, *d61-7*, and *d1-1/d61-7* double mutant after grain-filling stage

and a BR biosynthesis mutant *d2-1*) grown in a greenhouse. The results showed that the phenotype of *d1-1* was similar to that of known BR-deficient mutants, such as shortened second internodes and erect leaves (Oki et al. 2008). When these plants were grown in darkness, they showed constitutive photomorphogenic growth phenotype. These characteristics have been previously reported on some BR-deficient mutants (Yamamuro et al. 2000; Hong et al. 2002; Tanabe et al. 2005). Thus, the *d1* mutant and the BR-deficient mutants appear to share common developmental defects. In addition, transgenic plants expressing the constitutively active form of rice Gα/Q223L (Oki et al. 2005) showed hypersensitivity to 24-epiBL compared with the wild-type (Oki et al. 2009a), suggesting that rice Gα/Q223L may enhance the BR signaling pathway. *Arabidopsis* GCR1, a putative plant GPCR, acts as a positive regulator of BR signaling based on the decreased sensitivity of the *gcr1* mutant to exogenous BR treatment (Chen et al. 2004). This result is supportive of the conclusion that plant heterotrimeric G protein signaling is related to the BR signaling pathway in some way.

Involvement of rice Gα proteins in the BR cascade appears to be independent from that of the rice BR receptor, OsBRI1 (Oki et al. 2008). First, no apparent epistasis was observed in the length of internodes, the elongation pattern of internodes, the leaf morphology, and the number of cells in the leaf sheath between *d1-1* and *d61-7* (an OsBRI1 mutant). The phenotype of *d1-1/d61-7* double mutant was additive to each single mutant, *d1-1* and *d61-7*, respectively (Fig. 4). A similar relationship has also been observed between *gpa1* (an *Arabidopsis* Gα mutant) and *bri1* or *det2* (*Arabidopsis* BR deficient mutants) (Gao et al. 2008). Second, the feedback regulation of the expression of BR-biosynthetic genes with 24-epiBL was normal in *d1-1* as in T65, but it is impaired in a rice OsBRI1 mutant, *d61* (Yamamuro et al. 2000; Tanabe et al. 2005). Third, the amounts of BR-intermediates in *d1-1* were not different from those in T65, but they are higher in *d61* than in T65 (Yamamuro et al. 2000), probably to compensate for an impaired BR-signaling. The result suggests that the mechanism regulating the amounts of BR intermediates is not impaired in *d1-1* and that Gα may not be a part of the BRI1-mediated signaling system. Thus, rice Gα appears to be involved in the BR signaling pathway independently from a BRI1-mediated perception/transduction system. Possible functions of rice Gα in the BR signaling pathway were postulated and summarized in Fig. 5.

From the analysis of the *d1-1/d61-7* double mutant, the effects of plant Gα on cell proliferation and elongation were discussed (Oki et al. 2008). Rice Gα functions positively in cell proliferation, because the cell number in *d1-1* decreased

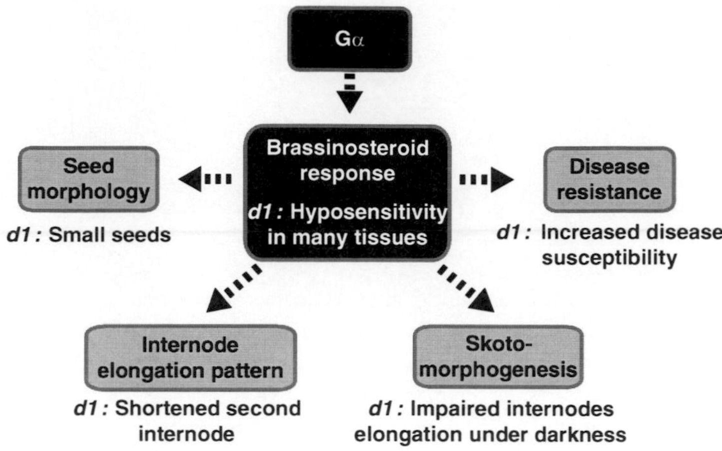

Fig. 5 Putative function of Gα in BR signaling pathway

relatively to that in T65. Similar function has also been reported for *Arabidopsis* Gα (Ullah et al. 2001; Chen et al. 2006). BRI1 is a positive regulator for cell proliferation in rice plant (Nakamura et al. 2006) and in *Arabidopsis* (Nakaya et al., 2002) as well. However, an additive effect on cell proliferation in the Gα and BRI1 double mutant was observed in rice (Oki et al. 2008) and *Arabidopsis* (Gao et al. 2008). Thus, the function of Gα in cell proliferation may be different from that of BRI1-mediated signaling. Rice Gα also functions positively in cell length, because the cell length in *d1-1* is slightly shorter than that in T65. The cell elongation induced by 24-epiBL was partially reduced in the lamina joint region of *d1-1*. The result supports the conclusion that rice Gα is a positive regulator for cell length in T65 background. OsBRI1 functions positively in controlling cell length as well, because the cell length in *d61-7* is shorter than that in T65. As a result, it was considered that cell length in the *d1-1/d61-7* double mutant results from a synergistic effect of Gα and OsBRI1. Thus, G protein signaling and BR signaling may cross-talk in regulating cell elongation.

4 Interdependency of Plant Heterotrimeric G Protein Signaling and Plant Hormone Signaling

Interdependency among plant hormones is an important concept and many examples have been reported. Among them, the most famous relationship is between GA and ABA (Achard et al. 2006; Razem et al. 2006). BR also works together with a wide variety of plant hormones such as auxin (Nakamura et al. 2003; Mouchel et al. 2006), GA (Shimada et al. 2006), ABA (Steber and McCourt 2001), and ET (Dingus et al. 2002; Shi et al. 2006). BR plays an important role in coordinating their functions in plants.

In *Arabidopsis*, abnormal responses to ABA in *gpa1* were observed in seed germination and stomatal response. The mutant is hypersensitive to ABA during seed germination (Pandey et al. 2006) and insensitive to ABA in the inhibition of stomatal opening (Wang et al. 2001). In stomata, the location of GPA1 in ABA signaling pathway is becoming clear now once the ABA receptors which interact with GPA1 have been found (Pandey et al. 2009). AtPLDα1 is another target of GPA1, and the activation of GPA1 leads to cancelation of the inactivated status of AtPLDα1 (Zhao and Wang 2004). AtPLDα1 is involved in both the promotion of stomatal closure and the inhibition of stomatal opening by ABA. These two processes are separately mediated by AtABI1 (protein phosphatase 2C) and GPA1 downstream of AtPLDα1 (Mishra et al. 2006). It was suggested that the ABA hypersensitivity of *gpa1* during the seed germination process may be brought about by the indirect effect of the BR response disruption (Ullah et al. 2002). However, the ABA hypersensitivity of *gpa1* may be explained by considering the function of the ABA receptors and GPA1. Study of the cross-talk between the ABA and the BR signaling will be necessary.

The *d1* mutant was originally identified as a GA signaling mutant based on its insensitivity to GA in the induction of α-amylase activity in aleurone cells (Ueguchi-Tanaka et al. 2000). However, the second leaf sheath of *d1* was elongated by GA. It was then suggested that the GA action was affected by BR signaling in *Arabidopsis* (Ullah et al. 2002). Given that decreased GA sensitivity of *gpa1* is a result of deficiency in BR action, decreased sensitivity to GA in aleurone cells of *d1* may be a result of an indirect effect of impaired BR action.

In addition, it has been reported that rice Gα functions as a key player in the enhancement of disease resistance against an avirulent race of blast fungus (Suharsono et al. 2002) and a bacterial pathogen, *Xanthomonas oryzae* pv. *oryzae* (Komatsu et al. 2004) in rice plants. The enhanced disease resistance is also induced by exogenous BR treatment of rice plants and it is called Brassinosteroid-mediated Disease Resistance (BDR) (Nakashita et al. 2003). Although the mechanism of BDR is unclear, the increased disease susceptibility in *d1* may be an indirect consequence of its aberrant BR response. In short, almost all of the features discovered in rice Gα mutants fit well to a BR deficient phenotype.

Among seven plant hormones, BR seems to be a signal which is closely related to the heterotrimeric G protein signaling in rice, while BR and ABA are related to this signaling pathway in *Arabidopsis*. In order to elucidate the functions of plant Gα, we will need to pay attention to plant species, tissues, developmental stages, growth conditions, and output events.

References

Achard P, Cheng H, De Grauwe L, Decat J, Schoutteten H, Moritz T, Van Der Straeten D, Peng J, Harberd NP (2006) Integration of plant responses to environmentally activated phytohormonal signals. Science 311:91–94

Ashikari M, Wu J, Yano M, Sasaki T, Yoshimura A (1999) Rice gibberellin-insensitive dwarf mutant gene *Dwarf 1* encodes the α-subunit of GTP-binding protein. Proc Natl Acad Sci USA 96:10284–10289

Assmann SM (2002) Heterotrimeric and unconventional GTP binding proteins in plant cell signaling. Plant Cell 14:S355–373

Chen JG, Gao Y, Jones AM (2006) Differential roles of *Arabidopsis* heterotrimeric G-protein subunits in modulating cell division in roots. Plant Physiol 141:887–897

Chen JG, Pandey S, Huang J, Alonso JM, Ecker JR, Assmann SM, Jones AM (2004) GCR1 can act independently of heterotrimeric G-protein in response to brassinosteroids and gibberellins in *Arabidopsis* seed germination. Plant Physiol 135:907–915

Coursol S, Fan LM, Le Stunff H, Spiegel S, Gilroy S, Assmann SM (2003) Sphingolipid signalling in *Arabidopsis* guard cells involves heterotrimeric G proteins. Nature 423:651–654

Dingus J, Tatum BS, Vaidyanathan G, Hildebrandt JD (2002) Purification of G protein βγ from bovine brain. Methods Enzymol 344:194–208

Fujisawa Y, Kato T, Ohki S, Ishikawa A, Kitano H, Sasaki T, Asahi T, Iwasaki Y (1999) Suppression of the heterotrimeric G protein causes abnormal morphology, including dwarfism, in rice. Proc Natl Acad Sci USA 96:7575–7580

Gao Y, Wang S, Asami T, Chen JG (2008) Loss-of-function mutations in the *Arabidopsis* heterotrimeric G-protein α subunit enhance the developmental defects of brassinosteroid signaling and biosynthesis mutants. Plant Cell Physiol 49:1013–1024

Hong Z, Ueguchi-Tanaka M, Shimizu-Sato S, Inukai Y, Fujioka S, Shimada Y, Takatsuto S, Agetsuma M, Yoshida S, Watanabe Y, Uozu S, Kitano H, Ashikari M, Matsuoka M (2002) Loss-of-function of a rice brassinosteroid biosynthetic enzyme, C-6 oxidase, prevents the organized arrangement and polar elongation of cells in the leaves and stem. Plant J 32: 495–508

Huang J, Taylor JP, Chen JG, Uhrig JF, Schnell DJ, Nakagawa T, Korth KL, Jones AM (2006) The plastid protein THYLAKOID FORMATION1 and the plasma membrane G-protein GPA1 interact in a novel sugar-signaling mechanism in *Arabidopsis*. Plant Cell 18:1226–1238

Ishikawa A, Iwasaki Y, Asahi T (1996) Molecular cloning and characterization of a cDNA for the β subunit of a G protein from rice. Plant Cell Physiol 37:223–228

Ishikawa A, Tsubouchi H, Iwasaki Y, Asahi T (1995) Molecular cloning and characterization of a cDNA for the α subunit of a G protein from rice. Plant Cell Physiol 36:353–359

Iwasaki Y, Kato H, Fujisawa Y, Oki K (2008) Rice heterotrimeric G protein signaling. In: Hirano H-Y, Hirai A, Sano Y, Sasaki T (eds) Rice Biology in the Genomics Era. Springer, Berlin, pp 135–148

Johnston CA, Taylor JP, Gao Y, Kimple AJ, Grigston JC, Chen JG, Siderovski DP, Jones AM, Willard FS (2007) GTPase acceleration as the rate-limiting step in *Arabidopsis* G protein-coupled sugar signaling. Proc Natl Acad Sci USA 104:17317–17322

Jones AM, Assmann SM (2004) Plants: the latest model system for G-protein research. EMBO Rep 5:572–578

Joo JH, Wang S, Chen JG, Jones AM, Fedoroff NV (2005) Different signaling and cell death roles of heterotrimeric G protein α and β subunits in the *Arabidopsis* oxidative stress response to ozone. Plant Cell 17:957–970

Kato C, Mizutani T, Tamaki H, Kumagai H, Kamiya T, Hirobe A, Fujisawa Y, Kato H, Iwasaki Y (2004) Characterization of heterotrimeric G protein complexes in rice plasma membrane. Plant J 38:320–331

Komatsu S, Yang G, Hayashi N, Kaku H, Umemura K, Iwasaki Y (2004) Alterations by a defect in a rice G protein α subunit in probenazole and pathogen-induced responses. Plant Cell Environm 27:947–957

Lapik YR, Kaufman LS (2003) The *Arabidopsis* cupin domain protein AtPirin1 interacts with the G protein α-subunit GPA1 and regulates seed germination and early seedling development. Plant Cell 15:1578–1590

Ma H, Yanofsky MF, Meyerowitz EM (1990) Molecular cloning and characterization of *GPA1*, a G protein α subunit gene from *Arabidopsis thaliana*. Proc Natl Acad Sci USA 87:3821–3825

Mason MG, Botella JR (2000) Completing the heterotrimer: isolation and characterization of an *Arabidopsis thaliana* G protein γ-subunit cDNA. Proc Natl Acad Sci USA 97:14784–14788

Mishra G, Zhang W, Deng F, Zhao J, Wang X (2006) A bifurcating pathway directs abscisic acid effects on stomatal closure and opening in *Arabidopsis*. Science 312:264–266

Mouchel CF, Osmont KS, Hardtke CS (2006) *BRX* mediates feedback between brassinosteroid levels and auxin signalling in root growth. Nature 443:458–461

Nakamura A, Higuchi K, Goda H, Fujiwara MT, Sawa S, Koshiba T, Shimada Y, Yoshida S (2003) Brassinolide induces *IAA5*, *IAA19*, and DR5, a synthetic auxin response element in *Arabidopsis*, implying a cross talk point of brassinosteroid and auxin signaling. Plant Physiol 133:1843–1853

Nakamura A, Fujioka S, Sunohara H, Kamiya N, Hong Z, Inukai Y, Miura K, Takatsuto S, Yoshida S, Ueguchi-Tanaka M, Hasegawa Y, Kitano H, Matsuoka M (2006) The role of *OsBRI1* and its homologous genes, *OsBRL1* and *OsBRL3*, in rice. Plant Physiol 140:580–590

Nakashita H, Yasuda M, Nitta T, Asami T, Fujioka S, Arai Y, Sekimata K, Takatsuto S, Yamaguchi I, Yoshida S (2003) Brassinosteroid functions in a broad range of disease resistance in tobacco and rice. Plant J 33:887–898

Nakaya M, Tsukaya H, Murakami N, Kato M (2002) Brassinosteroids control the proliferation of leaf cells of *Arabidopsis thaliana*. Plant Cell Physiol 43:239–244

Offermanns S (2000) Mammalian G-protein function in vivo: new insights through altered gene expression. Rev Physiol Biochem Pharmacol 140:63–133

Oki K, Fujisawa Y, Kato H, Iwasaki Y (2005) Study of the constitutively active form of the α subunit of rice heterotrimeric G proteins. Plant Cell Physiol 46:381–386

Oki K, Kitagawa K, Fujisawa Y, Kato H, Iwasaki Y (2009a) Function of the α subunit of heterotrimeric G protein in brassinosteroid response of rice plants. Plant Signal Behav 4:34–36

Oki K, Inaba N, Kitano H, Takahashi S, Fujisawa Y, Kato H, Iwasaki Y (2009b) Study of novel *d1* alleles, defective mutants of the α subunit of heterotrimeric G protein in rice. Genes Genet Syst 84(1):35–42

Oki K, Inaba N, Kitagawa K, Fujioka S, Kitano H, Fujisawa Y, Kato H, Iwasaki Y (2008) Function of the α subunit of rice heterotrimeric G protein in brassinosteroid signaling. Plant Cell Physiol 50:161–172

Pandey S, Assmann SM (2004) The *Arabidopsis* putative G protein-coupled receptor GCR1 interacts with the G protein α subunit GPA1 and regulates abscisic acid signaling. Plant Cell 16:1616–1632

Pandey S, Chen JG, Jones AM, Assmann SM (2006) G-protein complex mutants are hypersensitive to abscisic acid regulation of germination and postgermination development. Plant Physiol 141:243–256

Pandey S, Nelson DC, Assmann SM (2009) Two novel GPCR-type G proteins are abscisic acid receptors in *Arabidopsis*. Cell 136:136–148

Razem FA, Baron K, Hill RD (2006) Turning on gibberellin and abscisic acid signaling. Curr Opin Plant Biol 9:454–459

Shi YH, Zhu SW, Mao XZ, Feng JX, Qin YM, Zhang L, Cheng J, Wei LP, Wang ZY, Zhu YX (2006) Transcriptome profiling, molecular biological, and physiological studies reveal a major role for ethylene in cotton fiber cell elongation. Plant Cell 18:651–664

Shimada A, Ueguchi-Tanaka M, Sakamoto T, Fujioka S, Takatsuto S, Yoshida S, Sazuka T, Ashikari M, Matsuoka M (2006) The rice *SPINDLY* gene functions as a negative regulator of gibberellin signaling by controlling the suppressive function of the DELLA protein, SLR1, and modulating brassinosteroid synthesis. Plant J 48:390–402

Steber CM, McCourt P (2001) A role for brassinosteroids in germination in *Arabidopsis*. Plant Physiol 125:763–769

Suharsono U, Fujisawa Y, Kawasaki T, Iwasaki Y, Satoh H, Shimamoto K (2002) The heterotrimeric G protein α subunit acts upstream of the small GTPase Rac in disease resistance of rice. Proc Natl Acad Sci USA 99:13307–13312

Tanabe S, Ashikari M, Fujioka S, Takatsuto S, Yoshida S, Yano M, Yoshimura A, Kitano H, Matsuoka M, Fujisawa Y, Kato H, Iwasaki Y (2005) A novel cytochrome P450 is implicated in brassinosteroid biosynthesis via the characterization of a rice dwarf mutant, *dwarf11*, with reduced seed length. Plant Cell 17:776–790

Ueguchi-Tanaka M, Fujisawa Y, Kobayashi M, Ashikari M, Iwasaki Y, Kitano H, Matsuoka M (2000) Rice dwarf mutant *d1*, which is defective in the α subunit of the heterotrimeric G protein, affects gibberellin signal transduction. Proc Natl Acad Sci USA 97:11638–11643

Ullah H, Chen JG, Wang S, Jones AM (2002) Role of a heterotrimeric G protein in regulation of *Arabidopsis* seed germination. Plant Physiol 129:897–907

Ullah H, Chen JG, Young JC, Im KH, Sussman MR, Jones AM (2001) Modulation of cell proliferation by heterotrimeric G protein in *Arabidopsis*. Science 292:2066–2069

Wang L, Xu YY, Ma QB, Li D, Xu ZH, Chong K (2006) Heterotrimeric G protein α subunit is involved in rice brassinosteroid response. Cell Res 16:916–922

Wang S, Assmann SM, Fedoroff NV (2008) Characterization of the *Arabidopsis* heterotrimeric G protein. J Biol Chem 283:13913–13922

Wang XQ, Ullah H, Jones AM, Assmann SM (2001) G protein regulation of ion channels and abscisic acid signaling in *Arabidopsis* guard cells. Science 292:2070–2072

Warpeha KM, Lateef SS, Lapik Y, Anderson M, Lee BS, Kaufman LS (2006) G-protein-coupled receptor 1, G-protein Gα-subunit 1, and prephenate dehydratase 1 are required for blue light-induced production of phenylalanine in etiolated *Arabidopsis*. Plant Physiol 140:844–855

Warpeha KM, Upadhyay S, Yeh J, Adamiak J, Hawkins SI, Lapik YR, Anderson MB, Kaufman LS (2007) The GCR1, GPA1, PRN1, NF-Y signal chain mediates both blue light and abscisic acid responses in *Arabidopsis*. Plant Physiol 143:1590–1600

Weiss CA, Garnaat CW, Mukai K, Hu Y, Ma H (1994) Isolation of cDNAs encoding guanine nucleotide-binding protein β-subunit homologues from maize (ZGB1) and *Arabidopsis* (AGB1). Proc Natl Acad Sci USA 91:9554–9558

Yamamuro C, Ihara Y, Wu X, Noguchi T, Fujioka S, Takatsuto S, Ashikari M., Kitano H, Matsuoka M (2000) Loss of function of a rice brassinosteroid insensitive1 homolog prevents internode elongation and bending of the lamina joint. Plant Cell 12:1591–1606

Zhao J, Wang X (2004) *Arabidopsis* phospholipase Dα1 interacts with the heterotrimeric G-protein α-subunit through a motif analogous to the DRY motif in G-protein-coupled receptors. J Biol Chem 279:1794–1800

Auxin, Brassinosteroids, and G-Protein Signaling

Lei Wang and Kang Chong

Abstract Auxin is the earliest-discovered hormone in higher plants. It regulates many aspects of plant growth and development through its biosynthesis, polar transport, and signaling transduction. In the 1970s, brassinosteroids (BRs) were discovered from pollen of brassica plants and identified as functioning hormones. The function of BRs at the cellular level is mainly promoting cell elongation and differentiation. The signal transduction pathway of BRs is well known and is considered one of the clearest pathways in higher plants. Both auxin and BRs contribute to cell elongation and differentiation in plants. GTP-binding proteins are essential elements that mediate receptor and downstream members in cell signal transduction and are also involved in regulation of cell division and elongation. Only recently, the cross talk and communication among these three signaling pathways were explored. Cross talk between auxin and BRs depends on G-protein signaling, in part. This chapter summarizes the detailed cross talk mechanisms among the signal transduction pathways, the knowledge of which will help elucidate the nature of signaling transduction in higher plants.

1 Auxin Signaling

1.1 Auxin Physiological Functions in Higher Plants

As early as 1926, Frits Went obtained a diffusible growth-promoting factor from oat coleoptiles and thereafter named it auxin. Indole-3-acetic acid (IAA) is

K. Chong (✉)
Department of Plant Cellular and Molecular Biology, Ohio State University, Columbus 43210, OH, USA; Institute of Botany, Chinese Academy of Sciences, Beijing, 100093, China
e-mail: chongk@ibcas.ac.cn

L. Wang
Department of Plant Cellular and Molecular Biology, Ohio State University, Columbus 43210, OH, USA

S. Yalovsky et al. (eds.), *Integrated G Proteins Signaling in Plants*,
Signaling and Communication in Plants,
DOI 10.1007/978-3-642-03524-1_8,

the predominate auxin in plants and an indispensable plant hormone with a well-documented ability to regulate many aspects of plant growth and development (Teale et al. 2006). In 1937, Went and Thimann identified auxin as a phytohormone (Went and Thimann 1937). Two years later, IAA was used to create a first, true plant tissue culture (Sussex 2008). Whole physiological studies have identified the involvement of IAA in regulating cell elongation, apical dominance, and root growth and formation, for example. In plant tissue culture, the ratio of auxin and cytokinins in the medium orients the status of organ or tissue differentiation.

The physiological functions of auxin in dicots, represented by *Arabidopsis*, have been well addressed. Use of mutants lacking the ability to synthesize or perceive auxin and physiological treatment with IAA, synthetic auxins such as N-acetyl aspartate (NAA) and transport inhibitors such as N-(1-naphthyl) phthalamic acid (NPA) revealed auxin to play a pivotal role in regulating some developmental processes such as cell division, elongation, and differentiation. At the whole plant level, auxin plays critical roles in root formation, apical dominance, tropic response, and senescence (Jacobs and Ray 1976; Feldman 1985; Estelle 1992; Casimiro et al. 2001; Muday and DeLong 2001; Teale et al. 2006). The effects of auxin on plant growth and development depend highly on the type of auxin applied and its concentration. Although auxin is important to plant growth and development and it plays a central role in many physiological aspects, such as cell elongation and division, the mechanism of auxin action behind each process is not completely understood (Teale et al. 2006).

In contrast to dicots, in monocots, the response to auxin is less sensitive, and this knowledge has been used in the development of herbicides in agriculture. In monocots, even the underlying mechanisms of auxin signaling remain elusive (Nakamura et al. 2006a), although the physiological functions are similar to those in dicots. Wild-type rice plants treated with auxin will display multiple phenotypes such as increased number of adventitious roots and helix primary roots and delayed gravitropic response of roots (Nomura et al. 2004). *IAA* is a well-studied gene in *Arabidopsis* and serves as a negative regulator of auxin signaling. Gain-of-function of *OsIAA3*, an *IAA* gene homolog in rice, causes insensitivity to auxin in transgenic plants and confers multiple phenotypes, including short leaf blades, reduced crown root formation, and abnormal leaf formation (Nakamura et al. 2006b). Thus, auxin is also important for the development in monocot plants.

Auxin regulates many aspects of plant growth and development by altering the expression of diverse genes. However, through loss or gain of function of genes associated with auxin biosynthesis, polar transport and signaling transduction pathway of auxin will affect its action in higher plants.

1.2 Auxin Biosynthesis Pathway

IAA is the most abundant, naturally occurring auxin. Therefore, the main focus of auxin biosynthesis in higher plants is about biosynthesis of IAA. Plants produce active IAA both by de novo synthesis and by releasing IAA from conjugates (Bartel

1997; Schuller and Ludwig-Muller 2006). Multiple IAA de novo synthesis pathways have been characterized in genetic experiments and complimentary biochemical analysis, including both tryptophan-dependent and tryptophan-independent routes of IAA biosynthesis (Wright et al. 1991; Normanly et al. 1993). Tryptophan-dependent pathway for rapid auxin synthesis is a key to generating robust auxin gradients in response to environmental and developmental cues (Stepanova et al. 2008; Tao et al. 2008). Deficiency in tryptophan or its derivatives affects cell expansion during plant organogenesis, which suggests that tryptophan-dependent auxin biosynthesis also mediates the regulation of plant development (Jing et al. 2009). Probably, most of IAA is synthesized from a tryptophan-independent pathway, in which an intermediate in this pathway between anthranilate and tryptophan acts as precursor, rather than from the tryptophan-dependent pathway. Indole-3-glycerol phosphate in the *Arabidopsis* tryptophan biosynthetic pathway serves as a branch-point compound in the tryptophan-independent IAA de novo biosynthetic pathway (Ouyang et al. 2000).

Besides de novo auxin biosynthesis, free and active IAA also can be generated by hydrolyzing the conjugates of IAA. Actually, plants maintain most IAA in conjugated forms (Cohen and Bandurski 1982; Bartel 1997). Many conjugates have been identified, including conjugated carboxyl group of IAA to sugars, high-molecular-weight glycans, amino acids, and peptides (Cohen and Bandurski 1982). The different conjugates possibly perform different functions in plants for IAA destruction or to facilitate its transport (Aharoni and Yang 1983). Multiple enzymes have been characterized as modulating the hydrolysis process (Lopez-Bucio et al. 2005; Schuller and Ludwig-Muller 2006).

Both de novo biosynthesis and conjugate hydrolysis are important inputs to maintain free IAA level in higher plants. As well, methyl modulation of auxin plays an important role in regulating plant development and auxin homeostasis. *IAMT1 gene* encoding an IAA carboxyl methyltransferase is spatially and temporally regulated during the development of both rosette and cauline leaves (Qin et al. 2005).

1.3 Polar Auxin Transport

Plants employ a specialized delivery system to convey IAA from source to target tissues which was termed as polar auxin transport (PAT) (Swarup et al. 2000). PAT in plant cells is a unique and significant feature of the phytohormone auxin and is essential for normal plant growth and development (Blakeslee et al. 2005). It plays a crucial role in gravitropism, leaf vascular development, stabilization of phyllotactic patterning, and hypocotyl elongation in light-grown *Arabidopsis* plants (Bainbridge et al. 2008). In monocots, PAT plays important roles in regulation of leaf growth, control of tillering angle, and root development (Zhuang et al. 2006).

Auxin moves between plant cells through a combination of membrane diffusion and carrier-mediated transport (Kramer and Bennett 2006) with strict directionality mediated by specialized influx (such as AUXIN RESISTANT 1, AUX1) and efflux facilitators (such as PINs). PAT is mediated by a network involving the AUX1 influx facilitator and the PIN efflux facilitators. At the cellular level, directional auxin transport is primarily controlled by an efflux carrier complex that is characterized by the PIN-FORMED (PIN) family of proteins (Blakeslee et al. 2005), whereas AUX1 and its paralogs LIKE AUX1 (LAX1), LAX2, and LAX3 act to buffer the PIN-mediated patterning mechanism against environmental or developmental influences (Bainbridge et al. 2008). Besides the PIN proteins, plant orthologs of mammalian multidrug-resistance/P-glycoproteins (MDR/PGPs) function in auxin efflux (Titapiwatanakun and Murphy 2008). All these three families of cellular transport proteins, PIN-formed (PIN), P-glycoprotein (ABCB/PGP), and AUX1/LAX, can independently or coordinately transport auxin in plants (Titapiwatanakun and Murphy 2008). In addition, PAT can be modulated by many other plant hormones, such as cytokinin (Kuroha et al. 2006), giberrellin (Bjorklund et al. 2007), ethylene (Swarup et al. 2007; Negi et al. 2008), and brassinosteroids (Symons and Reid 2004; Symons et al. 2006). The polar localization of PIN and AUX1/LAX proteins is also regulated by many molecules. The localization of AUX1 depends on the activity of AUXIN RESISTANT 4 (AXR4), which encodes an endoplasmic reticulum (ER)-localized protein (Dharmasiri et al. 2006). Localization of AUX1 influx facilitator was modulated by OsARF-GTPase-activating protein (OsARF-GAP), which mediates the development of adventitious roots in rice (Zhuang et al. 2005). Localization of PIN1 and PIN7 depends on GNOM, which encodes guanine-nucleotide exchange factors for ADP-ribosylation factor GTPases (ARF-GEF), a regulator of vesicular trafficking that localizes to endosomes (Geldner et al. 2003). In addition, RHO OF PLANTS (ROP), a small class of GTPases, could be involved in PIN localization (Ellis and Miles 2001; Molendijk et al. 2004; Li et al. 2005). ARF-GTP/GDP, which is regulated by ARF-GEF and ARF-GAP, mediates auxin-dependent root architecture (Chong et al. 2006). Recently, PIN2 intracellular trafficking by promoting plasma membrane localization was found to depend on light, whereas vacuolar targeting for protein degradation is in the dark (Laxmi et al. 2008).

1.4 Auxin Signal Transduction in Higher Plants

Although the effects of auxin are thought to be dose dependent, with high and low doses eliciting different auxin responses (Teale et al. 2006), a framework of auxin signal transduction has been established. In this framework, TIR1 acts as an auxin receptor to respond to endogenously or exogenously applied auxin. This finding was considered as one of the most important advances in plant biology in recent years (Dharmasiri et al. 2005a; Dharmasiri et al. 2005b; Kepinski and Leyser 2005; Napier 2005; Lee et al. 2007). TIR1 encodes an F-box protein, one component of the SCF complex, which is well characterized in many organisms. SCF^{TIR1} specifically interacts with Aux/IAA1 proteins and covalently modifies them by ubiquitilation (Dharmasiri et al. 2005b; Kepinski and Leyser 2005). The interaction between TIR1 and Aux/IAA1 can be enhanced by auxin in a cell-free system (Dharmasiri et al.

2003). Studies of crystal structures showed the *Arabidopsis* TIR1-ASK1 complex free and in complexes with three different auxin compounds and an Aux/IAA substrate peptide. By filling in a hydrophobic cavity at the protein interface, auxin enhances the TIR1-substrate interactions by acting as a "molecular glue," which corresponds well to the physiological response of auxin (Tan et al. 2007).

The Aux/IAA family belongs to a group of early auxin-response genes. There are 29 Aux/IAA members with variation in amino acid identity ranging from 10 to 83% in *Arabidopsis* (Teale et al. 2006). Each individual Aux/IAA gene might have a set of nonessential functions, but they combine to perform essential regulatory functions (Dreher et al. 2006). Generally, Aux/IAA proteins are localized in the nucleus and contain four conserved domains (I–IV). Elimination of domain I does not affect its degradation, whereas domain II residues are required for the degradation of several *Arabidopsis thaliana* Aux/IAA proteins (Dreher et al. 2006). Domain III has a predicted ribbon–helix–helix DNA-binding domain found in bacterial transcriptional regulators (Ulmasov et al. 1997b, 1999). Aux/IAA proteins are able to associate with auxin response factor (ARF) proteins and affect their transcription activity (Ulmasov et al. 1997a). *Arabidopsis* genome contains 23 members of ARF genes. ARFs encode a kind of transcription factor that contains an amino-terminal B3-like DNA-binding domain that binds to an auxin-responsive element (ARE; TGTCTC) in the promoter of auxin-response genes in an auxin-independent manner (Ballas et al. 1993; Ulmasov et al. 1999). ARE-mediated transcription will be blocked by the interaction between Aux/IAA and ARFs (Ulmasov et al. 1997b). Until now, how TIR1 was promoted to interact with Aux/IAA after binding with auxin remained elusive. However, SCF^{TIR1} and the associated protein-degradation machinery, together with Aux/IAAs and ARFs, represent the full signal-transduction cascade from the auxin signal to gene expression, and these F-box proteins represent a new class of receptors (Teale et al. 2006).

Besides F-box proteins as auxin receptors and Aux/IAA-ARFs as their downstream components, alternative modes of auxin perception are represented by auxin-binding protein 1 (ABP1). ABP1 is a soluble, ER-located, dimeric glycoprotein that forms a β-jellyroll barrel carrying auxin in a central hydrophobic pocket (Woo et al. 2002). It binds the phytohormone auxin with high specificity and affinity (Braun et al. 2008). Constitutive expression of maize ABP1 in maize cell lines conferred the capacity to respond to auxin by increasing the cell size, which supports a role of ABP1 as an auxin receptor controlling plant growth (Jones et al. 1998). In addition, ABP1 was found to be required for organized cell elongation and division in *Arabidopsis* embryogenesis, which suggests that ABP1 mediates auxin-induced cell elongation and, directly or indirectly, cell division (Ullah et al. 2001). ABP1 is essential for the auxin control of cell division by acting at both the G1/S and G2/M checkpoints (David et al. 2007). Use of conditional repression of ABP1 to investigate its function during vegetative shoot development supports the model of ABP1 acting as a coordinator of cell division and expansion, with local auxin levels influencing ABP1 effectiveness (Braun et al. 2008). Although ABP1 mediates auxin signaling to regulate cell division and expansion, it can directly bind with auxin, but no direct downstream component of ABP1 has been characterized to regulate the expression of downstream genes (David et al. 2007; Bertosa et al. 2008; Braun et al. 2008). Given the almost instantaneous auxin response, ABP1 mediating gene expression may not be necessary for certain aspects of auxin signaling (Teale et al. 2006).

2 Brassinosteroids

2.1 *Physiological Functions of Brassinosteroids in Higher Plants*

BRs are naturally occurring plant steroids with structural similarities to insect and animal steroid hormones, including sex hormones, androgens, estrogens, and gestagen (progesterone), glucocorticoids, and mineral corticoids. BRs are C_{27}, C_{28}, and C_{29} steroids depending on their C-24 alkyl substituents (Mathur et al. 1998; Khripach et al. 2002; Oki et al. 2008). Since the discovery of the first brassinolide (BL, the most biologically active and naturally occuring form of the BRs) in the pollen of Western Rape, in 1979, and shown to be indispensable for plant growth and differentiation, more than 50 BL analogs have been identified (Oki et al. 2008). Either in whole plants or at cellular and molecular levels, the phenotype relation to BRs displayed diversity in many kinds of higher plants. Exogenously applied bioactive BRs increase the resistance of plants to a variety of stresses, including biotic and abiotic stress (Dhaubhadel et al. 1999, 2002; Abraham et al. 2003; Asami et al. 2003; Savenstrand et al. 2004; Kagale et al. 2007). In *Arabiopsis*, the model plant for dicots, mutants lacking the ability to synthesize or perceive BL display some severe phenotypes, such as dwarfed stature, reduced male fertility, round leaves, and photomorphogenic defects (Klahre et al. 1998; Neff et al. 1999; Yamamuro et al. 2000; Ellis and Miles 2001). In rice, the representative model plant for monocots, BRs are important to maintain optimal leaf angle and plant height (Baitsch et al. 2001; Kaneko et al. 2003; Wang et al. 2007; Gao et al. 2008; Malik et al. 2008), which are considered as important factors to obtain high yield grains. In addition, BRs were discovered to be involved in fruit development (Fujisawa et al. 2001) and regulation of grain filling (Yamamuro et al. 2000). Overall, BRs play essential and indispensable roles for normal plant growth and development in the whole life cycle.

2.2 *Brassinosteroid Synthesis Pathway*

BRs are widely distributed throughout reproductive and vegetative plant tissues and do not travel over long distances among different plant tissues (Symons et al. 2006). Thus, the local biosynthesis of BRs is critical for regulation of downstream signaling transduction. To date, many key enzymes involved in BR biosynthesis have been identified and characterized in different organisms. The BR biosynthetic pathways in the plant kingdom are relatively conserved and highly networked (Klahre et al. 1998; Bishop et al. 1999; Kim et al. 1999; Yamamuro et al. 2000; Shimada et al. 2001; Back et al. 2002; Nomura et al. 2004; Bancos et al. 2006; Bishopp et al. 2006; Hamberger and Bohlmann 2006; Ohnishi et al. 2006; He et al. 2007; Katsumata et al. 2008; Luo et al. 2008; Oki et al. 2008). Sterols, important membrane constituents, were recognized as precursors of BRs (Oki et al. 2008). The most biologically active C_{28} BR biosynthetic pathway mainly contains an early C-22 oxidation branch and two interconnected parallel routes, designated as the early and late C-6 oxidation pathways, respectively (Oki et al. 2008). Briefly, campesterol is first converted to campestanol in the early C-22 oxidation subpathway: (22*S*)-22-hydroxycampesterol→(22*S*,24*R*)-22-hydroxyergost-4-en-3-one→(22*S*,24*R*)-22-hydroxy-5α-ergostan-3-one→6-deoxocathasterone. Campestanol is the common precursor for both early and late C-6 oxidation pathways. In the early C-6 oxidation subpathway, campestanol is in turn converted to castasterone. In the late C-6 oxidation subpathway, campestanol is in turn converted to 6-deoxocathasterone. Finally, the intermediate molecules are converted to BL via early and late C-6 oxidation pathways (Oki et al. 2008). In addition, the early and late C-6 oxidation subpathways are connected at multiple steps via BR 6 oxidase (BR6ox) (Oki et al. 2008). Furthermore, using insect cell-expressed proteins (Ohnishi et al. 2006), both CYP90C1 and CYP90D1 were found to mediate a novel shortcut in BR biosynthesis via catalyzing C-23 hydroxylation of various 22-hydroxylated BRs. In addition, the biosynthesis of BRs is subject to light regulation and the feedback is suppressed by their signaling transduction (Ullah et al. 2002; Llorente et al. 2005; Bancos et al. 2006).

2.3 *Brassinosteroid Signal Transduction*

The emerging sketch of BR signal transduction diverges radically from the paradigms of animal steroid signaling, which generally involve the action of members of the nuclear receptor superfamily. In higher plants, BRs bind the extracellular domain of a small family of leucine-rich-repeat receptor-like kinases (LRR-RLKs) to activate intracellular signal transduction (Belkhadir et al. 2006). The first BR-insensitive mutant *bri1* was identified by genetic screens for mutants with reduced or abolished root growth inhibition in medium containing a high concentration of BL, the most active BR (Clouse 1996). Then, BRI1 was cloned by a map-based cloning strategy and discovered to encode a plasma membrane-localized LRR-RLK (Li and Chory 1997). Further biochemical and physiological analysis demonstrated that the extracellular domain of BRI1 was involved in the perception of BRs, whereas the intracellular domain was required for the initiation of the BR signaling cascade by phosphorylation (Lippert et al. 2000; Wang et al. 2001), which suggests that BRI1 functions as a receptor for BRs. In addition, two other *Arabidopsis*

BRI1-like LRR-RLKs, BRL1 and BRL3, play partially redundant roles with BRI1 (Cano-Delgado et al. 2004; Wei et al. 2008).

Another LRR-RLK, BRI1-Associated Receptor Kinase 1 (BAK1), distinct from BRI1, BRL1, or BRL3, was identified from *Arabidopsis* by a gain-of-function genetic screening for *bri1-5* suppressors and through a yeast two-hybrid investigation for proteins interacting with the BRI1 kinase domain (Li et al. 2002; Nam and Li 2002). It encodes an LRR type II RLK with five extracellular LRRs, a transmembrane domain following an intracellular kinase domain. Genetic, biochemical, and cellular analyzis demonstrated that BAK1 interacts with BRI1 both in vitro and in vivo (Li et al. 2001; Nam and Li 2002). Another family member of BAK1, BAK1-LIKE 1 (BKK1), functions redundantly with BAK1 in regulating BR signaling (He et al. 2007). Besides positively regulating a BR-dependent plant-growth pathway, BAK1 and BKK1 also negatively regulate a BR-independent cell-death pathway (He et al. 2007; Kemmerling et al. 2007). BRI1, BRL1, BRL3, BAK1, and BKK1 are involved in the early events in BR signal transduction. Interestingly, BRI1-interacting protein, BKI1, is able to prevent the activation of BRI1 by limiting the interaction of BRI1 with its proposed coreceptor, BAK1 (Wang et al. 2006a), which implies complex regulations are involved in early BR signaling events (see Fig. 1).

In downstream BR signal transduction, *Arabidopsis* Brassinosteroid-Insensitive 2 (BIN2) functions as a negative regulator, which BIN2 encodes a glycogen synthase kinase-3 (GSK3/SHAGGY-like kinase) (Li et al. 2001). As shown in Fig. 1, BIN2 catalyzes the phosphorylation of its two downstream components BZR1/BES1 (or BZR2) (Zhao et al. 2002) to inhibit their DNA binding and to

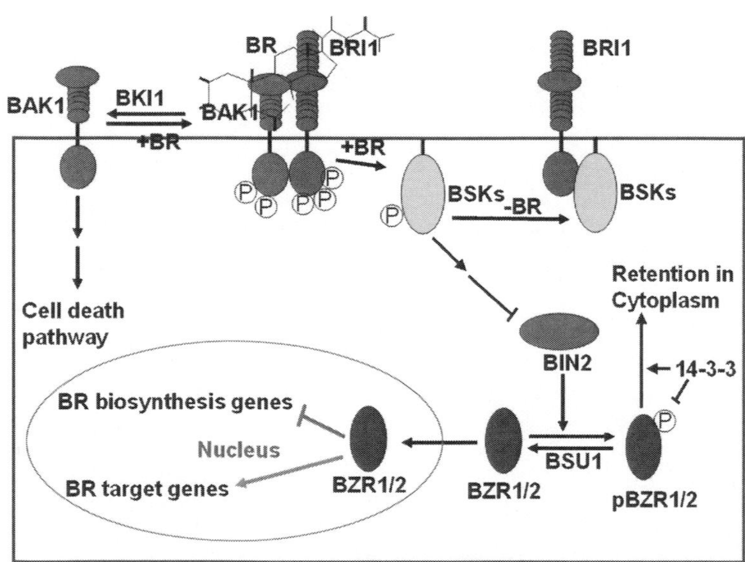

Fig. 1 Brassinosteroids signaling transduction pathway in higher plants

promote them binding to the 14-3-3 proteins for nuclear export (de Vries 2007; Ryu et al. 2007). BES1 shares 80% identity with BZR1 and belongs to the BZR1 family (Yin et al. 2002). BES1 and BZR1 provide a connection between the cytoplasmic BR response and the nucleus via functioning as transcription factors or through recruiting other transcriptional regulators (Yu et al. 2008). BR treatment induces rapid nuclear localization of BZR1/BZR2 through cell-surface receptors (BRI1 and BAK1) and a GSK3 kinase (BIN2) to initiate BR gene expression (Yin et al. 2002; He et al. 2005). BZR1 is a positive regulator in the BR signaling transduction pathway that mediates both downstream BR responses and feedback regulation of BR biosynthesis (He et al. 2005).

Very recently, in the area of BR signal transduction, the gap between early events and downstream gene regulation was filled partially by the knowledge of three identified homologous BR-signaling kinases (BSK1, BSK2, and BSK3) by a proteomic approach. Further genetic and transgenic analysis demonstrate that BSKs are the substrates of BRI1 kinase that activate downstream BR signal transduction (Tang et al. 2008). Drawing the whole BR signaling transduction network in higher plants has a long way to go. The knowledge of the cross talk with other plant signaling transduction pathways probably is beneficial to finally elucidate this network.

3 Physiological Functions of G-Protein Signaling in *Arabidopsis* and Rice

Heterotrimeric guanine nucleotide-binding proteins (G proteins) composed of Gα, Gβ, and Gγ subunits are important transducers of hormonal signals in organisms as evolutionarily distant as plants and humans (Assmann 2004). The G protein itself consists of three different subunits, α, β, and γ (Gα, Gβ, and Gγ, respectively), which form a heterotrimeric complex in the inactive state. The activating ligands, which associate with their specific G protein-coupled receptors (GPCRs), subsequently lead to the conversion of an inactive G protein to its active conformation. In this process, the GPCR acts as a guanine nucleotide exchange factor, causing Gα to exchange GDP for GTP. As a result, Gα–GTP separates from the Gβγ dimer, and both Gα–GTP and the Gβγ dimer separate from the receptor and can activate downstream effectors (Perfus-Barbeoch et al. 2004). Both Gα and Gβ are encoded by a single-copy gene in rice and *Arabidopsis*, designated GPA1 and AGB1 in *Arabidopsis* and RGA1 and RGB1 in rice, whereas Gγ has two copies – AGG1 and AGG2 in *Arabidopsis* and RGG1 and RGG2 in rice (Weiss et al. 1994; Ishikawa et al. 1996; Mason and Botella 2001; Trusov et al. 2007).

Studies in *Arabidopsis* have shown that plant G-protein signaling is important to many fundamental physiological processes such as cell proliferation, hormone perception, and ion-channel regulation (Ullah et al. 2001; Wang et al. 2001). The smaller repertoire of the heterotrimeric G-protein complex in plants offers a unique advantage over its counterpart in mammals for dissecting their roles in development (Chen 2008). In recent years, the diversity of signal transduction through plant G proteins has been revealed, thanks to the identification and mutation of genes in *Arabidopsis* and rice that encode specific G-protein components. These components include α, β, and γ subunits of the G-protein heterotrimer, possible heptahelical GPCRs, and regulator of G-protein signaling proteins (RGS). Knockdown of the α subunit of heterotrimeric G protein in rice caused abnormal morphology, such as dwarfism and the setting of small seeds (Fujisawa et al. 1999). The heterotrimeric G protein also was implicated in GA responses in oat and rice aleurone to regulate α-amylase gene expression (Jones et al. 1998; Ueguchi-Tanaka et al. 2000). These *gpa1* mutants display reduced cell division in both seedling leaves and stem (Ueguchi-Tanaka et al. 2000). In *Arabidopsis*, the GPA1 protein is required for the ABA inhibition of both stomata opening and inwardly rectifying K^+ channels (Wang et al. 2001). However, loss of function of the α subunit of heterotrimeric G protein in rice caused severe morphologic changes, but the loss of the sole function of GPA1 in *Arabidopsis* did not significantly change the morphology in light even though cell division was reduced; probably the reduced cell division was compensated by the increase of cell size (Ueguchi-Tanaka et al. 2000; Chen et al. 2006; Wang et al. 2006a; Gao et al. 2008). Dark-grown *gpa1* mutant seedlings display short hypocotyls and partially opened hooks (Ullah et al. 2001; Gao et al. 2008). In addition, the α subunit of heterotrimeric G protein may participate in pollen germination in *Arabidopsis* (Wu et al. 2007). Interestingly, loss of function of *AGB1*, which encodes a heterotrimeric G-protein β subunit, results in several striking phenotypes, including silique morphology, flower shape, inflorescence length and leaf shape (Lease et al. 2001). Recently, AGG1 and AGG2 were found to have distinct roles in plant development. AGG1-deficient but not AGG2-deficient mutants showed impaired resistance to necrotrophic pathogens. By contrast, both AGG1- and AGG2-deficient mutants are hypersensitive to auxin-mediated induction of lateral roots (Trusov et al. 2007). However, any of the G-γ single and double mutants did not display the distinctive traits observed in G-β-deficient mutants (such as reduced size of cotyledons, leaves, flowers, and siliques), which suggests that AGG1 together with AGG2 is not the functional equivalent of AGB1 (Trusov et al. 2007). Overall the heterotrimeric G protein plays versatile functions in higher plants, including cell division, plant defense, and stomatal opening.

Recently, the two GPCR-type G-proteins, GTG1 and GTG2, were reported to be ABA receptors in *Arabidopsis* (Pandey et al. 2009). GTG1 and GTG2 specifically bind ABA in vitro and mediate ABA responses in vivo. Both proteins interact with GPA1 and have intrinsic GTP-binding and GTPase activity. The GDP-bound form of the GTGs exhibits greater ABA binding than the GTP-bound form. GTPase activity of both proteins is inhibited by GPA1, and the *gpa1* null mutants exhibit ABA-hypersensitive phenotypes. Thus, the GDP-bound, instead of the GTP-bound form, actively relays the ABA signal for these novel G proteins with GPCR protein structure and distinguished function in hormone signal transduction.

Besides the heterotrimeric G protein, there is a branch of G proteins named small G proteins that include ROPs, RANs, and RAA1 in plant. The ROP family of small

GTPases has emerged as a versatile and pivotal regulator in plant signal transduction. The Yang's lab has reported a series of studies of ROP signaling in diverse processes ranging from cytoskeletal organization to hormone and stress responses (Hwang et al. 2008). The small GTPase Ran is the central element of a conserved signaling network that has a prominent role in mitotic regulation. The function of RAN in plant development is mediated by the cell cycle, and its novel role in meristem initiation is mediated by auxin signaling (Wang et al. 2006b). TaRAN1 from wheat is essential for all nucleocytoplasmic transport events and is associated with regulation of genome integrity and cell division in yeast systems (Wang et al. 2004). However, a new, small GTP-binding protein, RAA1, differs from typical small G-protein families in the sequence structure of its amino acids. OsRAA1, a 12-kD protein with GTP binding activity, is involved in regulation of the development of rice root systems, which is mediated by auxin. A positive feedback regulation mechanism of OsRAA1 to IAA metabolism may be involved in rice root development in nature. OsRAA1 may modulate root development mediated by the ubiquitin–proteasome pathway as a novel regulatory factor of the cell cycle from metaphase to anaphase (Ge et al. 2004; Han et al. 2005; Han et al. 2008).

4 Cross talk Between Signaling of Auxin, Brassinosteroids, and G Protein

4.1 Cross talk Between Auxin Signaling and Heterotrimeric G Protein

The G-protein mediated signaling pathway transduces multiple extracellular signals and plays versatile functions in higher plants. A few heterotrimeric G proteins mediated pathway in higher plants have been implicated. The signaling mediated by heterotrimeric G protein interacts with other plant signaling transduction pathways and study of potential involvement of heterotrimeric G protein in plant hormone signal transduction is increasing in interest. Among them, cross talk between auxin signaling transduction and heterotrimeric G-protein signaling merits attention. Early studies suggested that auxin promotes both GTPγS association with rice coleoptile membrane vesicles and GTP hydrolysis by those vesicles, because auxin activates a G-protein cycle (Zaina et al. 1990; Zaina et al. 1991). However, this direction was not followed up well (Hooley 1998). As is well known, auxin promotes cell division. Consistent with this observation, overexpression of the α subunit of heterotrimeric G protein in synchronized tobacco BY-2 cells increased the cell division (Ullah et al. 2001). The role of Gα in regulating cell division may result from an interaction with downstream proteins that share a conserved function (Assmann 2002); one example is PLA2, which is also activated by auxin in soybean cell cultures (Scherer 1994). More direct evidence was provided by Ullah and colleagues (Ullah et al. 2003). They found that heterotrimeric G-protein mutants with loss-of-function β subunit displayed hypersensitivity to auxin, whereas ectopic expression of the wild-type G α subunit phenocopies G β mutants, probably by sequestering the G β and γ subunits. Although overexpression of G β reduces auxin sensitivity, a constitutively active (Q222L) mutant of G α behaves like the wild type (Ullah et al. 2003). These data are consistent with a model in which G β-γ acts as a negative regulator of auxin-induced cell division. In addition, both AGG1- and AGG2-deficient mutants show hypersensitivity to auxin-mediated induction of lateral roots, which suggests that G β-γ1 and G β-γ2 synergistically inhibit auxin-dependent lateral root initiation (Trusov et al. 2007). Interestingly, the functions of the two differ in root response, with G β-γ1 acting within the central cylinder, attenuating acropetally transported auxin signaling, and G β-γ2 affecting the action of basipetal auxin and gravi-responsiveness within the epidermis and/or cortex (Trusov et al. 2007). Further evidence supporting the cross talk between auxin signaling and heterotrimeric G protein-mediated signaling may be discovered in more detail, especially, considering cell type and/or developmental stages.

For cross talk between auxin signaling and small G protein, ADP ribosylation factor (ARF)-GTPase is representative. Auxin signaling is modulated by polar transport mediated by the influx facilitator AUX1 and efflux facilitators PINs. Root development is affected by localization of AUX1, which is mediated by ARF-GAP through Golgi vesicle trafficking. As well, asymmetric distribution and activity for PIN1 is regulated by ARF-guanine-nucleotide exchange factor (GEF). For cell polarity determination, the activity of PIN2 is modulated by ROP2 depending on ARF1 located in Golgi stacks and the ER (Chong and Zhuang 2007). Therefore, the small G protein cross talks with auxin through its polar transport.

4.2 Cross talk Between Brassinosteroids Signaling Transduction and Heterotrimeric G Protein

As noted above, BR-mediated signaling transduction plays many fundamental roles in plant growth and development, as well as in biotic and abiotic stress responses. The partially overlapping functions between BR signaling and heterotrimeric G protein signaling suggest some cross talk between these two kinds of signaling. The first hint of this cross talk was suggested by *gpa1* mutants exhibiting a *rotundifolia*-like leaf shape when grown in light (Ullah et al. 2001). As previously known, *Rotundifolia* encodes cytochrome P450, which might be involved in BR synthesis (Kim et al. 1999). Investigation of the BR growth-promoting molecular mechanism in lamina joint rice cDNA microarray, containing 1,265 genes, revealed a novel BL-induced gene, designated OsBLE2, which have a much weaker response to BL rice α subunit mutant than in control plants. This evidence further suggests that heterotrimeric G protein may be a component of BR signaling (Yang et al. 2003). Driven by these implications, the detailed physiological investigation of heterotrimeric G protein was performed using null mutation of rice α subunit of heterotrimeric G protein. The reduced sensitivity of *d1* mutant plants (caused by the null mutation of *RGA*, the DK22 mutant, an allele of *d1* whose recurrent parent is Nipponbare) to 24-epibrassinolide (24-epiBL) was discovered by root elongation inhibition assay, lamina inclination assay, and coleoptile elongation analysis (Wang et al. 2006a). This result was further confirmed and thoroughly investigated with another rice Gα-defective mutant, T65d1, whose recurrent parent is T65, and rice BR-deficient mutants (three BR receptor mutants *d61-1*, *d61-2*, and *d61-7* and with one BR biosynthesis mutant *d2-1* (Oki et al. 2008). The T65d1 mutant exhibited decreased sensitivity to 24-epiBL in many aspects, such as growth inhibition of seminal roots, adventitious roots, and aerial parts (Oki et al. 2008). In addition, the T65d1 mutant displayed constitutive photomorphogenic growth phenotypes in darkness, which is not observed in DK22, possibly because of the difference in the recurrent parents (Oki et al. 2008). However, the T65d1 mutant and *d61-7* did not show epistasis on analysis of the T65d1/*d61-7* double mutant (Oki et al. 2008). Furthermore, *d61-1* and wild-type plants responded similarly to 24-epiBL in transcriptional expression of RGA (Wang et al. 2006a). Finally, the 24-epiBL feedback regulation of the expression of BR-biosynthetic genes in T65d1 as in the wild type (Oki et al. 2008). All these results suggest that Gα-mediated signaling is not directly connected to BR perception mediated by OsBRI1. Whether the signaling is through another plasmamembrane-located BR receptor, such as OsBRL1 or OsBRL3, is still unclear so far. Similarly in *Arabidopsis*, genetic evidence suggests that loss of function of GPA1 enhances the developmental defects of *bri1-5*, a weak allele of a BR receptor mutant, and *det2-1*, a BR-deficient mutant in *Arabidopsis* (Gao et al. 2008). G protein- and BR-mediated pathways may converge to modulate cell proliferation in a cell/tissue-specific manner (Gao et al. 2008). However, GPA1- and BR-mediated signaling pathways may be through their interaction with other plant hormones, because *gpa1* mutants display altered sensitivity to multiple hormones.

5 Future Prospects

Auxin and BR-mediated signaling and heterotrimeric G protein-mediated signaling play overlapping and distinct roles during plant growth and development, which suggests some cross talk between these signaling transduction pathways. In some cases, the kinds of cross talks are at developmental stage or are organ specific. Until now, little has been known about the underlying mechanism of the cross talk. The following are some issues that should be addressed: the relation between auxin and BR signaling; the coordinated signalings from auxin and BRs on G-protein signaling in living plant cells; and the relation between the signalings and vascular trafficking to regulate cell elongation and division. Final elucidation of these issues will help in understanding the function of this signaling transduction in physical action. Besides auxin and BRs cross talk, cross talk exists between heterotrimeric

G protein and other plant hormones, such as gibberellins (Fujisawa et al. 1999). Whether the latter cross talk interacts directly or through auxin or BR signaling to cross talk with heterotrimeric G protein needs further investigation.

References

Abraham E, Rigo G, Szekely G, Nagy R, Koncz C, Szabados L (2003) Light-dependent induction of proline biosynthesis by abscisic acid and salt stress is inhibited by brassinosteroid in *Arabidopsis*. Plant Mol Biol 51:363–372

Aharoni N, Yang SF (1983) Auxin-induced ethylene production as related to auxin metabolism in leaf discs of tobacco and sugar beet. Plant Physiol 73:598–604

Asami T, Nakano T, Nakashita H, Sekimata K, Shimada Y, Yoshida S (2003) The influence of chemical genetics on plant science: shedding light on functions and mechanism of action of brassinosteroids using biosynthesis inhibitors. J Plant Growth Regul 22:336–349

Assmann SM (2002) Heterotrimeric and unconventional GTP binding proteins in plant cell signaling. Plant Cell 14:S355–373

Assmann SM (2004) Plant G proteins, phytohormones, and plasticity: three questions and a speculation. Sci STKE 2004:re20

Back TG, Janzen L, Pharis RP, Yan Z (2002) Synthesis and bioactivity of C-2 and C-3 methyl ether derivatives of brassinolide. Phytochemistry 59:627–634

Bainbridge K, Guyomarc'h S, Bayer E, Swarup R, Bennett M, Mandel T, Kuhlemeier C (2008) Auxin influx carriers stabilize phyllotactic patterning. Genes Dev 22:810–823

Baitsch D, Sandu C, Brandsch R, Igloi GL (2001) Gene cluster on pAO1 of *Arthrobacter nicotinovorans* involved in degradation of the plant alkaloid nicotine: cloning, purification, and characterization of 2, 6-dihydroxypyridine 3-hydroxylase. J Bacteriol 183:5262–5267

Ballas N, Wong LM, Theologis A (1993) Identification of the auxin-responsive element, AuxRE, in the primary indoleacetic acid-inducible gene, PS-IAA4/5, of pea (*Pisum sativum)*. J Mol Biol 233:580–596

Bancos S, Szatmari AM, Castle J, Kozma-Bognar L, Shibata K, Yokota T, Bishop GJ, Nagy F, Szekeres M (2006) Diurnal regulation of the brassinosteroid-biosynthetic CPD gene in *Arabidopsis*. Plant Physiol 141:299–309

Bartel B (1997) Auxin biosynthesis. Annu Rev Plant Physiol Plant Mol Biol 48:51–66

Belkhadir Y, Wang X, Chory J (2006) Brassinosteroid signaling pathway. Sci STKE 2006:cm4

Bertosa B, Kojic-Prodic B, Wade RC, Tomic S (2008) Mechanism of auxin interaction with Auxin Binding Protein (ABP1): a molecular dynamics simulation study. Biophys J 94:27–37

Bishop GJ, Nomura T, Yokota T, Harrison K, Noguchi T, Fujioka S, Takatsuto S, Jones JD, Kamiya Y (1999) The tomato DWARF enzyme catalyses C-6 oxidation in brassinosteroid biosynthesis. Proc Natl Acad Sci USA 96:1761–1766

Bishopp A, Mahonen AP, Helariutta Y (2006) Signs of change: hormone receptors that regulate plant development. Development 133:1857–1869

Bjorklund S, Antti H, Uddestrand I, Moritz T, Sundberg B (2007) Cross-talk between gibberellin and auxin in development of Populus wood: gibberellin stimulates polar auxin transport and has a common transcriptome with auxin. Plant J 52:499–511

Blakeslee JJ, Peer WA, Murphy AS (2005) Auxin transport. Curr Opin Plant Biol 8:494–500

Braun N, Wyrzykowska J, Muller P, David K, Couch D, Perrot-Rechenmann C, Fleming AJ (2008) Conditional repression of AUXIN BINDING PROTEIN1 reveals that it coordinates cell division and cell expansion during postembryonic shoot development in *Arabidopsis* and tobacco. Plant Cell 20:2746–2762

Cano-Delgado A, Yin Y, Yu C, Vafeados D, Mora-Garcia S, Cheng JC, Nam KH, Li J, Chory J (2004) BRL1 and BRL3 are novel brassinosteroid receptors that function in vascular differentiation in *Arabidopsis*. Development 131:5341–5351

Casimiro I, Marchant A, Bhalerao RP, Beeckman T, Dhooge S, Swarup R, Graham N, Inze D, Sandberg G, Casero PJ, Bennett M (2001) Auxin transport promotes *Arabidopsis* lateral root initiation. Plant Cell 13:843–852

Chen JG (2008) Heterotrimeric G-proteins in plant development. Front Biosci 13:3321–3333

Chen JG, Gao Y, Jones AM (2006) Differential roles of *Arabidopsis* heterotrimeric G-protein subunits in modulating cell division in roots. Plant Physiol 141:887–897

Chong K, Zhuang X (2007) ARF-GTPase as a molecular switch for polar auxin transport mediated by vesicle trafficking in root development. Plant Signal Behav 2:101–102

Clouse SD (1996) Plant hormones: brassinosteroids in the spotlight. Curr Biol 6:658–661

Cohen JD, Bandurski RS (1982) Chemistry and physiology of the bound auxins. Annu Rev Plant Physiol 33:403–430

David KM, Couch D, Braun N, Brown S, Grosclaude J, Perrot-Rechenmann C (2007) The auxin-binding protein 1 is essential for the control of cell cycle. Plant J 50:197–206

de Vries SC (2007) 14–3-3 proteins in plant brassinosteroid signaling. Dev Cell 13:162–164

Dharmasiri N, Dharmasiri S, Estelle M (2005a) The F-box protein TIR1 is an auxin receptor. Nature 435:441–445

Dharmasiri N, Dharmasiri S, Jones AM, Estelle M (2003) Auxin action in a cell-free system. Curr Biol 13:1418–1422

Dharmasiri N, Dharmasiri S, Weijers D, Lechner E, Yamada M, Hobbie L, Ehrismann JS, Jurgens G, Estelle M (2005b) Plant development is regulated by a family of auxin receptor F box proteins. Dev Cell 9:109–119

Dharmasiri S, Swarup R, Mockaitis K, Dharmasiri N, Singh SK, Kowalchyk M, Marchant A, Mills S, Sandberg G, Bennett MJ, Estelle M (2006) AXR4 is required for localization of the auxin influx facilitator AUX1. Science 312:1218–1220

Dhaubhadel S, Chaudhary S, Dobinson KF, Krishna P (1999) Treatment with 24-epibrassinolide, a brassinosteroid, increases the basic thermotolerance of *Brassica napus* and tomato seedlings. Plant Mol Biol 40:333–342

Dhaubhadel S, Browning KS, Gallie DR, Krishna P (2002) Brassinosteroid functions to protect the translational machinery and heat-shock protein synthesis following thermal stress. Plant J 29:681–691

Dreher KA, Brown J, Saw RE, Callis J (2006) The *Arabidopsis* Aux/IAA protein family has diversified in degradation and auxin responsiveness. Plant Cell 18:699–714

Ellis BE, Miles GP (2001) Plant biology. One for all? Science 292:2022–2023

Estelle M (1992) The plant hormone auxin: insight in sight. Bioessays 14:439–444

Feldman LJ (1985) Root gravitropism. Physiol Plant 65:341–344

Fujisawa Y, Kato H, Iwasaki Y (2001) Structure and function of heterotrimeric G proteins in plants. Plant Cell Physiol 42:789–794

Fujisawa Y, Kato T, Ohki S, Ishikawa A, Kitano H, Sasaki T, Asahi T, Iwasaki Y (1999) Suppression of the heterotrimeric G protein causes abnormal morphology, including dwarfism, in rice. Proc Natl Acad Sci USA 96:7575–7580

Gao Y, Wang S, Asami T, Chen JG (2008) Loss-of-function mutations in the *Arabidopsis* heterotrimeric G-protein alpha subunit enhance the developmental defects of brassinosteroid signaling and biosynthesis mutants. Plant Cell Physiol 49:1013–1024

Ge L, Chen H, Jiang JF, Zhao Y, Xu ML, Xu YY, Tan KH, Xu ZH, Chong K (2004) Over-expression of OsRAA1 causes pleiotropic phenotypes in transgenic rice plants, including altered leaf, flower, and root development and root response to gravity. Plant Physiol 135:1502–1513

Geldner N, Anders N, Wolters H, Keicher J, Kornberger W, Muller P, Delbarre A, Ueda T, Nakano A, Jurgens G (2003) The Arabidopsis GNOM ARF-GEF mediates endosomal recycling, auxin transport, and auxin-dependent plant growth. Cell 112:219–230

Hamberger B, Bohlmann J (2006) Cytochrome P450 mono-oxygenases in conifer genomes: discovery of members of the terpenoid oxygenase superfamily in spruce and pine. Biochem Soc Trans 34:1209–1214

Han Y, Wang X, Jiang J, Xu Y, Xu Z, Chong K (2005) Biochemical character of the purified OsRAA1, a novel rice protein with GTP-binding activity, and its expression pattern in *Oryza sativa*. J Plant Physiol 162:1057–1063

Han Y, Cao H, Jiang J, Xu Y, Du J, Wang X, Yuan M, Wang Z, Xu Z, Chong K (2008) Rice ROOT ARCHITECTURE ASSOCIATED1 binds the proteasome subunit RPT4 and is degraded in a D-box and proteasome-dependent manner. Plant Physiol 148:843–855

He JX, Gendron JM, Sun Y, Gampala SS, Gendron N, Sun CQ, Wang ZY (2005) BZR1 is a transcriptional repressor with dual roles in brassinosteroid homeostasis and growth responses. Science 307:1634–1638

He K, Gou X, Yuan T, Lin H, Asami T, Yoshida S, Russell SD, Li J (2007) BAK1 and BKK1 regulate brassinosteroid-dependent growth and brassinosteroid-independent cell-death pathways. Curr Biol 17:1109–1115

Hooley R (1998) Plant hormone perception and action: a role for G-protein signal transduction? Philos Trans R Soc Lond B Biol Sci 353:1425–1430

Hwang JU, Vernoud V, Szumlanski A, Nielsen E, Yang Z (2008) A tip-localized RhoGAP controls cell polarity by globally inhibiting Rho GTPase at the cell apex. Curr Biol 18:1907–1916

Ishikawa A, Iwasaki Y, Asahi T (1996) Molecular cloning and characterization of a cDNA for the beta subunit of a G protein from rice. Plant Cell Physiol 37:223–228

Jacobs M, Ray PM (1976) Rapid auxin-induced decrease in free space pH and its relationship to auxin-induced growth in maize and pea. Plant Physiol 58:203–209

Jing Y, Cui D, Bao F, Hu Z, Qin Z, Hu Y (2009) Tryptophan deficiency affects organ growth by retarding cell expansion in *Arabidopsis*. Plant J 57:511–521

Jones HD, Smith SJ, Desikan R, Plakidou-Dymock S, Lovegrove A, Hooley R (1998) Heterotrimeric G proteins are implicated in gibberellin induction of a-amylase gene expression in wild oat aleurone. Plant Cell 10:245–254

Kagale S, Divi UK, Krochko JE, Keller WA, Krishna P (2007) Brassinosteroid confers tolerance in *Arabidopsis thaliana* and *Brassica napus* to a range of abiotic stresses. Planta 225:353–364

Kaneko M, Itoh H, Inukai Y, Sakamoto T, Ueguchi-Tanaka M, Ashikari M, Matsuoka M (2003) Where do gibberellin biosynthesis and gibberellin signaling occur in rice plants? Plant J 35:104–115

Katsumata T, Hasegawa A, Fujiwara T, Komatsu T, Notomi M, Abe H, Natsume M, Kawaide H (2008) *Arabidopsis* CYP85A2 catalyzes lactonization reactions in the biosynthesis of 2-deoxy-7-oxalactone brassinosteroids. Biosci Biotechnol Biochem 72:2110–2117

Kemmerling B, Schwedt A, Rodriguez P, Mazzotta S, Frank M, Qamar SA, Mengiste T, Betsuyaku S, Parker JE, Mussig C, Thomma BP, Albrecht C, de Vries SC, Hirt H, Nurnberger T (2007) The BRI1-associated kinase 1, BAK1, has a brassinolide-independent role in plant cell-death control. Curr Biol 17:1116–1122

Kepinski S, Leyser O (2005) The *Arabidopsis* F-box protein TIR1 is an auxin receptor. Nature 435:446–451

Khripach VA, Zhabinskii VN, Konstantinova OV, Antonchick AP, Schneider B (2002) Synthesis of [26–2H(3)]brassinosteroids. Steroids 67:587–595

Kim GT, Tsukaya H, Saito Y, Uchimiya H (1999) Changes in the shapes of leaves and flowers upon overexpression of cytochrome P450 in *Arabidopsis*. Proc Natl Acad Sci USA 96:9433–9437

Klahre U, Noguchi T, Fujioka S, Takatsuto S, Yokota T, Nomura T, Yoshida S, Chua NH (1998) The *Arabidopsis* DIMINUTO/DWARF1 gene encodes a protein involved in steroid synthesis. Plant Cell 10:1677–1690

Kramer EM, Bennett MJ (2006) Auxin transport: a field in flux. Trends Plant Sci 11:382–386

Kuroha T, Ueguchi C, Sakakibara H, Satoh S (2006) Cytokinin receptors are required for normal development of auxin-transporting vascular tissues in the hypocotyl but not in adventitious roots. Plant Cell Physiol 47:234–243

Laxmi A, Pan J, Morsy M, Chen R (2008) Light plays an essential role in intracellular distribution of auxin efflux carrier PIN2 in *Arabidopsis thaliana*. PLoS ONE 3:e1510

Lease KA, Wen J, Li J, Doke JT, Liscum E, Walker JC (2001) A mutant *Arabidopsis* heterotrimeric G-protein beta subunit affects leaf, flower, and fruit development. Plant Cell 13:2631–2641

Lee JJ, Woodward AW, Chen ZJ (2007) Gene expression changes and early events in cotton fibre development. Ann Bot 100:1391–1401

Li J, Chory J (1997) A putative leucine-rich repeat receptor kinase involved in brassinosteroid signal transduction. Cell 90:929–938

Li J, Nam KH, Vafeados D, Chory J (2001) BIN2, a new brassinosteroid-insensitive locus in *Arabidopsis*. Plant Physiol 127:14–22

Li J, Wen J, Lease KA, Doke JT, Tax FE, Walker JC (2002) BAK1, an Arabidopsis LRR receptor-like protein kinase, interacts with BRI1 and modulates brassinosteroid signaling. Cell 110:213–222

Li L, Xu J, Xu ZH, Xue HW (2005) Brassinosteroids stimulate plant tropisms through modulation of polar auxin transport in *Brassica* and *Arabidopsis*. Plant Cell 17:2738–2753

Lippert E, Jacques Y, Hermouet S (2000) Positive regulation of human T cell activation by Gi2 proteins and interleukin-8. J Leukoc Biol 67:742–748

Llorente F, Alonso-Blanco C, Sanchez-Rodriguez C, Jorda L, Molina A (2005) ERECTA receptor-like kinase and heterotrimeric G protein from *Arabidopsis* are required for resistance to the necrotrophic fungus *Plectosphaerella cucumerina*. Plant J 43:165–180

Lopez-Bucio J, Hernandez-Abreu E, Sanchez-Calderon L, Perez-Torres A, Rampey RA, Bartel B, Herrera Estrella L (2005) An auxin transport independent pathway is involved in phosphate stress-induced root architectural alterations in *Arabidopsis*. Identification of BIG as a mediator of auxin in pericycle cell activation. Plant Physiol 137:681–691

Luo M, Tan K, Xiao Z, Hu M, Liao P, Chen K (2008) Cloning and expression of two sterol C-24 methyltransferase genes from upland cotton (*Gossypium hirsuturm* L.). J Genet Genomics 35:357–363

Malik MR, Wang F, Dirpaul JM, Zhou N, Hammerlindl J, Keller W, Abrams SR, Ferrie AM, Krochko JE (2008) Isolation of an embryogenic line from non-embryogenic Brassica napus cv. Westar through microspore embryogenesis. J Exp Bot 59:2857–2873

Mason MG, Botella JR (2001) Isolation of a novel G-protein gamma-subunit from *Arabidopsis thaliana* and its interaction with G beta. Biochim Biophys Acta 1520:147–153

Mathur J, Molnar G, Fujioka S, Takatsuto S, Sakurai A, Yokota T, Adam G, Voigt B, Nagy F, Maas C, Schell J, Koncz C, Szekeres M (1998) Transcription of the Arabidopsis CPD gene, encoding a steroidogenic cytochrome P450, is negatively controlled by brassinosteroids. Plant J 14:593–602

Molendijk AJ, Ruperti B, Palme K (2004) Small GTPases in vesicle trafficking. Curr Opin Plant Biol 7:694–700

Muday GK, DeLong A (2001) Polar auxin transport: controlling where and how much. Trends Plant Sci 6:535–542

Nakamura A, Umemura I, Gomi K, Hasegawa Y, Kitano H, Sazuka T, Matsuoka M (2006a) Production and characterization of auxin-insensitive rice by overexpression of a mutagenized rice IAA protein. Plant J 46:297–306

Nakamura A, Nakajima N, Goda H, Shimada Y, Hayashi K, Nozaki H, Asami T, Yoshida S, Fujioka S (2006b) *Arabidopsis* Aux/IAA genes are involved in brassinosteroid-mediated growth responses in a manner dependent on organ type. Plant J 45:193–205

Nam KH, Li J (2002) BRI1/BAK1, a receptor kinase pair mediating brassinosteroid signaling. Cell 110:203–212

Napier RM (2005) TIRs of joy: new receptors for auxin. Bioessays 27:1213–1217

Neff MM, Nguyen SM, Malancharuvil EJ, Fujioka S, Noguchi T, Seto H, Tsubuki M, Honda T, Takatsuto S, Yoshida S, Chory J (1999) BAS1: A gene regulating brassinosteroid levels and light responsiveness in *Arabidopsis*. Proc Natl Acad Sci USA 96:15316–15323

Negi S, Ivanchenko MG, Muday GK (2008) Ethylene regulates lateral root formation and auxin transport in *Arabidopsis thaliana*. Plant J 55:175–187

Nomura T, Jager CE, Kitasaka Y, Takeuchi K, Fukami M, Yoneyama K, Matsushita Y, Nyunoya H, Takatsuto S, Fujioka S, Smith JJ, Kerckhoffs LH, Reid JB, Yokota T (2004) Brassinosteroid deficiency due to truncated steroid 5alpha-reductase causes dwarfism in the lk mutant of pea. Plant Physiol 135:2220–2229

Normanly J, Cohen JD, Fink GR (1993) *Arabidopsis thaliana* auxotrophs reveal a tryptophan-independent biosynthetic pathway for indole-3-acetic acid. Proc Natl Acad Sci USA 90:10355–10359

Ohnishi T, Szatmari AM, Watanabe B, Fujita S, Bancos S, Koncz C, Lafos M, Shibata K, Yokota T, Sakata K, Szekeres M, Mizutani M (2006) C-23 hydroxylation by Arabidopsis CYP90C1 and CYP90D1 reveals a novel shortcut in brassinosteroid biosynthesis. Plant Cell 18:3275–3288

Oki K, Inaba N, Kitagawa K, Fujioka S, Kitano H, Fujisawa Y, Kato H, Iwasaki Y (2008) Function of the alpha subunit of rice heterotrimeric G protein in brassinosteroid signaling. Plant Cell Physiol 50:161–172

Ouyang J, Shao X, Li J (2000) Indole-3-glycerol phosphate, a branchpoint of indole-3-acetic acid biosynthesis from the tryptophan biosynthetic pathway in *Arabidopsis thaliana*. Plant J 24:327–333

Pandey S, Nelson DC, Assmann SM (2009) Two novel GPCR-type G proteins are abscisic acid receptors in *Arabidopsis*. Cell 136:136–148

Perfus-Barbeoch L, Jones AM, Assmann SM (2004) Plant heterotrimeric G protein function: insights from *Arabidopsis* and rice mutants. Curr Opin Plant Biol 7:719–731

Qin G, Gu H, Zhao Y, Ma Z, Shi G, Yang Y, Pichersky E, Chen H, Liu M, Chen Z, Qu LJ (2005) An indole-3-acetic acid carboxyl methyltransferase regulates *Arabidopsis* leaf development. Plant Cell 17:2693–2704

Ryu H, Kim K, Cho H, Park J, Choe S, Hwang I (2007) Nucleocytoplasmic shuttling of BZR1 mediated by phosphorylation is essential in *Arabidopsis* brassinosteroid signaling. Plant Cell 19:2749–2762

Savenstrand H, Brosche M, Strid A (2004) Ultraviolet-B signalling: *Arabidopsis* brassinosteroid mutants are defective in UV-B regulated defence gene expression. Plant Physiol Biochem 42:687–694

Scherer GF (1994) Phospholipid signalling by phospholipase A2 in plants. The role of mastoparan and lysophospholipids as weak 'auxin-like' agonists. Symp Soc Exp Biol 48:229–242

Schuller A, Ludwig-Muller J (2006) A family of auxin conjugate hydrolases from *Brassica rapa*: characterization and expression during clubroot disease. New Phytol 171:145–157

Shimada Y, Fujioka S, Miyauchi N, Kushiro M, Takatsuto S, Nomura T, Yokota T, Kamiya Y, Bishop GJ, Yoshida S (2001) Brassinosteroid-6-oxidases from Arabidopsis and tomato catalyze multiple C-6 oxidations in brassinosteroid biosynthesis. Plant Physiol 126:770–779

Stepanova AN, Robertson-Hoyt J, Yun J, Benavente LM, Xie DY, Dolezal K, Schlereth A, Jörgens G, Alonso JM (2008) HYPERLINK "http://www.ncbi.nlm.nih.gov/pubmed/18394997?ordinalpos=5&itool=EntrezSystem2.PEntrez.Pubmed.Pubmed_ResultsPanel.Pubmed_DefaultReportPanel.Pubmed_RVDocSum" TAA1-mediated auxin biosynthesis is essential for hormone crosstalk and plant development. Cell 133:177–191

Sussex IM (2008) The scientific roots of modern plant biotechnology. Plant Cell 20:1189–1198

Swarup R, Marchant A, Bennett MJ (2000) Auxin transport: providing a sense of direction during plant development. Biochem Soc Trans 28:481–485

Swarup R, Perry P, Hagenbeek D, Van Der Straeten D, Beemster GT, Sandberg G, Bhalerao R, Ljung K, Bennett MJ (2007) Ethylene upregulates auxin biosynthesis in *Arabidopsis* seedlings to enhance inhibition of root cell elongation. Plant Cell 19:2186–2196

Symons GM, Reid JB (2004) Brassinosteroids do not undergo long-distance transport in pea. Implications for the regulation of endogenous brassinosteroid levels. Plant Physiol 135:2196–2206

Symons GM, Davies C, Shavrukov Y, Dry IB, Reid JB, Thomas MR (2006) Grapes on steroids. Brassinosteroids are involved in grape berry ripening. Plant Physiol 140:150–158

Tan X, Calderon-Villalobos LI, Sharon M, Zheng C, Robinson CV, Estelle M, Zheng N (2007) Mechanism of auxin perception by the TIR1 ubiquitin ligase. Nature 446:640–645

Tang W, Kim TW, Oses-Prieto JA, Sun Y, Deng Z, Zhu S, Wang R, Burlingame AL, Wang ZY (2008) BSKs mediate signal transduction from the receptor kinase BRI1 in *Arabidopsis*. Science 321:557–560

Tao Y, Ferrer JL, Ljung K, Pojer F, Hong F, Long JA, Li L, Moreno JE, Bowman ME, Ivans LJ, Cheng Y, Lim J, Zhao Y, Ballaré CL, Sandberg G, Noel JP, Chory J (2008) HYPERLINK "http://www.ncbi.nlm.nih.gov/pubmed/18394996?ordinalpos=6&itool=EntrezSystem2.PEntrez.Pubmed.Pubmed_ResultsPanel.Pubmed_DefaultReportPanel.Pubmed_RVDocSum" Rapid synthesis of auxin via a new tryptophan-dependent pathway is required for shade avoidance in plants. Cell 133:164–176

Teale WD, Paponov IA, Palme K (2006) Auxin in action: signalling, transport and the control of plant growth and development. Nat Rev Mol Cell Biol 7:847–859

Titapiwatanakun B, Murphy AS (2008) Post-transcriptional regulation of auxin transport proteins: cellular trafficking, protein phosphorylation, protein maturation, ubiquitination, and membrane composition. J Exp Bot 60:1093–1107

Trusov Y, Rookes JE, Tilbrook K, Chakravorty D, Mason MG, Anderson D, Chen JG, Jones AM, Botella JR (2007) Heterotrimeric G protein gamma subunits provide functional selectivity in G betagamma dimer signaling in *Arabidopsis*. Plant Cell 19:1235–1250

Ueguchi-Tanaka M, Fujisawa Y, Kobayashi M, Ashikari M, Iwasaki Y, Kitano H, Matsuoka M (2000) Rice dwarf mutant d1, which is defective in the alpha subunit of the heterotrimeric G protein, affects gibberellin signal transduction. Proc Natl Acad Sci USA 97:11638–11643

Ullah H, Chen JG, Wang S, Jones AM (2002) Role of a heterotrimeric G protein in regulation of *Arabidopsis* seed germination. Plant Physiol 129:897–907

Ullah H, Chen JG, Young JC, Im KH, Sussman MR, Jones AM (2001) Modulation of cell proliferation by heterotrimeric G protein in *Arabidopsis*. Science 292:2066–2069

Ullah H, Chen JG, Temple B, Boyes DC, Alonso JM, Davis KR, Ecker JR, Jones AM (2003) The beta-subunit of the *Arabidopsis* G protein negatively regulates auxin-induced cell division and affects multiple developmental processes. Plant Cell 15:393–409

Ulmasov T, Hagen G, Guilfoyle TJ (1997a) ARF1, a transcription factor that binds to auxin response elements. Science 276:1865–1868

Ulmasov T, Hagen G, Guilfoyle TJ (1999) Activation and repression of transcription by auxin response factors. Proc Natl Acad Sci USA 96:5844–5849

Ulmasov T, Murfett J, Hagen G, Guilfoyle TJ (1997b) Aux/IAA proteins repress expression of reporter genes containing natural and highly active synthetic auxin response elements. Plant Cell 9:1963–1971

Wang L, Xu YY, Ma QB, Li D, Xu ZH, Chong K (2006a) Heterotrimeric G protein alpha subunit is involved in rice brassinosteroid response. Cell Res 16:916–922

Wang S, Narendra S, Fedoroff N (2007) Heterotrimeric G protein signaling in the *Arabidopsis* unfolded protein response. Proc Natl Acad Sci USA 104:3817–3822

Wang X, Xu W, Xu Y, Chong K, Xu Z, Xia G (2004) Wheat RAN1, a nuclear small G protein, is involved in regulation of cell division in yeast. Plant Sci 167:1183–1190

Wang X, Xu Y, Han Y, Bao S, Du J, Yuan M, Xu Z, Chong K (2006b) Overexpression of RAN1 in rice and *Arabidopsis* alters primordial meristem, mitotic progress, and sensitivity to auxin. Plant Physiol 140:91–101

Wang XQ, Ullah H, Jones AM, Assmann SM (2001) G protein regulation of ion channels and abscisic acid signaling in *Arabidopsis* guard cells. Science 292:2070–2072

Wei Q, Zhou W, Hu G, Wei J, Yang H, Huang J (2008) Heterotrimeric G-protein is involved in phytochrome A-mediated cell death of *Arabidopsis* hypocotyls. Cell Res 18:949–960

Weiss CA, Garnaat CW, Mukai K, Hu Y, Ma H (1994) Isolation of cDNAs encoding guanine nucleotide-binding protein beta-subunit homologues from maize (ZGB1) and Arabidopsis (AGB1). Proc Natl Acad Sci USA 91:9554–9558

Went FW, Thimann KV (1937) Phytohormones. MacMillan, New York

Woo EJ, Marshall J, Bauly J, Chen JG, Venis M, Napier RM, Pickersgill RW (2002) Crystal structure of auxin-binding protein 1 in complex with auxin. EMBO J 21:2877–2885

Wright AD, Sampson MB, Neuffer MG, Michalczuk L, Slovin JP, Cohen JD (1991) Indole-3-acetic acid biosynthesis in the mutant maize orange pericarp, a tryptophan auxotroph. Science 254:998–1000

Wu Y, Xu X, Li S, Liu T, Ma L, Shang Z (2007) Heterotrimeric G-protein participation in Arabidopsis pollen germination through modulation of a plasma membrane hyperpolarization-activated Ca^{2+}-permeable channel. New Phytol 176:550–559

Yamamuro C, Ihara Y, Wu X, Noguchi T, Fujioka S, Takatsuto S, Ashikari M, Kitano H, Matsuoka M (2000) Loss of function of a rice brassinosteroid insensitive1 homolog prevents internode elongation and bending of the lamina joint. Plant Cell 12:1591–1606

Yang G, Matsuoka M, Iwasaki Y, Komatsu S (2003) A novel brassinolide-enhanced gene identified by cDNA microarray is involved in the growth of rice. Plant Mol Biol 52:843–854

Yin Y, Wang ZY, Mora-Garcia S, Li J, Yoshida S, Asami T, Chory J (2002) BES1 accumulates in the nucleus in response to brassinosteroids to regulate gene expression and promote stem elongation. Cell 109:181–191

Yu HY, Seo JA, Kim JE, Han KH, Shim WB, Yun SH, Lee YW (2008) Functional analyses of heterotrimeric G protein G alpha and G beta subunits in *Gibberella zeae*. Microbiology 154:392–401

Zaina S, Reggiani R, Bertani A (1990) Preliminary evidence for involvement of GTP-binding protein(s) in auxin signal transduction in rice (*Oryza sativa* L.) coleoptile. J Plant Physiol 136:653–658

Zaina S, Mapelli S, Reggiani R, Bertani A (1991) Auxin and GTPase activity in membranes from aerobic and anaerobic rice coleoptile. J Plant Physiol 138:760–762

Zhao J, Peng P, Schmitz RJ, Decker AD, Tax FE, Li J (2002) Two putative BIN2 substrates are nuclear components of brassinosteroid signaling. Plant Physiol 130:1221–1229

Zhuang X, Xu Y, Chong K, Lan L, Xue Y, Xu Z (2005) OsAGAP, an ARF-GAP from rice, regulates root development mediated by auxin in *Arabidopsis*. Plant Cell Environ 28:147–156

Zhuang X, Jiang J, Li J, Ma Q, Xu Y, Xue Y, Xu Z, Chong K (2006) Over-expression of OsAGAP, an ARF-GAP, interferes with auxin influx, vesicle trafficking and root development. Plant J 48:581–591

Heterotrimeric G-Proteins and Cell Division in Plants

Jin-Gui Chen

Abstract The heterotrimeric GTP-binding protein (G-protein) complex is a conserved signaling module found in all eukaryotes. G-proteins function as molecular switches to regulate diverse signal transduction pathways. Although in contrast to its counterpart in mammals, the repertoire of G-protein complex in plants is much simpler, G-proteins play important roles in plant development, hormonal signaling, and biotic and abiotic stress responses. Gene expression and protein localization studies, pharmacological analysis, and genetic characterization demonstrated that G-proteins are critical modulators of plant cell division. Many of these studies have been concentrated on the model plant *Arabidopsis thaliana*, and the role of G-proteins in cell division has been best characterized in hypocotyls, rosette leaves, and roots. However, little is known about the upstream and downstream components coupled to G-proteins in the regulation of cell division. Future studies are expected to reveal the molecular mechanism through which G-proteins exert their modulatory roles in plant cell division.

1 Introduction

The heterotrimeric GTP-binding proteins (G-proteins) consist of three subunit, Gα, Gβ, and Gγ. G-proteins function as molecular switches to regulate diverse signal transduction pathways (Gilman 1987). In the classical G-protein signaling paradigm, ligand binding to the cell surface 7-transmembrane (7TM) G protein-coupled receptor (GPCR) activates the GPCR by inducing a conformational change. The activated GPCR acts as a Guanine Nucleotide Exchange Factor (GEF) to prompt the exchange of GDP for GTP in the Gα and subsequent dissociation of Gβγ dimer

J.-G. Chen
Department of Botany, University of British Columbia, Vancouver, BC V6T 1Z4, Canada
e-mail: jingui@interchange.ubc.ca

S. Yalovsky et al. (eds.), *Integrated G Proteins Signaling in Plants*,
Signaling and Communication in Plants,
DOI 10.1007/978-3-642-03524-1_9, © Springer-Verlag Berlin Heidelberg 2010

from the activated Gα (GTP-bound). The GTP-bound Gα and freely released Gβγ can then interact with downstream target proteins, termed as effectors, rendering specific cellular responses. Eventually, the intrinsic GTPase activity of the Gα allows the bound GTP to be hydrolyzed into GDP, and the G-proteins return to an inactive state. The GTPase activity of the Gα can be accelerated by the Regulator of G-protein Signaling (RGS) proteins which possess GTPase Accelerating Protein (GAP) activity for Gα (Neubig and Siderovski 2002).

The heterotrimeric G-protein complex is a conserved signaling module found in all eukaryotes, including plants (Assmann 2002; Temple and Jones 2007; Ding et al. 2008). In this chapter, I begin with a brief overview of known G-protein subunits and component proteins (referred to proteins that interact with G-proteins) in plants. Then, I specifically focus on the role of G-proteins in plant cell division. For more details about other aspects of plant G-proteins, readers are referred to recent review articles and references therein (Assmann 2002; Jones 2002; Jones and Assmann 2004; Perfus-Barbeoch et al. 2004; Temple and Jones 2007; Ding et al. 2008; Chen 2008).

2 Heterotrimeric G-Proteins and Component Proteins in Plants

The key components of the classical heterotrimeric G-protein signaling complex, including Gα, Gβ, Gγ, putative GPCRs, and RGS have all been found in plants. However, in contrast to its counterpart in mammals, the repertoire of G-protein complex is much simpler in plants. In model plant *Arabidopsis*, there are one canonical Gα (GPA1) (Ma et al. 1990), one canonical Gβ (AGB1) (Weiss et al. 1994), and two canonical Gγ (AGG1 and AGG2) (Mason and Botella 2000, 2001). Several dozens of proteins are predicted to contain 7TM domain (Moriyama et al. 2006; Gookin et al. 2008), a structural hallmark of classical GPCRs. Four proteins have been proposed as GPCRs, including GCR1 (Josefsson and Rask 1997; Plakidou-Dymock et al. 1998; Chen et al. 2004; Pandey and Assmann 2004), GCR2 (Liu et al. 2007), GTG1, and GTG2 (Pandey et al. 2009). Each of these four proteins has been shown to physically interact with the sole *Arabidopsis* Gα, GPA1, but only GCR1 is predicted to possess 7TM domain. Although GCR2 was initially predicted to possess 7TM, subsequent studies suggested that GCR2 lacks 7TM and is likely a plant homolog of bacterial lanthionine synthetase (Gao et al. 2007; Johnston et al. 2007b; Illingworth et al. 2008). GTG1 and GTG2 were predicted to have 9TM, not 7TM (Pandey et al. 2009). Moreover, GTG1 and GTG2 display GTPase activity that is inhibited by GTP-bound GPA1. Therefore, GTG1 and GTG2 do not appear to be classical GPCRs. There is a single RGS protein (AtRGS1) in *Arabidopsis* (Chen et al. 2003). AtRGS1 is an unusual RGS because it contains an *N*-terminal 7TM, characteristic of GPCRs, but its GAP activity has been shown to be required for its function in G-protein-coupled sugar signaling pathway (Johnston et al. 2007a).

Four proteins have been shown to physically bind the sole *Arabidopsis* Gα (GPA1), including AtPirin1 (Lapik and Kaufman 2003), PLDα1 (Zhao and Wang 2004), PD1 (Warpeha et al. 2006), and THF1 (Huang et al. 2006), and are candidate effectors of GPA1. AtPirin1 is a member of the cupin protein superfamily which has diverse biological functions (Lapik and Kaufman 2003). PLDα1 is a major isoform of phospholipase D in *Arabidopsis* (Zhao and Wang 2004). PD1 is a cytosolic prephenate dehydratase (Warpeha et al. 2006). THF1 is a plastid protein localized to both the outer plastid membrane and the stroma and does not share significant sequence with any known protein (Huang et al. 2006).

Similar to *Arabidopsis*, rice contains one canonical Gα (RGA1) (Ishikawa et al. 1995; Seo et al. 1995), one canonical Gβ (RGB1) (Ishikawa et al. 1996) and two canonical Gγ (RGG1 and RGG2) (Kato et al. 2004). No component protein has been found to physically interact with G-proteins in rice. However, RGA1 is found to function genetically with the small GTPase Rac and act upstream of Rac in disease resistance (Suharsono et al. 2002).

The heterotrimeric G-protein subunits and component proteins have also been found in pea which contains two Gα (PsGα1 and PsGα2), one Gβ (PsGβ), and two Gγ (PsGγ1 and PsGγ2) (Misra et al. 2007). Moreover, PsGα1, PsGβ, and PsGγ1 all physically bind PsGPCR, a homolog of *Arabidopsis* GCR1. It was also found that PsGα1 and PsGβ bind PsPLCδ, an isoform of phospholipase C (Misra et al. 2007).

Although the protein sequence homologs of G-protein subunits and component proteins have been found in maize, tomato, tobacco, soybean, spinach, lotus, lupin, wild oat, and alfalfa (Assmann 2002), not all three G-protein subunits have been identified and relatively little is known about the function of G-proteins in these plant species. So far, the formation of G-protein heterotrimer has only been demonstrated in *Arabidopsis* (Ullah et al. 2003; Chen et al. 2006b; Adjobo-Hermans et al. 2006; Wang et al. 2008), rice (Kato et al. 2004) and pea (Misra et al. 2007). Table 1 summarizes the known G-protein subunits and component proteins in *Arabidopsis*, rice, and pea and their proposed roles in cell division.

3 Gene Expression and Protein Localization Studies Support a Role of G-Proteins in Cell Division

3.1 Gα Expression and Localization

If G-proteins play an important role in regulating cell division, one would expect that G-proteins are expressed in actively dividing tissues, such as shoot apical meristem (SAM), root apical meristem (RAM), leaf primordia, and lateral root primordia. When the first Gα was cloned in *Arabidopsis* (Ma et al. 1990), the expression patterns of *GPA1* across various tissues and organs became available. The transcript of *GPA1* could be detected in both vegetative and floral tissues and organs (Ma et al. 1990). Western blot analysis using a polyclonal serum raised

Table 1 Heterotrimeric G-protein subunits and component proteins in *Arabidopsis*, rice, and pea and their proposed roles in cell division

Description		Roles in cell division	References
Arabidopsis heterotrimeric G-protein subunits			
GPA1	The sole *Arabidopsis* heterotrimeric G-protein α subunit	Promotes advance of the cell cycle from G1 to G2 phase in tobacco BY-2 cells; positively regulates axial cell division in hypocotyls; positively regulates cell division in rosette leaves; GTP-bound GPA1 positively regulates cell division whereas intact heterotrimer negatively regulates cell division in the primary root; acts through AGB1 to positively regulate lateral root formation; promotes cell division in stomatal development	Ma et al. (1990); Ullah et al. (2001, 2003); Chen et al. (2003, 2006a); Zhang et al. (2008)
AGB1	The sole *Arabidopsis* heterotrimeric G-protein β subunit	Positively regulates axial cell division and negatively regulates circumferential cell division in hypocotyls; likely positively regulates cell division in rosette leaves; intact heterotrimer negatively regulates cell division in the primary root; acts downstream of GPA1 to negatively regulate lateral root formation; inhibits cell division in stomatal development	Weiss et al. (1993); Ullah et al. (2003); Chen et al. (2006a); Zhang et al. (2008)
AGG1	*Arabidopsis* heterotrimeric G-protein γ subunit 1	Negatively regulates lateral root formation	Mason and Botella (2000); Trusov et al. (2007)
AGG2	*Arabidopsis* heterotrimeric G-protein γ subunit 2	Negatively regulates lateral root formation	Mason and Botella (2001); Trusov et al. (2007)
Arabidopsis heterotrimeric G-protein component proteins			
GCR1	G-protein-coupled receptor 1; binds GPA1; predicted to contain 7TM	Promotes cell division in tobacco BY-2 cells through an increase in phosphatidylinositol-specific phospholipase C activity and an elevation of inositol 1,4,5-trisphosphate levels	Josefsson and Rask (1997); Plakidou-Dymock et al. (1998); Colucci et al. (2002); Apone et al. (2003); Chen et al. (2004); Pandey and Assmann (2004)

GCR2	G-protein-coupled receptor 2; binds GPA1; likely a plant homolog of bacterial lanthionine synthetase	Unknown	Liu et al. (2007); Gao et al. (2007); Johnston et al. (2007b); Illingworth et al. (2008)
GTG1	A GPCR-type G-protein; binds GPA1; binds GTP and possesses GTPase activity; predicted to contain 9TM	Unknown	Pandey et al. (2009)
GTG2	A GPCR-type G-protein; binds GPA1; binds GTP and possesses GTPase activity; predicted to contain 9TM	Unknown	Pandey et al. (2009)
AtRGS1	The sole *Arabidopsis* Regulator of G-protein Signaling (RGS); preferentially binds GTP-bound GPA1; contains an N-terminal 7TM domain and a C-terminal RGS box	Negatively regulates cell division in the primary root	Chen et al. (2003, 2006a)
AtPirin1	A member of the cupin protein superfamily; binds GPA1	Unknown	Lapik and Kaufman (2003)
PLDα1	A major isoform of phospholipase D in *Arabidopsis*; binds GPA1	Unknown	Zhao and Wang (2004)
PD1	A cytosolic prephenate dehydratase; binds GPA1	Unknown	Warpeha et al. (2006)
THF1	A plastid protein localized to both the outer plastid membrane and the stroma; binds GPA1; does not share significant sequence with any known proteins	Unknown	Huang et al. (2006)
SGB1	A Golgi-localized hexose transporter; interacts genetically with AGB1 in regulation of hypocotyl cell division	Positively regulates axial cell division in hypocotyls	Wang et al. (2006a)
Rice heterotrimeric G-protein subunits			
RGA1	The sole rice heterotrimeric G-protein α subunit	Acts independently of BRI1 to positively regulate cell division in the leaf sheaths	Ishikawa et al. (1995); Seo et al. (1995); Ashikari et al. (1999); Fujisawa et al. (1999); Ueguchi-Tanaka et al. (2000); Oki et al. (2009)
RGB1	The sole rice heterotrimeric G-protein β subunit	Unknown	Ishikawa et al. (1996)

(*continued*)

Table 1 (continued)

Description		Roles in cell division	References
RGG1	Rice heterotrimeric G-protein γ subunit 1	Unknown	Kato et al. (2004)
RGG2	Rice heterotrimeric G-protein γ subunit 2	Unknown	Kato et al. (2004)
Rice heterotrimeric G-protein component protein			
Rac	A small GTPase; interacts genetically with and acts downstream of RGA1 in disease resistance	Unknown	Suharsono et al. (2002)
Pea heterotrimeric G-protein subunits			
PsGα1	Pea heterotrimeric G-protein α subunit 1	Unknown	Misra et al. (2007)
PsGα2	Pea heterotrimeric G-protein α subunit 2	Unknown	Misra et al. (2007)
PsGβ	The sole pea heterotrimeric G-protein β subunit	Unknown	Misra et al. (2007)
PsGγ1	Pea heterotrimeric G-protein γ subunit 1	Unknown	Misra et al. (2007)
PsGγ2	Pea heterotrimeric G-protein γ subunit 2	Unknown	Misra et al. (2007)
Pea heterotrimeric G-protein component proteins			
PsGPCR	Highly similar to *Arabidopsis* GCR1; binds PsGα1, PsGβ, and PsGγ1; predicted to contain 7TM	Unknown	Misra et al. (2007)
PsPLCδ	An isoform of phospholipase C; binds PsGα1 and PsGβ	Unknown	Misra et al. (2007)

against a 15-amino acid peptide sequence unique to the *C*-terminus of GPA1 indicated that GPA1 protein is expressed widely in all tissues and organs examined, including roots, inflorescence stems, rosette leaves, cauline leaves, and siliques (Weiss et al. 1993). Moreover, the level of GPA1 is higher in immature organs than in mature organs, suggesting that the expression of GPA1 may be developmentally regulated and that GPA1 may have a more important role in actively growing tissues and organs. Immunolocalization study revealed more details of the site of GPA1 expression. GPA1 is found to be strongly expressed in SAM, RAM, lateral root apical meristem, and the leaf primordia (Weiss et al. 1993). GPA1 also appeared to be expressed strongly in pericycle, which is potentially meristematic and the site of lateral root initiation (Weiss et al. 1993).

A translational fusion between *GPA1* genomic DNA and *β-glucuronidase* (*GUS*) was used to examine the expression of *GPA1* in various tissues and organs (Huang et al. 1994). This 4.1 kb genomic DNA fragment contains 3.2 kb of upstream nontranscribed region and 0.9 kb of the *GPA1* 5′ transcribed region, including the first two exons, the first two introns, and part of the third exon, and is believed to contain all regulatory elements controlling *GPA1* expression. Results from the analysis of *GPA1::GUS* translational fusion reporter lines are largely consistent with those of immunolocalization study (Weiss et al. 1993; Huang et al. 1994). In both cases, *GPA1* and its gene product are expressed strongly in actively dividing tissues, which supports the hypothesis that GPA1 is involved in the regulation of cell division (Ma et al. 1990). Consistent with this view, at subcellular level, it has been found that GPA1-cyan fluorescent protein (CFP), a fusion protein, is preferentially expressed in the cell plate of dividing cells in *Arabidopsis* suspension cells (Chen et al. 2003; Chen et al. 2006a).

3.2 *Gβ Expression and Localization*

After Gβ was cloned from *Arabidopsis* and maize (Weiss et al. 1994), the information regarding the expression patterns of Gβ in various tissues and organs also became available. Like Gα, the transcript of *AGB1* could be detected in both vegetative and floral tissues and organs (Weiss et al. 1994). More detailed information about the possible sites of *AGB1* expression was revealed using the *AGB1* promoter::GUS transcriptional reporter lines in *Arabidopsis*. Two different genomic DNA sequences upstream of *AGB1* translation initiation site were used as putative *AGB1* promoter (Chen et al. 2006c; Anderson and Botella 2007). Some differences in expressions have been observed. For example, by using a larger genomic sequence, which also contains the entire coding region of the *AGB1*'s neighboring gene, *ASK12*, it was found that the promoter of *AGB1* was also active in the root cap of young seedlings (Anderson and Botella 2007). In both cases, however, it was found that the promoter of *AGB1* is active widely across various tissues and organs (Chen et al. 2006c; Trusov et al. 2007; Anderson and Botella 2007). In particular, the promoter of *AGB1* appeared to be active in the hypocotyls, young leaves, primary roots, and lateral roots, the organs that display cell division defects in *agb1* mutants (discussed below). Western blot and immunolocalization study, like that used for GPA1, may help provide more details about the expression and localization of AGB1. However, no antibodies specific for AGB1 have been reported, although the antibodies for Gβ in rice (Kato et al. 2004), tobacco (Peskan and Oelmüller 2000), and pea (Misra et al. 2007) have been reported. By using anti-tobacco Gβ serum, it was found that in young tissues, the Gβ protein level was relatively high, while it declined substantially during later stages of leaf development (Peskan and Oelmüller 2000), supporting a role of Gβ in leaf cell division. It has not been tested if any of these antibodies can cross-react with Gβ in *Arabidopsis*. At subcellular level, by using AGB1-yellow fluorescent protein (YFP), a fusion protein reporter, it has been shown that AGB1 is preferentially expressed in the cell plates of newly dividing cells (Chen et al. 2006a).

3.3 Gene Expression and Location Studies of Gγ Subunits and Component Proteins

The gene expression and protein location studies of Gγ subunits and component proteins also support a role of G-proteins in cell division. For example, the expression of Gγ has been studied using northern blot (Mason and Botella 2000, 2001) and the promoter::GUS transcriptional reporter lines in *Arabidopsis* (Chen et al. 2006c; Trusov et al. 2007). The promoters of *AGG1* and *AGG2* are active in SAM, leaves, and roots. Interestingly, the expression profiles of *AGG1* and *AGG2* appeared to overlap *AGB1* expression (Trusov et al. 2007). The promoter of the sole *RGS* gene in *Arabidopsis*, *AtRGS1*, was found to be predominantly expressed in the SAM and RAM (Chen et al. 2003), where *GPA1* is also expressed (Weiss et al. 1993, Huang et al. 1994). By using AtRGS1-green fluorescent protein (GFP), a fusion protein reporter, it was shown that AtRGS1 is accumulated at the nascent cell plate of dividing cells (Chen et al. 2003). Collectively, the analyzes of the expression and localization of G-protein subunit and component proteins support a role of G-proteins in cell division.

4 Pharmacological Analysis Supports a Role of G-Proteins in Cell Division

Gα functions as a molecular switch: Gα is activated when it binds GTP and inactivated when its bound GTP is hydrolyzed to GDP. Therefore, molecules that can change the activation status of Gα have been used as an activator or an inhibitor for Gα and are instrumental for studying the function of G-proteins.

Some commonly used activators for Gα include mastoparan, cholera toxin, and aluminum tetrafluoride (AlF_4^-). Mastoparan is a peptide that mimics the activating domain of GPCRs. A synthetic version of this peptide, Mas7, has been generally used as an activator of Gα. Cholera toxin, which is produced by the bacterium that causes cholera, is an enzyme that catalyzes the transfer of ADP ribose from intracellular NAD^+ to the Gαs (stimulatory Gα). Because this ADP ribosylation alters the Gα so that it can no longer hydrolyze its bound GTP, addition of cholera toxin causes Gα to remain in an active state. AlF_4^- binding to the GDP-bound Gα causes the G-protein to adopt an activated conformation capable of signaling to downstream effectors.

Pertussis toxin is one of the commonly used inhibitors for Gα. Pertussis toxin catalyzes the ADP ribosylation of Gαi (inhibitory Gα), thereby blocking Gαi activation by preventing GDP release from Gαi subunits. It should be noted that although plant Gα subunits contain a conserved sequence that serves as a potential site for ADP-ribosylation by cholera toxin, plant Gα subunits do not appear to contain the *C*-terminal cysteine ribosylation site for modification by pertussis toxin (Temple and Jones 2007). Therefore, the results of pertussis toxin sensitivity should be interpreted with cautions.

These activators and inhibitors for G-protein have been used in plant cell cultures to study the role of G-proteins in cell division. It was found that Mas7, but not its inactive analog Mas17, can stimulate cell division in the presence of auxin in tobacco BY-2 cells whereas pertussis toxin appeared to inhibit auxin-induced cell division (Ullah et al. 2001; Jones et al. 2003b). Pertussis toxin has also been shown to inhibit auxin-induced cell division in tobacco cell line cv Virginia Bright Italia-0 (VBI-0) (Campanoni and Nick 2005). On the other hand, AlF_4^- stimulates cell division in the presence or absence of auxin (Campanoni and Nick 2005). Taken together, pharmacological studies using G-protein activators and inhibitors implied that activation of Gα normally promotes cell division in plants.

5 Genetic Characterization Provides Direct Evidence that G-Proteins Play a Modulatory Role in Cell Division

Gene expression and protein location studies and pharmacological analysis provided evidence that G-proteins may have a role in regulating cell division. The direct evidence supporting the role of G-proteins in cell division had been lacking till the loss-of-function mutants of G-protein subunits became available. Characterization of these mutants as well as transgenic lines overexpressing G-protein subunits and component proteins provided convincing evidence that G-proteins are critical modulators of cell proliferation. The role of G-proteins in plant cell division has been best characterized in the hypocotyls, rosette leaves, and roots of model plant, *Arabidopsis*.

5.1 *Heterotrimeric G-Proteins and Hypocotyl Cell Division*

5.1.1 G-protein Subunit Mutants Display Hypocotyl Cell Division Phenotype

The first loss-of-function alleles of Gα in plants were isolated in *Arabidopsis*, in 2001 (Ullah et al. 2001). The *gpa1* mutants displayed multiple morphological phenotypes (Ullah et al. 2001, 2003; Jones et al. 2003a). One of the characteristic morphological traits observed in *gpa1* mutants was the short hypocotyl phenotype in etiolated seedlings (Ullah et al. 2001). Because the number of cells in the hypocotyl epidermis and cortex is predetermined during embryogenesis and little cell division occurs in the epidermis or cortex in dark- or light-grown wild-type *Arabidopsis* seedlings (Gendreau et al. 1997), the short hypocotyl of *gpa1* mutants could be due to a defect in either cell division (predetermined during embryogenesis) or cell elongation (determined postembryogenesis), or in both. By counting the number of cells and measuring the length of cells in a single cell file longitudinally from the base (hypocotyl/root junction) to the top (lateral to cotyledons) of a hypocotyl, it was determined that the short hypocotyl of *gpa1* etiolated seedlings was due to the decrease in cell number (Ullah et al. 2001). From the base to the top of a hypocotyl, wild-type *Arabidopsis* seedlings have about 20 cells in a single cell file longitudinally. *gpa1* mutants had only about 12 cells in a single cell file (Ullah et al. 2001), suggesting that cell division is inhibited in the hypocotyls of *gpa1* mutants. Consistent with the view that GPA1 regulates hypocotyl cell division, ectopic expression of *GPA1* conferred increased, ectopic cell division in the hypocotyl epidermal cells of *Arabidopsis* seedlings (Ullah et al. 2001).

Later on, it was found that this hypocotyl cell division phenotype is also shared by *agb1* mutants (Ullah et al. 2003). *agb1* mutants have short hypocotyls due to a decrease in axial cell division. However, unlike *gpa1* mutants, circumferential cell division was increased in *agb1* mutants (Ullah et al. 2003), suggesting that AGB1 may negatively regulate circumferential cell division in hypocotyls.

5.1.2 SGB1 Functions Downstream of AGB1 to Regulate Hypocotyl Cell Division

Because *agb1* mutants have short hypocotyls due to reduced cell division, *agb1* mutants were mutagenized through transformation with activation tagging vector to facilitate a screen for genetic repressors for *agb1*. One mutant which restored the short hypocotyl phenotype of *agb1* to wild-type length was obtained. This mutant was designated as *suppressor of G-protein β1* (*sgb1*) (Wang et al. 2006a). Moreover, microscopic analysis revealed that overexpression of *SGB1* restored the hypocotyl cell division phenotype of *agb1* mutants. These results suggested that SGB1 is genetically coupled to AGB1 and likely acts downstream of AGB1 to regulate cell division in the hypocotyls. Therefore, SGB1 represents the first candidate effector for G-proteins in the cell division pathway. Consistent with a role of *SGB1* in cell division, *SGB1* is predominantly expressed in tissues with active cell division, including SAM, RAM, and lateral root primordia. The relationship between *AGB1* and *SGB1* in regulating hypocotyl cell division is supported by the characterization of *sgb1* loss-of-function mutants and the epistatic analysis of *agb1* and *sgb1* mutations. Similar to *agb1* mutant, etiolated *sgb1* mutant seedlings have short hypocotyls than the wild type. No additive or synergistic phenotype was observed between *agb1* and *sgb1* mutations (Wang et al. 2006a).

Molecular cloning revealed that *SGB1* encodes a member of large superfamily of known and putative hexose transporters (Wang et al. 2006a). Indeed, heterologous expression in oocytes in conjunction with 2-deoxy D-glucose uptake assays support the predicted hexose transport activity of SGB1. Because AGB1 is predominantly localized at the plasmamembrane whereas SGB1 is Golgi-localized, it remains elusive the mechanism through which AGB1 acts with SGB1 to regulate cell division.

5.1.3 The Relationship Between G-Proteins and Brassinosteroids in Regulating Cell Division

Gao et al. (2008) used the *Arabidopsis* hypocotyl as a model system to study the relationship between GPA1 and BRI1, a receptor for brassinosteroids (BRs), and found that the short hypocotyl phenotype of *gpa1* mutants was enhanced by loss-of-function mutations in *BRI1*. BRs are best known for their roles in promoting cell elongation. As expected, the epidermal cells in the hypocotyls of *bri1* mutants are much shorter than those in wild-type (Gao et al. 2008). Similarly, the hypocotyl epidermal cells were also much shorter in BR biosynthesis mutants, such as *det2-1* and *dwf4-102*. Interestingly, it was found that the hypocotyl of *bri1* etiolated seedling also consisted of fewer cells in a single cell file longitudinally from the base to the top of a hypocotyl, and the hypocotyl of *gpa1 bri1* double mutants contained even fewer cells than *gpa1* and *bri1* single mutants. These results suggested that GPA1 and BRI1 likely act in parallel pathways to regulate cell division in hypocotyls.

The view that GPA1 and BRI1 may act independently to regulate cell division is also supported by the work in rice. The rice Gα (RGA1) mutant, *d1*, was originally identified as a dwarf mutant and was regarded as a gibberellin (GA) signaling mutant because *d1* mutant showed reduced sensitivity to GA in the induction of α-amylase expression and activity and in the promotion of internode elongation (Ashikari et al. 1999; Fujisawa et al. 1999; Ueguchi-Tanaka et al. 2000). Recently, it has been found that similar to *Arabidopsis* Gα mutant (Ullah et al. 2002), *d1* mutants also displayed reduced sensitivity to BR (Wang et al. 2006b; Oki et al., 2009), raising the possibility that G-proteins may be involved in BR signaling pathway. It was found that the dwarfism of rice Gα mutant (*T65d1*) was enhanced by mutation in rice *BRI1* (*d61-7*) (Oki et al. 2009). Moreover, the number of cells in the leaf sheaths of both *T65d1* and *d61-7* decreased significantly compared with that in wild-type, and a further reduction in cell number was observed in *T65d1 d61-7* double mutants. Because mutations in *Gα* and *BRI1* are not epistatic to each other to regulate cell division in *Arabidopsis* and rice, these results suggested that Gα may not be directly coupled to BRI1-mediated signaling pathway to regulate the cell division.

5.2 Heterotrimeric G-Proteins and Leaf Cell Division

5.2.1 G-Protein Subunit Mutants Display Defects in Cell Division in Leaves

Another characteristic morphological phenotype found in *gpa1* and *agb1* mutants is the formation of round-shaped rosette leaves (Ullah et al. 2001; Lease et al. 2001; Ullah et al. 2003; Chen et al. 2006a). Microscopic analysis revealed that the rosette leaves of *gpa1* mutants consisted of fewer but larger epidermal cells than that of wild-type (Ullah et al. 2001). The increase in epidermal cell size was interpreted as a compensating effect on the reduction in cell division (Ullah et al. 2001). Therefore, GPA1 appears to positively regulate the cell division in rosette leaves. By using *cyc1At-CDB::GUS* as a mitotic reporter, Ullah et al. (2001) found that the expression of this mitotic reporter was reduced in developing leaves of *gpa1* mutants, supporting the view that GPA1 regulates cell division in leaves.

5.2.2 G-Protein Regulates Cell Division During Stomatal Development

Stomatal complex on the surface of shoot epidermis is developed through a series of asymmetric divisions and ends with a symmetric one (Bergmann and Sack 2007). Recently, Zhang et al. (2008) found that G-proteins also regulate stomatal density on the epidermis of *Arabidopsis* cotyledons. *gpa1* displayed a reduction in stomatal density whereas overexpression of a constitutively active form of GPA1 (GPA1^{QL}) resulted in an increase in stomatal density. These results suggested that GPA1 promotes cell division in stomatal development. Interestingly, it was found the stomatal density was increased in the *agb1* mutants and decreased in transgenic lines overexpressing *AGB1*, suggesting that AGB1 inhibits cell division in stomatal development. Therefore, it appeared that GPA1 and AGB1 may have antagonistic effect on cell division in stomatal development. Epistatic analysis indicated that the stomatal density in *gpa1 agb1* double mutants was higher than wild-type but lower than *agb1* single mutant, supporting the view that GPA1 and AGB1 may function antagonistically in the same pathway or in different pathways to regulate stomatal density.

By examining the number of the primary, secondary, and tertiary stomatal complexes on the cotyledon epidermis, it was found that *gpa1* mutants have increased number of the primary stomatal complex but lack the secondary and tertiary stomatal complexes compared with wild type (Zhang et al. 2008). In contrast, *agb1* mutants have more tertiary stomatal complexes and fewer primary and secondary stomatal complexes. However, the one-celled spacing rule (stomata normally do not directly contact each other) which guides stomatal complex development was not affected either in *gpa1* or in *agb1* mutants. These results suggested that G-proteins mainly regulate the frequency of amplifying divisions occurring in the meristemoids (intermediate precursor cells that undergo one or more amplifying asymmetric divisions that regenerate the meristemoid and increase the number of larger daughter cells capable of founding the next generation in the lineage). Consistent with this view, the number of meristemoids was decreased in *gpa1* mutants and increased in *agb1* mutants (Zhang et al. 2008).

Because stomatal density is mainly determined by the frequency of asymmetric cell division and because heterotrimeric G-proteins are important regulators of asymmetric cell division in other organism, such as flies and worms (Gönczy 2002), these findings raised the possibility that G-proteins may regulate asymmetric cell divisions in plants. The asymmetric cell division in flies and worms are largely determined by spindle positioning during mitosis. There is, however, no evidence supporting a similar role of G-proteins in plant cell division.

5.3 Heterotrimeric G-Proteins and Root Cell Division

5.3.1 G-Protein Subunit Mutants Display Root Cell Division Phenotype

Loss-of-function mutants of *GPA1* and *AGB1* both displayed phenotypes in roots. *gpa1* mutants have normal length of primary roots but have fewer lateral roots whereas *agb1* mutants have longer primary roots and produced more lateral roots compared with wild type (Ullah et al. 2003; Chen et al. 2006a). These root phenotypes have become the key morphological traits to analyze the role of G-proteins in cell division.

Because primary root growth reflects a combined effect of cell division in the cell division zone and cell elongation in the cell elongation zone, a longer primary root as observed in *agb1* mutants could be due to alternation in cell division, cell elongation, or both. A defect in cell elongation in the primary root can be assessed by measuring the cell length in the root maturation zone, because cells in this region have reached their final length which reflects if an alternation in cell elongation has occurred (in cell elongation zone). A defect in cell division in the primary root can be assessed by measuring the cell production rate, which is often expressed as the root growth rate divided by the root cortex cell length in the root maturation zone. By measuring these parameters, it was determined that the longer primary root of *agb1* mutants was due to increased cell production (Chen et al. 2006a). No significant difference in the length of mature cortex cells was observed between wild-type and *agb1* mutants. Therefore, AGB1 is considered to function as a negative regulator of cell division in the primary roots.

Lateral root formation involves cell cycle activation or reentry in the pericycle founder cells and the subsequent establishment of a new meristem (Malamy and Benfey 1997; Casimiro et al. 2003; Malamy 2005; de Smet et al. 2006; Fukaki et al. 2007). In *Arabidopsis*, the first lateral root is formed in the root maturation zone at a position that corresponds with the region where pericycle cells progress via the S to G2 phase (Beeckman et al. 2001). Then, other lateral roots occur in an acropetal pattern (Dubrovsky et al. 2006). Because *gpa1* mutants produced fewer lateral roots whereas *agb1* produced more lateral roots, these results suggested that G-proteins regulate either cell cycle activation or reentry in the pericycle cells or the establishment of a new meristem (e.g., protrusion of the lateral root primordia, a process requires both cell division and cell elongation), or both. Cell division in the pericycle founder cells can be indirectly measured by counting the number of lateral root primordia and lateral roots. By measuring the number, position, and developmental stage of lateral root primordia in *gpa1* and *agb1* mutants, it was found that the number of lateral root primordia was decreased in *gpa1* mutants but increased in *agb1* mutants (Ullah et al. 2003), suggesting that GPA1 and AGB1 primarily regulate cell division during lateral root formation.

As discussed above, both GPA1 and AGB1 are expressed in the RAM and the lateral root primordia, which are consistent with a role of G-proteins in regulating root division in the primary root and during lateral root formation.

Characterization of loss-of-function Gγ mutants and mutants and transgenic lines of G-protein component proteins also support a role of G-proteins in regulating root cell division. Both the *agg1* and *agg2* mutants produced more lateral roots than did the wild type, and an additive effect was observed in *agg1 agg2* double mutants (Trusov et al. 2007), suggesting that Gγ subunits, likely functioning in the form of Gβγ dimers, negatively regulate the lateral root formation. On the other hand, loss-of-function mutations in the sole *RGS* in *Arabidopsis*, *AtRGS1*, resulted in longer primary roots due to increase root production (Chen et al. 2003). Because AtRGS1 preferentially interacts with the GTP-bound form of GPA1 and accelerates the GTPase activity of GPA1 (Chen et al. 2003), in the *Atrgs1* mutant, more GPA1 is believed to stay in the GTP-bound form due to the slow GTPase activity of GPA1 and the loss of GAP activity of AtRGS1. Therefore, these results supported that the GTP-bound GPA1 normally promotes cell division in the primary root. Consistent with this view, overexpression of a constitutively active form of GPA1, GPA1^{QL}, conferred longer primary roots than wild type due to increased cell division (Chen et al. 2003).

5.3.2 Differential Roles of G-Protein Subunits in Regulating Root Cell Division

The different primary root and lateral root phenotypes in *gpa1* and *agb1* mutants suggested that the Gα (GPA1) and Gβ (AGB1) subunits may have differential roles in regulating root cell division. In the classical model of G-protein signaling, activation of the Gα releases the sequestration of Gβγ dimer by Gα, therefore an opposite phenotype, such as number of lateral roots of Gα and Gβ mutants, is generally interpreted to mean that the Gβ subunit is the predominant factor of the heterotrimer regulating the cellular process of this given phenotype. Therefore, during lateral root formation, AGB1 plays more important role than GPA1 in the modulation of the cell cycle activation or reentry in the pericycle founder cells.

Double mutants analysis indicated that *agb1* is epistatic to *gpa1* in lateral root formation (Chen et al. 2006a), suggesting that AGB1 likely acts downstream of GPA1 to negatively regulate the lateral root formation.

While *agb1* mutants produced longer primary root due to increased cell production, *gpa1* mutants have wild-type length of primary root. This made a direct assignment of individual G-protein subunits in regulating cell division in primary root ambiguous. Chen et al. (2006a) generated transgenic lines overexpressing *GPA1* and *AGB1* in *agb1* and *gpa1* mutant backgrounds, respectively, and examined cell production rate in the RAM and counted the number of lateral roots in these lines. The major advantage of overexpressing Gα (GPA1) in the loss-of-function Gβ (AGB1) mutant background is to allow a direct examination of the role of Gα because Gβ is no longer available for recruitment by Gα to form the heterotrimer. On the other hand, overexpression of Gβ in the loss-of-function Gα mutant background allows a direct test of the role of Gβ because the sequestration of Gβ by Gα is eliminated. By using these transgenic lines as well as the *gpa1 agb1*, *Atrgs1 gpa1*, and *Atrgs1 agb1* double mutants, it was confirmed that AGB1 is the predominant factor of the G-protein complex regulating lateral root formation and that AGB1 works downstream of GPA1 (Chen et al. 2006a). In the RAM, interestingly, it was found that the formation of the heterotrimer is required to modulate cell division and that the heterotrimer functions as an attenuator whereas the activated form of GPA1 (GTP-bound) positively modulates cell division (Chen et al. 2006a).

5.3.3 Relationship Between G-Proteins and Auxin in Regulation of Root Cell Division

Little is know about how G-proteins regulate cell division in roots. Auxin is the most important plant hormone controlling cell division in roots. However, *gpa1* and *agb1* do not appear to be typical auxin mutants. Moreover, both *gpa1* and *agb1* mutants could still respond to exogenous auxin (Ullah et al. 2003). These results suggested that G-proteins are not directly coupled to auxin signaling pathways. Recent studies using *agg1* and *agg2* mutants revealed a possible mechanism through which G-proteins participate in auxin pathways (Trusov et al. 2007). It was found that both *agg1* and *agg2* mutants are hypersensitive to auxin in lateral root formation. It was proposed that AGG1 is coupled to AGB1 to attenuate acropetally transported auxin signaling whereas AGG2 is coupled to AGB1 to attenuate basipetally transported auxin signaling. Consistent with this view, the promoter of *AGG1* is active in the central cylinder whereas the promoter of *AGG2* is active in the epidermis and cortex. Further, in auxin-induced adventitious root formation in hypocotyls, *agg1* displayed hypersensitive responses whereas *agg2* mutants had wild-type responses. Because adventitious root formation is mainly dependent on auxin transport in the hypocotyl stele, these results are consistent with a role of AGG1 attenuating auxin transport in the stele (acropetal auxin transport in roots). On the other hand, both basipetal and acropetal auxin transports play their roles in gravitropism response with the basipetal auxin transport playing a major role, because inhibition of basipetal auxin transport blocks gravity response in roots while inhibition of acropetal auxin transport only partially reduces it (Rashotte et al. 2000). Therefore, if the heterotrimeric G-proteins indeed modulate lateral root formation through regulation of auxin polar transport, one would expect that the G-protein mutants display defects in gravitropism response. Indeed, *agb1*, *agg1*, and *agg2* single mutants and *agg1 agg2* double mutant all displayed reduced sensitivity to gravistimulation compared with the wild type (Trusov et al. 2007). Although a direct measurement of auxin transport in the roots of G-protein subunit mutants has not been performed, these results provide evidence that G-proteins regulate lateral root formation by modulating auxin polar transport and that AGG1 and AGG2 provide functional selectivity in the AGB1–AGG1/AGG2 dimer-mediated signaling. AGG1 and AGG2 were also found to provide functional selectivity in other AGB1-mediated pathways. For example, *agg1* mutants are hypersensitive to glucose and the osmotic agent mannitol during seed germination whereas *agg2* mutants were only affected by glucose (Trusov et al. 2007).

6 G-Protein Subunits May Target Different Nuclear Stage in Regulating Cell Division

Gene expression and protein location studies, pharmacological analysis, and genetic characterization all supported that G-proteins play a modulatory role in cell division. What could be the precise nuclear stage at which G-proteins exert their modulatory role? Overexpression of GPA1 in synchronized tobacco BY-2 cells shortened the G1 phase of the cell cycle and promoted the formation of nascent cell plates (Ullah et al. 2001), suggesting that GPA1 may promote cell division at the G1 or G1-to-S transition of the cell cycle.

Because formation of the lateral root results from cell cycle activation or reentry in the pericycle founder cells (Malamy and Benfey 1997; Casimiro et al. 2003; Malamy 2005; de Smet et al. 2006; Fukaki et al. 2007) whereas AGB1 negatively regulates lateral root formation, the exact nuclear stage at which AGB1 experts its modulatory role in root cell division remains unclear. Because AGB1 acts downstream of GPA1 and does not require GPA1 to inhibit lateral root formation (Ullah et al. 2003; Chen et al. 2006a), AGB1 may not necessarily target at the same phase of cell cycle as that of GPA1. For example, it has been proposed that AGB1 may target the G2-to-M transition of the cell cycle (Ding et al. 2008).

7 Concluding Remarks

The regulation of cell division is one of the conserved functions of the heterotrimeric G-proteins in eukaryotes (Knoblich 2001; Bellaiche and Gotta 2005; Wilkie and Kinch 2005; Yu et al. 2006). Gene expression and protein location studies, pharmacological analysis, and genetic characterization supported that G-proteins are important modulators of plant cell division. In addition to hypocotyls, rosette leaves, and roots, mutations in Gα and Gβ also resulted in alternations in organ shapes in flower and fruit in *Arabidopsis* (Lease et al. 2001; Ullah et al. 2003). Similarly, loss-of-function mutations or suppression of Gα altered seed shape in rice (Ashikari et al. 1999; Fujisawa et al. 1999), and suppression of Gβ in tobacco altered the shapes and development of floral organs (Peskan-Berghöfer et al. 2005). It remains unclear if alternations in these morphological traits in Gα and Gβ mutants or transgenic lines are also primarily caused by defects in cell division.

In classical G-protein signaling paradigm, G-proteins are coupled to upstream GPCRs and switch on-or-off-specific cellular processes through interaction with downstream effectors. As discussed above, several candidate GPCRs (e.g., GCR1, GTG1, and GTG2) and Gα-interacting proteins (e.g., AtPirin1, PD1, PLDα1, and THF1) have been identified. However, none of these G-protein component proteins has been unequivocally demonstrated to couple to Gα or Gβγ dimer to regulate cell division. The only putative effector coupled to Gβγ to regulate cell division is SGB1 (Wang et al. 2006a). SGB1 genetically interacts with AGB1 to regulate cell division in hypocotyls, but the physical interaction between AGB1 and SGB1 has yet to be established.

There is evidence indicating that GCR1 may have a role in regulating cell division. For example, overexpression of *Arabidopsis* GCR1 in tobacco BY-2 cells caused an increase in thymidine incorporation and in the mitotic index of aphidicolin synchronized cells (Colucci et al. 2002). Furthermore, it was found that GCR1-enhanced thymidine incorporation depends on an increase in phosphatidylinositol-specific phospholipase C activity and an elevation of inositol 1,4,5-trisphosphate levels in the cells (Apone et al. 2003). Because overexpression of GPA1 resulted in a similar effect on phosphatidylinositol-specific phospholipase C activity and inositol 1,4,5-trisphosphate levels, it was proposed that phosphatidylinositol-specific phospholipase C may be an effector for GPA1 and inositol 1,4,5-trisphosphate acts as a second messenger in this process (Apone et al. 2003). However, *gcr1* mutants do not appear to display any cell division phenotypes as that observed in *gpa1* mutants, including cell division defects in hypocotyls, rosette leaves, and roots. Furthermore, although the promoter of *GCR1* was shown to be active in hypocotyls and young leaves, it was not active in the root tip (Chen et al. 2004), the site at which both *GPA1* and *AGB1* are expressed. Therefore, GPA1 does not appear to be activated by and coupled to GCR1 to regulate cell division in these organs.

GTG1 and GTG2 are proposed to function as GPCR-type G-proteins (Pandey et al. 2009). Because GTP-bound GPA1 inhibits the GTPase activity of GTG1 and GTG2, it does not appear that GTG1 and GTG2 function as classical GPCRs (e.g., with GEF activity). Because no cell division defects, similar to that found in *gpa1* or *agb1* mutants, have been reported in *gtg1* and *gtg2* single and double mutants, it remains unclear if GTG1 and GTG2 and G-proteins are coupled to regulate cell division.

Similarly, none of the four GPA1-interacting proteins identified so far, including AtPirin1, PLDα1, PD1, and THF1, appears to function as an effector for GPA1 to regulate cell division. On the other hand, there are two canonical Gγ subunits in *Arabidopsis* (Mason and Botella 2000, 2001). Gγ is generally thought to be required for the proper function of Gβ. Therefore, it was predicted that mutants lacking both *AGG1* and *AGG2* in *Arabidopsis* phenocopy *agb1* mutant. Indeed, *agg1* and *agg2* mutants share the root phenotype with *agb1* mutants (Trusov et al. 2007). However, neither *agg1* nor *agg2* mutants displayed *agb1*-characteristic cell division defects in hypocotyls and rosette leaves (Trusov et al. 2008) although overexpression of a truncated Gγ lacking the isoprenylation motif resulted in a phenotype similar to *gpa1* and *agb1* mutants (Chakravorty and Botella 2007), raising the possibility that there exists unknown elements in the G-protein signaling complex. Taken together, future studies are required to uncover the upstream components (possibly GPCRs) and downstream effectors of G-proteins in the regulation of cell division.

References

Adjobo-Hermans MJ, Goedhart J, Gadella TW Jr (2006) Plant G protein heterotrimers require dual lipidation motifs of Gα and Gγ and do not dissociate upon activation. J Cell Sci 119:5087–5097

Anderson DJ, Botella JR (2007) Expression analysis and subcellular localization of the *Arabidopsis thaliana* G-protein β-subunit AGB1. Plant Cell Rep 26:1469–14680

Apone F, Alyeshmerni N, Wiens K, Chalmers D, Chrispeels MJ, Colucci G (2003) The G-protein-coupled receptor GCR1 regulates DNA synthesis through activation of phosphatidylinositol-specific phospholipase C. Plant Physiol 133:571–579

Ashikari M, Wu J, Yano M, Sasaki T, Yoshimura A (1999) Rice gibberellin-insensitive dwarf mutant gene *Dwarf 1* encodes the α-subunit of GTP-binding protein. Proc Natl Acad Sci USA 96:10284–10289

Assmann SM (2002) Heterotrimeric and unconventional GTP binding proteins in plant cell signaling. Plant Cell (Suppl) 14:S355–S373

Beeckman T, Burssens S, Inzé D (2001) The peri-cell-cycle in *Arabidopsis*. J Exp Bot 52:403–411

Bellaiche Y, Gotta M (2005) Heterotrimeric G proteins and regulation of size asymmetry during cell division. Curr Opin Cell Biol 17:658–663

Bergmann DC, Sack FD (2007) Stomatal development. Annu Rev Plant Biol 58:163–181

Campanoni P, Nick P (2005) Auxin-dependent cell division and cell elongation. 1-naphthaleneacetic acid and 2, 4-dichlorophenoxyacetic acid activate different pathways. Plant Physiol 137:939–948

Casimiro I, Beeckman T, Graham N, Bhalerao R, Zhang H, Casero P, Sandberg G, Bennett MJ (2003) Dissecting *Arabidopsis* lateral root development. Trends Plant Sci 8:165–171

Chakravorty D, Botella JR (2007) Over-expression of a truncated *Arabidopsis thaliana* heterotrimeric G protein γ subunit results in a phenotype similar to α and β subunit knockouts. Gene 393:163–170

Chen JG (2008) Heterotrimeric G proteins in plant development. Front Biosci 13:3321–3333

Chen JG, Willard FS, Huang J, Liang J, Chasse SA, Jones AM, Siderovski DP (2003) A seven-transmembrane RGS protein that modulates plant cell proliferation. Science 301:1728–1731

Chen JG, Pandey S, Huang JR, Alonso JM, Ecker JR, Assmann SM, Jones AM (2004) GCR1 can act independently of heterotrimeric G-protein in response to brassinosteroids and gibberellins in *Arabidopsis* seed germination. Plant Physiol 135:907–915

Chen JG, Gao Y, Jones AM (2006a) Differential roles of *Arabidopsis* heterotrimeric G-protein subunits in modulating cell division in roots. Plant Physiol 141:887–897

Chen JG, Ullah H, Temple B, Liang J, Guo J, Alonso JM, Ecker JR, Jones AM (2006b) RACK1 mediates multiple hormone responsiveness and developmental processes in *Arabidopsis*. J Exp Bot 57:2697–2708

Chen Z, Hartmann HA, Wu MJ, Friedman EJ, Chen JG, Pulley M, Schulze-Lefert P, Panstruga R, Jones AM (2006c) Expression analysis of the AtMLO gene family encoding plant-specific seven-transmembrane domain proteins. Plant Mol Biol 60:583–597

Colucci G, Apone F, Alyeshmerni N, Chalmers D, Chrispeels MJ (2002) GCR1, the putative *Arabidopsis* G protein-coupled receptor gene is cell cycle-regulated, and its overexpression abolishes seed dormancy and shortens time to flowering. Proc Natl Acad Sci USA 99:4736–4741

de Smet I, Vanneste S, Inzé D, Beeckman T (2006) Lateral root initiation or the birth of a new meristem. Plant Mol Biol 60:871–887

Ding L, Chen JG, Jones AM, Assmann SM (2008) Heterotrimeric G protein-coupled signaling in higher plants. In: Yang Z (Ed) Intracellular signalling in plants. Blackwell Publishing, Oxford. Annu Plant Rev 33:30–63

Dubrovsky JG, Gambetta GA, Hernandez-Barrera A, Shishkova S, Gonzalez I (2006) Lateral root initiation in *Arabidopsis*: developmental window, spatial patterning, density and predictability. Ann Bot 97:903–915

Fujisawa Y, Kato T, Ohki S, Ishikawa A, Kitano H, Sasaki T, Asahi T, Iwasaki Y (1999) Suppression of the heterotrimeric G protein causes abnormal morphology, including dwarfism, in rice. Proc Natl Acad Sci USA 96:7575–7580

Fukaki H, Okushima Y, Tasaka M (2007) Auxin-mediated lateral root formation in higher plants. Int Rev Cytol 256:111–137

Gao Y, Zeng Q, Guo J, Cheng J, Ellis BE, Chen JG (2007) Genetic characterization reveals no role for the reported ABA receptor, GCR2, in ABA control of seed germination and early seedling development in *Arabidopsis*. Plant J 52:1001–1013

Gao Y, Wang S, Asami T, Chen JG (2008) Loss-of-function mutations in the *Arabidopsis* heterotrimeric G-protein α subunit enhance the developmental defects of brassinosteroid signaling and biosynthesis mutants. Plant Cell Physiol 49:1013–1024

Gendreau E, Traas J, Desnos T, Grandjean O, Caboche M, Hofte H (1997) Cellular basis of hypocotyl growth in *Arabidopsis thaliana*. Plant Physiol 114:295–305

Gilman AG (1987) G proteins: transducers of receptor-generated signals. Annu Rev Biochem 56:615–649

Gönczy P (2002) Mechanisms of spindle positioning: focus on flies and worms. Trends Cell Biol 12:332–339

Gookin TE, Kim J, Assmann SM (2008) Whole proteome identification of plant candidate G-protein coupled receptors in *Arabidopsis*, rice, and poplar: computational prediction and in-vivo protein coupling. Genome Biol 9:R120

Huang H, Weiss CA, Ma H (1994) Regulated expression of the *Arabidopsis* G-protein α-subunit gene GPA1. Int J Plant Sci 155:3–14

Huang J, Taylor JP, Chen JG, Uhrig JF, Schnell DJ, Nakagawa T, Korth KL, Jones AM (2006) The plastid protein THYLAKOID FORMATION1 and the plasma membrane G-protein GPA1 interact in a novel sugar signaling mechanism in *Arabidopsis*. Plant Cell 18:1226–1238

Illingworth CJR, Parkes KE, Snell CR, Mullineaux PM, Reynolds CA (2008) Criteria for confirming sequence periodicity identified by Fourier transform analysis: application to GCR2, a candidate plant GPCR. Biophys Chem 133:28–35

Ishikawa A, Tsubouchi H, Iwasaki Y, Asahi T (1995) Molecular cloning and characterization of a cDNA for the α subunit of a G protein from rice. Plant Cell Physiol 36:353–359

Ishikawa A, Iwasaki Y, Asahi T (1996) Molecular cloning and characterization of a cDNA for the β subunit of a G protein from rice. Plant Cell Physiol 37:223–228

Johnston CA, Taylor JP, Gao Y, Kimple AJ, Grigston JC, Chen JG, Siderovski DP, Jones AM, Willard FS (2007a) GTPase acceleration as the rate-limiting step in *Arabidopsis* G protein-coupled sugar signaling. Proc Natl Acad Sci USA 104:17317–17322

Johnston CA, Temple BR, Chen JG, Gao Y, Moriyama EN, Jones AM, Siderovski DP, Willard FS (2007b) Comment on 'A G Protein-Coupled Receptor Is a Plasma Membrane Receptor for the Plant Hormone Abscisic Acid'. Science 318:914c

Jones AM (2002) G-protein-coupled signaling in *Arabidopsis*. Curr Opin Plant Biol 5:402–407

Jones AM, Assmann SM (2004) Plants: The latest model system for G-protein research. EMBO Rep 5:572–578

Jones AM, Ecker JR, Chen JG (2003a) A reevaluation of the role of the heterotrimeric G protein in coupling light responses in *Arabidopsis*. Plant Physiol 131:1623–1627

Jones AM, Ullah H, Chen JG (2003b) Dual pathways for auxin regulation of cell division and expansion. In: Nagata T, Hasezawa S, Inze D (Eds.) Tobacco BY-2 cells. Biotechnol Agric For 53:181–191

Josefsson LG, Rask L (1997) Cloning of a putative G-protein-coupled receptor from *Arabidopsis thaliana*. Eur J Biochem 249:415–420

Kato C, Mizutani T, Tamaki H, Kumagai H, Kamiya T, Hirobe A, Fujisawa Y, Kato H, Iwasaki Y (2004) Characterization of heterotrimeric G protein complexes in rice plasma membrane. Plant J 38:320–331

Knoblich JA (2001) Asymmetric cell division during animal development. Nat Rev Mol Cell Biol 2:11–20

Lapik YR, Kaufman LS (2003) The *Arabidopsis* cupin domain protein AtPirin1 interacts with the G protein α-subunit GPA1 and regulates seed germination and early seedling development. Plant Cell 15:1578–1590

Lease KA, Wen J, Li J, Doke JT, Liscum E, Walker JC (2001) A mutant *Arabidopsis* heterotrimeric G-protein b subunit affects leaf, flower, and fruit development. Plant Cell 13:2631–2641

Liu X, Yue Y, Li B, Nie Y, Li W, Wu WH, Ma L (2007) A G protein-coupled receptor is a plasma membrane receptor for the plant hormone abscisic acid. Science 315:1712–1716

Ma H, Yanofsky MF, Meyerowitz EM (1990) Molecular cloning and characterization of GPA1, a G protein α subunit gene from *Arabidopsis thaliana*. Proc Natl Acad Sci USA 87:3821–3825

Malamy JE (2005) Intrinsic and environmental response pathways that regulate root system architecture. Plant Cell Environ 28:67–77

Malamy JE, Benfey PN (1997) Organization and cell differentiation in lateral roots of *Arabidopsis thaliana*. Development 124:33–44

Mason MG, Botella JR (2000) Completing the heterotrimer: isolation and characterization of an *Arabidopsis thaliana* G protein γ-subunit cDNA. Proc Natl Acad Sci USA 97:14784–14788

Mason MG, Botella JR (2001) Isolation of a novel G-protein γ-subunit from *Arabidopsis thaliana* and its interaction with Gβ. Biochim Biophys Acta 1520:147–153

Misra S, Wu Y, Venkataraman G, Sopory SK, Tuteja N (2007) Heterotrimeric G-protein complex and G-protein-coupled receptor from a legume (*Pisum sativum*): role in salinity and heat stress and cross-talk with phospholipase C. Plant J 51:656–669

Moriyama EN, Strope PK, Opiyo SO, Chen Z, Jones AM (2006) Mining the *Arabidopsis thaliana* genome for highly-divergent seven transmembrane receptors. Genome Biol 7:R96

Neubig RR, Siderovski DP (2002) Regulators of G-protein signalling as new central nervous system drug targets. Nat Rev Drug Discov 1:187–197

Oki K, Inaba N, Kitagawa K, Fujioka S, Kitano H, Fujisawa Y, Kato H, Iwasaki Y (2009) Function of the α subunit of rice heterotrimeric G protein in brassinosteroid signaling. Plant Cell Physiol 50:161–172

Pandey S, Assmann SM (2004) The *Arabidopsis* putative G-protein coupled receptor GCR1 interacts with the G protein α subunit GPA1 and regulates abscisic acid signaling. Plant Cell 16:1616–1632

Pandey S, Nelson DC, Assmann SM (2009) Two novel GPCR-type G proteins are abscisic acid receptors in *Arabidopsis*. Cell 136:136–148

Perfus-Barbeoch L, Jones AM, Assmann SM (2004) Plant heterotrimeric G protein function: insights from *Arabidopsis* and rice mutants. Curr Opin Plant Biol 7:719–731

Peskan T, Oelmüller R (2000) Heterotrimeric G-protein beta-subunit is localized in the plasma membrane and nuclei of tobacco leaves. Plant Mol Biol 42:915–922

Peskan-Berghöfer T, Neuwirth J, Kusnetsov V, Oelmüller R (2005) Suppression of heterotrimeric G-protein beta-subunit affects anther shape, pollen development and inflorescence architecture in tobacco. Planta 220:737–746

Plakidou-Dymock S, Dymock D, Hooley R (1998) A higher plant seven transmembrane receptor that influences sensitivity to cytokinins. Curr Biol 8:315–324

Rashotte AM, Brady SR, Reed RC, Ante SJ, Muday GK (2000) Basipetal auxin transport is required for gravitropism in roots of *Arabidopsis*. Plant Physiol 122:481–490

Seo HS, Kim HY, Jeong JY, Lee SY, Cho MJ, Bahk JD (1995) Molecular cloning and characterization of RGA1 encoding a G protein α subunit from rice (*Oryza sativa* L. IR-36). Plant Mol Biol 27:1119–1131

Suharsono U, Fujisawa Y, Kawasaki T, Iwasaki Y, Satoh H, Shimamoto K (2002) The heterotrimeric G protein α subunit acts upstream of the small GTPase Rac in disease resistance of rice. Proc Natl Acad Sci USA 99:13307–13312

Temple BRS, Jones AM (2007) The plant heterotrimeric G-protein complex. Annu Rev Plant Biol 58:249–266

Trusov Y, Rookes JE, Tilbrook K, Chakravorty D, Mason MG, Anderson D, Chen JG, Jones AM, Botella JR (2007) Heterotrimeric G protein γ subunits provide functional selectivity in $G\beta\gamma$ dimer signaling in *Arabidopsis*. Plant Cell 19:1235–1250

Trusov Y, Zhang W, Assmann SM, Botella JR (2008) $G\gamma1 + G\gamma2 \neq G\beta$: heterotrimeric G protein $G\gamma$-deficient mutants do not recapitulate all phenotypes of $G\beta$-deficient mutants. Plant Physiol 147:636–649

Ueguchi-Tanaka M, Fujisawa Y, Kobayashi M, Ashikari M, Iwasaki Y, Kitano H, Matsuoka M (2000) Rice dwarf mutant d1, which is defective in the α subunit of the heterotrimeric G protein, affects gibberellin signal transduction. Proc Natl Acad Sci USA 97:11638–11643

Ullah H, Chen JG, Young JC, Im KH, Sussman MR, Jones AM (2001) Modulation of cell proliferation by heterotrimeric G protein in *Arabidopsis*. Science 292:2066–2069

Ullah H, Chen JG, Wang S, Jones AM (2002) Role of a heterotrimeric G protein in regulation of *Arabidopsis* seed germination. Plant Physiol 129:897–907

Ullah H, Chen JG, Temple B, Boyes DC, Alonso JM, Davis KR, Ecker JR, Jones AM (2003) The β-subunit of the *Arabidopsis* G protein negatively regulates auxin-induced cell division and affects multiple developmental processes. Plant Cell 15:393–409

Wang HX, Weerasinghe RR, Perdue TD, Cakmakci NG, Taylor JP, Marzluff WF, Jones AM (2006a) A golgi-localized hexose transporter is involved in heterotrimeric G protein-mediated early development in *Arabidopsis*. Mol Biol Cell 17:4257–4269

Wang L, Xu YY, Ma QB, Li D, Xu ZH, Chong K (2006b) Heterotrimeric G protein α subunit is involved in rice brassinosteroid response. Cell Res 16:916–922

Wang S, Assmann SM, Fedoroff NV (2008) Characterization of the *Arabidopsis* heterotrimeric G protein. J Biol Chem 283:13913–13922

Warpeha KM, Lateef SS, Lapik Y, Anderson M, Lee BS, Kaufman LS (2006) G-protein-coupled receptor 1, G-protein $G\alpha$-subunit 1, and prephenate dehydratase 1 are required for blue light-induced production of phenylalanine in etiolated *Arabidopsis*. Plant Physiol 140:844–855

Weiss CA, Huang H, Ma H (1993) Immunolocalization of the G protein α subunit encoded by the GPA1 gene in *Arabidopsis*. Plant Cell 5:1513–1528

Weiss CA, Garnaat CW, Mukai K, Hu Y, Ma H (1994) Isolation of cDNAs encoding guanine nucleotide-binding protein β-subunit homologues from maize (ZGB1) and *Arabidopsis* (AGB1). Proc Natl Acad Sci USA 91:9554–9558

Wilkie TM, Kinch L (2005) New roles for Gα and RGS proteins: communication continues despite pulling sisters apart. Curr Biol 15:R843–R854

Yu F, Kuo CT, Jan YN (2006) Drosophila neuroblast asymmetric cell division: recent advances and implications for stem cell biology. Neuron 51:13–20

Zhang L, Hu G, Cheng Y, Huang J (2008) Heterotrimeric G protein α and β subunits antagonistically modulate stomatal density in *Arabidopsis thaliana*. Dev Biol 324:68–75

Zhao J, Wang X (2004) *Arabidopsis* phospholipase Dα1 interacts with the heterotrimeric G-protein α-subunit through a motif analogous to the DRY motif in G-protein-coupled receptors. J Biol Chem 279:1794–1800

Heterotrimeric G Protein Regulation of Stomatal Movements

Sarah E. Nilson and Wei Zhang

Abstract Fine-tuned regulation of stomatal aperture size is key to the survival of land plants. This chapter discusses the roles of heterotrimeric G proteins in the regulation of stomatal movements and ion channel activities of guard cells. Evidence implicating heterotrimeric G protein function in light-induced stomatal opening, ABA-induced stomatal closure, and pathogen-induced stomatal movements is described from early pharmacological experiments to phenotypic studies of *Arabidopsis* heterotrimeric G protein mutants.

1 Introduction

Among the earliest identified and most extensively studied functions of heterotrimeric G proteins in plants are their roles in the regulation of stomatal movements. Stomata, which are found on the aerial surfaces of plants, consist of two guard cells surrounding a pore or stoma. The pore allows gas exchange with the atmosphere; through the pore, CO_2 enters the leaf for photosynthesis and water evaporates, driving the transport of water from the roots to the shoots. Land plants are faced with the challenge of balancing CO_2 acquisition with desiccation avoidance; guard cells, by regulating the size of the stomatal pore, are critical to maintaining this balance. This chapter reviews the primary guard cell signal transduction pathways that pertain to G proteins, describes the main methods used for assaying stomatal function, and discusses the pharmacological and molecular genetic evidence which has identified heterotrimeric G protein subunits as key players in stomatal aperture regulation. A summary table of the guard-cell-related phenotypes of the heterotrimeric G proteins mutants (Table 1) and an integrated

S.E. Nilson (✉) and W. Zhang
Biology Department, Penn State University, University Park, PA, 16802-5301, USA
e-mail: sen130@psu.edu; wuz12@psu.edu

S. Yalovsky et al. (eds.), *Integrated G Proteins Signaling in Plants*,
Signaling and Communication in Plants,
DOI 10.1007/978-3-642-03524-1_10,

Table 1 Stomatal regulation and density-related phenotypes of *Arabidopsis* heterotrimeric G protein mutants

Allele	Ecotype	Ion channel activities	Stomatal movements	Excised leaf water loss	Drought tolerance	Stomatal density
gpa1-1 *gpa1-2*	Ws, Ws	Insensitive to ABA inhibition of inward K^+ channels and pH-independent ABA activation of slow anion channels Wang et al. (2001) Insensitive to S1P inhibition of inward K^+ channels and S1P promotion of slow anion channels Coursol et al. (2003)	Insensitive to ABA inhibition of opening Wang et al. (2001) Wild-type ABA promotion of closure Wang et al. (2001) Insensitive to S1P promotion of closure and inhibition of opening Coursol et al. (2003) Reduced extracellular calmodulin-induced closure Chen et al. (2004)	Increased water loss from excised leaves Wang et al. (2001)	Not phenotyped	Not phenotyped
gpa1-3 *gpa1-4*	Col Col	*gpa1-4*: hyposensitive to ABA inhibition of inward K^+ channels Fan et al. (2008) Insensitive towards flg22-induced inhibition of inward K^+ channels Zhang et al. (2008b)	*gpa1-4*: hyposensitive to ABA inhibition of opening Fan et al. (2008) *gpa1-4*: wild-type ABA-induced closure *gpa1-3*: insensitive to PA inhibition of opening Mishra et al. (2006) Insensitive towards flg22-induced inhibition of opening Zhang et al. (2008b)	Allele not specified: Reduced water loss from detached cotyledons Zhang et al. (2008a) *gpa1-3*: Increased water loss from detached leaves Mishra et al. (2006)	Allele not specified: Reduced seedling dessication under low humidity conditions Zhang et al. (2008a)	Allele not specified: Reduced cotyledon stomatal density Zhang et al. (2008a)

agb1-1 *agb1-2*	Col Col	Hyposensitive to ABA inhibition of inward K^+ channels and pH-independent activation of outward anion channels Fan et al. (2008)	Hyposensitive to ABA inhibition of opening Fan et al. (2008) Wild-type ABA promotion of closure Fan et al. (2008)	Allele not specified: Increased water loss from cotyledons Zhang et al. (2008a)	Allele not specified: Increased seedling dessication under low humidity conditions Zhang et al. (2008a)	Allele not specified: Increased cotyledon stomatal density Zhang et al. (2008a)
agg1-1c	Ws allele introgressed into Col	Wild-type ABA inhibition of inward K^+ channels Trusov et al. (2008)	Wild-type ABA inhibition of opening and promotion of closure Trusov et al. (2008)	Not phenotyped	Not phenotyped	Not phenotyped
agg2-1 *agg2-2*	Col	Wild-type ABA inhibition of inward K^+ channels Trusov et al. (2008)	Wild-type ABA inhibition of opening and promotion of closure Trusov et al. (2008)	Not phenotyped	Not phenotyped	Not phenotyped
gcr1-3 *gcr1-4*	Ws Col	Not phenotyped	Hypersensitive to ABA and S1P inhibition of stomatal opening and ABA and S1P promotion of closure Pandey and Assmann (2004)	Reduced water loss from excised leaves Pandey and Assmann, (2004)	Plants exhibit drought tolerance Pandey and Assmann (2004)	Not phenotyped
rgs1-1 *rgs1-2*	Col Col	Wild-type ABA inhibition of inward K^+ channels for current amplitude but channels show accelerated voltage activation kinetics Fan et al. (2008)	Not phenotyped	Not phenotyped	Not phenotyped	*rgs1-2*: Increased stomatal density in cotyledons Zhang et al. (2008a)
gtg1gtg2 (double mutant)	Ws	Not phenotyped	Wild-type ABA inhibition of stomatal opening, hyposensitive to ABA promotion of closure Pandey et al. (2009)	Not phenotyped	Not phenotyped	Not phenotyped

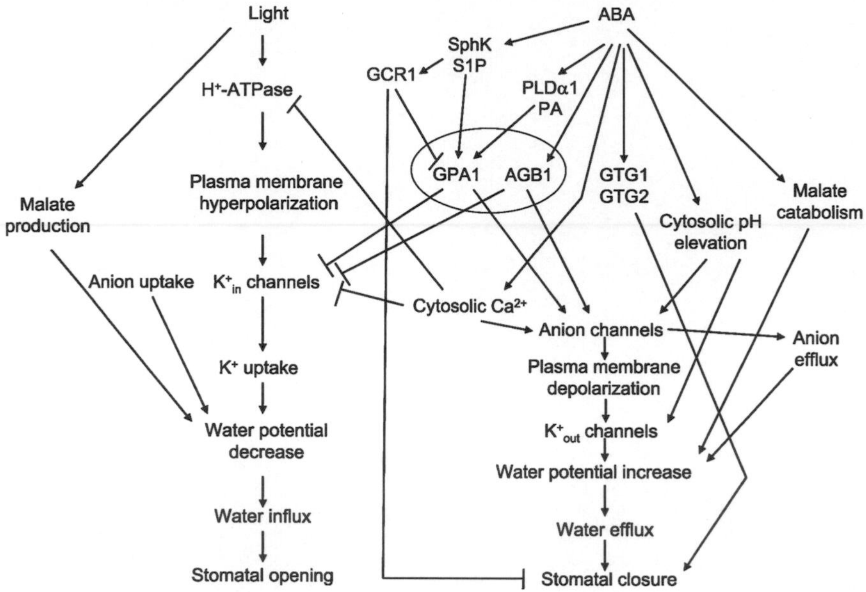

Fig. 1 Integrated model of G protein regulation of ABA-induced stomatal closure and light-induced stomatal opening. *Arrow* indicates positive regulation while *open square* indicates negative regulation

model of G protein regulation of abscisic acid (ABA)-induced stomatal movements (Fig. 1) are provided.

2 Mechanisms of Stomatal Movements

Guard cells are able to finely regulate the size of stomatal apertures in response to an array of environmental signals including CO_2 concentration, humidity, ABA, red and blue light, pathogen infection, and wounding (Assmann 1993; Schroeder et al. 2001; Melotto et al. 2006; Shimazaki et al. 2007; Underwood et al. 2007; Zhang et al. 2008b; Zhao et al. 2008). Stomatal movements are driven by turgor changes within the guard cells. Because of the radial arrangement of cellulose microfibrils in guard cell walls, an increase in cell turgor, originating from water influx into the guard cells, results in stomatal opening while water efflux from the cell, and the subsequent loss of turgor, facilitates stomatal closure (MacRobbie 1998; Schroeder et al. 2001). The turgor changes are initiated by the influx and efflux of solutes, including K^+ and anions, by malate synthesis, catabolism, and export, and, under some conditions, by the production and transport of soluble sugars (MacRobbie 1998; Fan et al. 2004; Shimazaki et al. 2007).

Guard cells integrate numerous signals and have become a model cell type for plant signal transduction (Assmann 1993; Schroeder et al. 2001). Mature guard cells lack plasmodesmata, so solute movements must occur across the plasma membrane (Willmer and Sexton 1979; Pandey et al. 2007). Ion channels, including inward- and outward-rectifying K^+ channels, anion channels, and Ca^{2+} permeable channels, are capable of mediating large ion fluxes across the cell membrane (Hille 2001) and are important components of stomatal movements and G-protein regulation (Wang et al. 2001; Coursol et al. 2003; Fan et al. 2008; Zhang et al. 2008b). Two of the best characterized pathways that involve heterotrimeric G proteins are light-induced stomatal opening and ABA promotion of stomatal closure.

2.1 Light-Induced Stomatal Opening

Guard cell exposure to light, especially blue light, activates plasma membrane H^+-ATPases, which results in H^+ efflux from the cell and hyperpolarization of the cell membrane (Shimazaki et al. 2007). This hyperpolarization opens voltage-gated inward-rectifying K^+ channels that mediate the influx of K^+ into the cell. K^+ influx, along with anion influx and malate^{2-} production from starch metabolism reduces the cell water potential (Shimazaki et al. 2007; Lee et al. 2008). Water moves into the guard cell to compensate for the reduction in water potential, resulting in an increase in cell turgor and stomatal opening (MacRobbie 1983; Talbott and Zeiger 1998; Schroeder et al. 2001; Pandey et al. 2007).

2.2 ABA Promotion of Closure

In response to drought stress, ABA concentration increases in the leaves (Davies and Zhang 1991; Wilkinson and Davies 2002). ABA induces closure of open stomata and inhibits light-induced opening of closed stomata. ABA stimulates stomatal closure, in part, by activating plasma membrane channels that are permeable to Ca^{2+}, causing an increase in cytosolic Ca^{2+}. The influx of Ca^{2+} also stimulates the release of Ca^{2+} from intracellular stores (Gilroy et al. 1990; Staxen et al. 1999). Elevated cytosolic Ca^{2+} levels can activate both slow and rapid anion channels which facilitate anion loss from the cell and plasma membrane depolarization (Schroeder and Hagiwara 1989; Hedrich et al. 1990); however, Ca^{2+}-independent activation of anion channels has also been observed (Levchenko et al. 2005; Marten et al. 2007). Depolarization of the membrane stimulates outward-rectifying K^+ channels and K^+ efflux from the cell. The net loss of K^+ and anions from the cell increases the water potential and drives water loss from the cell, turgor reduction, and stomatal closure. ABA-induced cytosolic Ca^{2+} elevations also inhibit the plasma membrane H^+-ATPase and inward K^+ channel activation, thereby inhibiting the opening of closed stomata (Schroeder and Hagiwara 1989; McAinsh et al. 1990; Lemtiri-Chlieh and MacRobbie 1994; Kinoshita et al. 1995; Schroeder et al. 2001; Pandey et al. 2007).

3 Measuring Stomatal Movements and Ion Channel Activities

3.1 *Electrophysiological Studies of Guard Cells*

Ion channel activities can be measured in guard cell protoplasts or isolated membrane patches using the patch clamp technique. (Schroeder et al. 1987; Wu and Assmann 1994; Forestier et al. 1998; Zhang et al. 2008c). This electrophysiological method can measure either net ion fluxes in whole cells or activities of single ion channels in isolated membrane patches (Neher et al. 1978; Kornreich 2007). The type of ion channel studied is controlled by the ionic composition and concentration of the patch pipette solution and the bath solution (the solution surrounding the protoplasts). Additionally, two-electrode voltage clamp is an electrophysiological technique that can be used to measure guard cell ion channels in epidermal peels or intact plants (Blatt 1992; Roelfsema et al. 2001). Electrophysiology experiments using *Vicia faba*, *Commelina communis*, and *Arabidopsis thaliana* protoplasts (both mutant and wild type) in combination with G-protein modulators and hormones have greatly contributed to our understanding of the downstream targets of G protein regulation of stomatal movements.

3.2 *Stomatal Aperture Assays*

Stomatal aperture assays allow the measurement of stomatal movements in response to hormonal, environmental, or pharmacological treatments. Typically, excised leaves are incubated in a buffer to induce stomatal closure or opening, a treatment is applied (e.g., ABA, blue light, or a G-protein-modifying drug), and after a period of incubation the epidermal strips are peeled off the leaves with forceps. For some species, particularly *Commelina* and *Vicia,* for which the epidermis can be peeled off easily, the isolated epidermal strips, rather than the intact leaves, can be directly subjected to the incubation solution. After treatment, the peels are examined under a microscope and stomatal apertures are measured with an ocular micrometer. Alternatively, digital photographs of peels can be taken and apertures measured on the photographs using image analysis software. By comparing apertures from treated and untreated samples, stomatal movements can be assessed. This technique is particularly effective for species in which the epidermis can be easily separated from the leaf. These studies are best performed blind to avoid inadvertent bias in measurement.

3.3 *Whole-Leaf/Plant Measures of Stomatal Function*

The implication of G-protein involvement in ABA-regulated stomatal movements has led researchers to conduct water loss assays from excised leaves and rosettes of

G-protein mutants by weighing the samples periodically following excision (Wang et al. 2001; Pandey and Assmann 2004; Zhang et al. 2008a). While the excision of plant tissue does result in wilting and water loss, the physiological validity of these assays is limited given that the leaves and/or rosette and, therefore, the stomata are separated from the roots, a primary source of ABA. Nonuniform humidity and temperatures can also confound results. An alternative approach to measuring whole-leaf changes in stomatal functioning is to measure stomatal conductance, an indicator of how open stomata are using a gas exchange system. Gas exchange systems allow for rates of photosynthesis and stomatal conductance to be assessed on whole plants or intact leaves under tightly controlled environmental conditions. Recent technological advances, such as the availability of a whole-plant *Arabidopsis* gas exchange chamber, will no doubt contribute to our understanding of G-protein regulation of stomatal function and whole-plant water status in the future.

4 G-protein Regulation of Stomatal Movements

Evidence suggesting that G-proteins regulate stomatal movements and ion channel activities was first obtained in the early 1990s using electrophysiological and pharmacological methods mainly applied to broad bean, *V. faba* (Fairley-Grenot and Assmann 1991; Wu and Assmann 1994). With the sequencing of the *Arabidopsis* genome and identification of G protein encoding genes, *Arabidopsis* quickly became the model system for the study of G-protein function in plants. The development of guard cell protoplast isolation and patch clamping techniques suitable for *Arabidopsis* (*Arabidopsis* guard cells are considerably smaller than those of *V. faba*) (Pei et al. 1997) and the acquisition and characterization of mutants lacking functional genes for heterotrimeric G-proteins allowed direct testing and confirmation of the role of heterotrimeric G-proteins in the regulation of stomatal movements (Wang et al. 2001).

4.1 Early Pharmacological Studies

Early and ongoing, mammalian studies have demonstrated that nonhydrolysable forms of GTP and GDP can be used to manipulate heterotrimeric G-protein function in cells. G-proteins can be constitutively activated upon binding to GTPγS or inactivated upon binding to GDPβS. Additionally, two pharmacological agents which function as ADP-ribosyltransferases, cholera toxin and pertussis toxin, can lock G proteins that contain an ADP-ribosylation site in either an active or an inactive state, respectively (Gilman 1987). The first evidence of G-protein regulation of ion channels was found in *V. faba* guard cell protoplasts, when G protein activity modulators were combined with whole-cell electrophysiological recordings of K^+ channels (Fairley-Grenot and Assmann 1991). Inward-rectifying

K^+ channels were found to be activated by GDPβS and inhibited by GTPγS. Additionally, both cholera and pertussis toxins were found to inhibit inward K^+ channels (Fairley-Grenot and Assmann 1991). No significant effects of GTPγS, GDPβS, or bacterial toxins were found on outward K^+ channels of *V. faba*. Synthetic mastoparan toxin (mas7), which essentially mimics ligand binding to a GPCR and activates heterotrimeric G proteins in mammalian species, inhibited inward K^+ but not outward K^+ channels in *Vicia* guard cell protoplasts (Armstrong and Blatt 1995). Application of GDPβS blocked the mas7 attenuation of inward K^+ channels, further supporting a role for heterotrimeric G proteins in ion channel regulation in plants (Armstrong and Blatt 1995). It is important to note, however, that mastoparan can activate MAP kinase signaling independent of Gα or Gβ, suggesting that mastoparan may not always specifically modulate G-protein signaling in *Arabidopsis* (Miles et al. 2004).

Regulation of inward K^+ channels by pharmacological modulations of G proteins was also found in single-channel recordings from isolated membrane patches, showing that a cytosolic signal transduction cascade was not required for ion channel regulation by G proteins, i.e., that regulation can occur via a membrane delimited pathway (Wu and Assmann 1994). However, electrophysiological studies using both calcium chelators and G-protein activity modifiers also suggested a role for cytosolic Ca^{2+} in G-protein regulation of inward K^+ channels. Thus, G-protein regulation of plant ion channels may occur via both membrane delimited pathways and cytosolic pathways involving secondary messengers (Fairley-Grenot and Assmann 1991; Kelly et al. 1995).

In epidermal peel experiments, microinjection of GTPγS into guard cells of *C. communis L* somewhat promoted light-induced stomatal opening (Lee et al. 1993), supporting a regulatory role for plant G-proteins in stomatal movements, although one inconsistent with the observed effects on K^+ channel activity. Also, application of G-protein antagonists (mas17 or GP Ant-2) to *C. communis* epidermal peels was shown to inhibit promotion of stomatal opening by 1 mM auxin as well as promotion of stomatal closure by 100 nM ABA (Cousson and Vavasseur 1998a, b). While these early studies strongly suggested the involvement of heterotrimeric G proteins in the regulation of ion channels and stomatal movements, it was the cloning of G-protein subunit genes and functional analyzes of T-DNA insertional mutants that unequivocally identified heterotrimeric G proteins as regulators of stomatal movements.

4.2 Arabidopsis Heterotrimeric G-Protein Genes

The *Arabidopsis* genome contains only one canonical Gα subunit, encoded by *GPA1*, and one Gβ subunit, encoded by *AGB1* (Ma et al. 1990; Weiss et al. 1994). Two Gγ subunits encoded by *AGG1* and *AGG2* have also been identified (Mason and Botella 2001). Northern analysis showed expression of *GPA1* in a mixed vegetative tissue sample that included both leaf and root tissue

(Ma et al. 1990). Promoter GUS analysis shows *GPA1* expression throughout the plant body, with stronger expression in developing tissues rather than in mature tissues (Huang et al. 1994). *GPA1* transcript can also be amplified from cDNA obtained from guard cell protoplast RNA (Wang et al. 2001). Microarray analysis also indicates *GPA1* expression in guard cell protoplasts (Yang et al. 2008). Western blotting and immunolocalization experiments indicate that GPA1 protein is present in all plant organs but mature dry seeds (Weiss et al. 1993; Huang et al. 1994; Pandey et al. 2006) and is more highly expressed in seedling roots as opposed to seedling shoots (Chen et al. 2006a). Like *GPA1*, has ubiquitous expression *AGB1* throughout the plant (Weiss et al. 1994; Anderson and Botella 2007) and is also expressed in guard cell protoplasts (Yang et al. 2008). *AGB1* promoter::GUS transgenic lines show strong GUS activity in the guard cells of cotyledons and seedling leaves (Anderson and Botella 2007; Trusov et al. 2007). RNA blots show *AGG1* and *AGG2* expression throughout the plant. Promoter GUS fusions show guard cell expression of *AGG2* but not *AGG1* (Mason and Botella 2000, 2001; Trusov et al. 2007); however, microarray analysis of guard cell protoplasts indicates that *AGG1* is expressed in guard cells (Yang et al. 2008)

All heterotrimeric G protein subunits have been shown to associate with the plasma membrane. Immunolocalization (Weiss et al. 1997) and transient expression assays (Adjobo-Hermans et al. 2006; Chen et al. 2006a, Wang et al. 2008) show GPA1 at the plasma membrane. While more than 60% of leaf GPA1 protein is associated with the plasma membrane (Wang et al. 2007), GPA1 also has been immunolocalized to the Golgi apparatus and the endoplasmic reticulum (Weiss et al. 1997). Transgenic *Arabidopsis* overexpressing an AGB1–GFP reporter fusion shows plasma membrane and nuclear localization of AGB1 in leaf epidermal cells (Anderson and Botella 2007) and the plasma membrane localization of 35S::YFP-AGB1 in *Arabidopsis* suspension cells (Chen et al. 2006a). In transient mesophyll protoplast expression assays, AGB1 localizes to the cytoplasm, but plasma membrane localization of AGB1 can be observed when AGB1 is coexpressed with either AGG1 or AGG2 (Adjobo-Hermans et al. 2006, Wang et al. 2008). Protein fractionation experiments also show that AGB1 is associated with the plasma membrane, endoplasmic reticulum, and nuclei (Obrdlik et al. 2000; Peškan and Oelmüller 2000; Wang et al. 2007). AGG1 and AGG2 both show plasma membrane localization in leaf epidermal cells in YFP-tagged AGG1 and AGG2 *Arabidopsis* lines (Zeng et al. 2007). Additionally, AGG1 also localizes to the Golgi apparatus (Zeng et al. 2007). Transient expression experiments in protoplasts show plasma membrane association for AGG1 and AGG2 (Adjobo-Hermans et al. 2006; Wang et al. 2008). Although Adjobo-Hermans et al. (2006) found that the plasma membrane localization of AGG1 requires AGB1 coexpression, this requirement was not observed by Wang et al. (2008). FRET experiments using *Arabidopsis* (Wang et al. 2008) and cowpea (Adjobo-Hermans et al. 2006) mesophyll cell protoplasts transfected with GPA1, AGB1, and AGG1 show interaction among the subunits and suggest that the heterotrimer exists at the plasma membrane. To date, no heterotrimeric G protein subcellular localization studies have been performed in guard cells or guard cell protoplasts.

In addition to the genes encoding heterotrimeric G protein subunits in *Arabidopsis*, one regulator of G protein signaling gene, *RGS1* (Chen et al. 2003), and several putative G-protein-coupled receptors (GPCRs), one of which is GCR1, have been identified (Josefsson and Rask 1997; Pandey and Assmann 2004; Moriyama et al. 2006, Gookin et al. 2008). *RGS1* is expressed in root and shoot meristems (Chen et al. 2003) and in guard cell protoplasts (Fan et al. 2008). RGS1-GFP localizes to the plasma membrane of *Arabidopsis* suspension cells (Chen et al. 2003). *GCR1* transcript can be detected throughout the plant body, including guard cell protoplasts (Pandey and Assmann 2004). Both western blotting (Pandey and Assmann 2004) and GFP localization studies (Humphrey and Botella 2001) show that GCR1 associates with cell membranes. The gene expression patterns and subcellular localizations of additional candidate GPCRs, predicted using bioinformatics (Moriyama et al. 2006; Gookin et al. 2008) and biochemical interaction assays with GPA1 (Gookin et al. 2008), have not been studied.

Two novel GPCR-like G proteins, GTG1 and GTG2, have been identified in *Arabidopsis* that localize at the plasma membrane (Pandey et al. 2009). The *GTG* genes are widely expressed in the plant according to RT-PCR and promoter::GUS analysis, including expression in guard cells (Pandey et al. 2009). It should be noted that another gene, *GCR2*, has been reported to function as both an ABA receptor and a GPCR and to regulate stomatal movements in response to ABA (Liu et al. 2007). However, these conclusions have been challenged; phenotypic discrepancies and erroneous topology predictions indicate it is neither an ABA receptor nor a GPCR (Gao et al. 2007; Guo et al. 2008; Illingworth et al. 2008).

4.3 G protein Regulation of ABA Inhibition of Light-Induced Stomatal Opening

Using two null mutants of *gpa1*, Wang et al. (2001) found the first definitive evidence for heterotrimeric G protein regulation of stomatal movements. ABA inhibition of light-induced stomatal opening was diminished in the mutants compared with wild type, Ws. Consistent with the aperture data, patch clamp experiments showed that *gpa1* mutants are insensitive to ABA inhibition of inward K^+ channels suggesting that GPA1 is a positive regulator of ABA inhibition of stomatal opening (Wang et al. 2001). Recently, it has been found that the *Arabidopsis* gene *SPHK1* encodes a sphingosine kinase (SphK) which is involved in ABA-induced stomatal closure and inhibition of stomatal opening (Worrall et al. 2008). ABA inhibits stomatal opening in part via activation of SphK, resulting in the production of spingosine-1-phosphate (S1P) which propagates the signal via GPA1 (Coursol et al. 2003). Application of S1P or the similar molecule phytosphingosine (Coursol et al. 2005) inhibited stomatal opening in wild-type plants, but not in *gpa1* mutants. Also, *gpa1* mutants are insensitive to S1P inhibition of inward K^+ channels (Coursol et al. 2003). Since S1P elicits elevation of cytosolic Ca^{2+} concentration in guard cells (Ng et al. 2001) and Ca^{2+} inhibits inward K^+ channels (Schroeder and Hagiwara 1989), this Ca^{2+}-elevation may be initiated by G-protein activation (Wang et al. 2001; Coursol et al. 2003).

Phospholipase Dα1 (PLDα1) is an additional downstream component of ABA regulation of stomatal movements. Activation of PLDα1 by ABA results in phosphatidic acid (PA) production, which inhibits stomatal opening (Jacob et al. 1999; Zhang et al. 2004). Application of PA to *gpa1* and wild-type epidermal peels inhibited stomatal opening in wild type, but not in *gpa1*, suggesting that GPA1 is a downstream component of PA-induced inhibition of opening (Mishra et al. 2006). Interestingly, PLDα1 has been shown to interact with GPA1 and stimulate GTPase activity of GPA1 and GPA1–GDP can bind PLDα1 and attenuate its activity suggesting a possible negative feedback mechanism in ABA signaling (Zhao and Wang 2004, Mishra et al. 2006).

In mammalian studies of Gα regulation of inward K^+ channels, Gα-GDP inhibits G-protein-coupled inwardly rectifying K^+ (GIRK) channels, and upon activation of the G protein, the free Gβγ dimer activates GIRK channels (Riven et al. 2006). If plant heterotrimeric G proteins regulate ion channels in a similar manner, it would be expected that *gpa1* and *agb1* mutants would show different phenotypes. In *Arabidopsis*, however, electrophysiology and stomatal aperture assays of *gpa1*, *agb1*, and *gpa1agb1* mutants all show statistically identical phenotypes: in the absence of ABA, there is no alteration in basal K^+ currents in any of these mutants, while in the presence of ABA, inhibition of inward K^+ currents and light-induced stomatal opening is attenuated similarly in all of these mutants (Fan et al. 2008). Taken together, the data suggest that plant heterotrimeric G proteins regulate K^+ channels differently than animal heterotrimeric G proteins. This idea is also supported by electrophysiological studies of Gγ mutants. According to the mammalian G protein paradigm, Gβγ always functions as a dimer; however, mutants of the two identified Gγ genes in *Arabidopsis*, *agg1* and *agg2*, show wild-type ABA-induced stomatal closure, ABA inhibition of stomatal opening, and ABA inhibition of inward K^+ channels (Trusov et al. 2008). Given the known lack of extensive sequence similarity among mammalian Gγs, it is possible that additional *Arabidopsis* Gγ(s), which would function with AGB1 in the regulation of stomatal movements, exist but have not yet been identified (Mason and Botella 2001; Trusov et al. 2008).

Of the putative GPCRs that have been identified in *Arabidopsis*, only *gcr1*, *gtg1*, and *gtg2* mutants have been phenotyped for stomatal aperture regulation. *gcr1* mutants, unlike *gpa1* or *agb1* mutants, are hypersensitive to ABA and S1P inhibition of stomatal opening (Pandey and Assmann 2004). GPA1 and GCR1 have been shown to interact in yeast and in coimmunoprecipitation from plant tissue, leading to the hypothesis that GCR1 is a negative regulator of GPA1 (Pandey and Assmann 2004). GTG1 and GTG2 are membrane-localized proteins, have both GPCR-like topologies and intrinsic GTPase activity, bind ABA, and interact with GPA1. While it is tempting to postulate that these ABA receptors activate heterotrimeric G proteins, which then propagate the ABA signal, resulting in inhibition of K^+ channels, analysis of the *gtg1gtg2* mutants shows wild-type ABA inhibition of opening (Pandey et al. 2009). Characterization of GTG regulation of ion channels and analysis of *gtggpa1* double mutants will, no doubt, help elucidate this novel component of G protein signaling in guard cells. *rgs1* mutants have also been assessed for altered ion channel regulation. The mutants did not show alterations in the magnitude of ABA inhibition of inward K^+ currents; however, the voltage-activation kinetics of inward K^+ channels were accelerated, suggesting a function for RGS in the regulation of channel response to signals (Fan et al. 2008).

4.4 G Protein Regulation of ABA Promotion of Stomatal Closure

Analysis of *Arabidopsis* heterotrimeric G protein mutants has also suggested a role for G proteins in the regulation of ABA promotion of stomatal closure. The above-mentioned *gtg1gtg2* double mutants show hyposensitivity in ABA promotion of stomatal closure (and in all other, nonstomatal ABA responses that were assayed; Pandey et al., 2009). ABA activation of slow anion channels during stomatal closure can occur via at least two pathways, one of which is dependent on the G protein heterotrimer, while the other involves cytosolic pH. Thus, *gpa1* mutants show reduced ABA activation of slow anion channels when cytosolic pH is strongly buffered. However, under weak pH buffering, ABA activation of anion efflux channels is identical in wild type and *gpa1* mutants (Wang et al. 2001). The functional redundancy of these two pathways for ABA activation of anion channels likely explains the lack of an ABA promotion of closure phenotype in the *gpa1* mutants, unless cytosolic pH is similarly clamped (Wang et al. 2001). These identical phenomena are seen in assays of *agb1* guard cells (Fan et al. 2008).

Like ABA, S1P promotes stomatal closure (Ng et al. 2001). *gcr1* mutants show hypersensitivity toward both ABA and S1P in promotion of stomatal closure, consistent with the idea that GCR1 functions as a negative regulator of GPA1-mediated ABA and S1P signaling (Pandey and Assmann 2004). Ion channel activity in *gcr1* mutants has not yet been assessed. However, it is known that *gpa1* mutants are insensitive to both S1P promotion of stomatal closure and S1P activation of slow anion channels (Coursol et al. 2003). The effect of S1P on anion channels, unlike that of ABA, is obligately dependent on GPA1 (Coursol et al., 2003).

S1P induces an elevation in cytosolic Ca^{2+} concentration (Ng et al. 2001), and elevated cytosolic Ca^{2+} is sufficient to activate slow anion channels (Schroeder and Hagiwara 1989), therefore G proteins may mediate S1P response via cytosolic Ca^{2+} signals (Coursol et al. 2003). Both ABA and S1P induce cytosolic Ca^{2+} transients (Hetherington 1990; Schroeder and Hagiwara 1990; McAinsh et al. 1992; Allen et al. 2001; Ng et al. 2001). Experiments in which cytosolic Ca^{2+} transients are stimulated in guard cells have found evidence for two distinct mechanisms for Ca^{2+}-dependent stomatal closure, a rapid "Ca^{2+} reactive" response which contributes to the closure response itself, and a sustained "Ca^{2+} programmed" response which contributes to the maintenance of closed stomata after cessation of cytosolic

Ca^{2+} elevation (Allen et al. 2001). S1P application to stomata induces Ca^{2+} transients and rapid and temporary stomatal closure (Ng et al. 2001), presumably invoking the "Ca^{2+} reactive" response and not the "Ca^{2+} programmed" response (Allen et al. 2001). Since G proteins function downstream of S1P, they are similarly implicated in the "Ca^{2+} reactive" response.

One mechanism by which ABA promotes an increase in cytosolic Ca^{2+} is through the production of H_2O_2 which activates the plasma membrane Ca^{2+}-permeable channels that mediate Ca^{2+} influx (Pei et al. 2000). Extracellular calmodulin can also promote stomatal closure and triggers H_2O_2 production and cytosolic Ca^{2+} increase (Chen et al. 2004). H_2O_2 generation, in response to extracellular calmodulin, is attenuated in the *gpa1* mutants (Chen et al. 2004), consistent with the observations that stomatal closure induced by extracellular calmodulin is impaired in *gpa1* mutants and heightened in plants overexpressing GPA1 (Chen et al. 2004).

4.5 G Protein Regulation of Pathogen-Induced Stomatal Movements

Stomata play the critical roles in regulating plant water status and photosynthetic carbon assimilation; however, their pores also can serve as convenient entry points for plant bacterial pathogens (Melotto et al. 2006; Underwood et al. 2007; Melotto et al. 2008). It has been shown that guard cells function in plant innate immunity by closing stomata and inhibiting stomatal opening in response to bacterial pathogens or their elicitors, such as the flg22 peptide derived from flagellin, a pathogen-associated molecular pattern (PAMP) (Melotto et al. 2006; Zhang et al. 2008b). The *gpa1* mutant shows insensitivity in flg22-induced inhibition of stomatal opening and inhibition of inward K^+ channels, implicating G proteins in PAMP signaling (Zhang et al. 2008b). Interestingly, G proteins are not the only shared signaling components between PAMP and ABA signaling pathways: nitric oxide, H_2O_2, and the kinase OST1 all function in both ABA- and PAMP-induced stomatal closure (Melotto et al. 2006). In addition, Ca^{2+}-permeable channels (CNGC2/DND1) facilitate Ca^{2+} influx and act as an upstream signal component for NO production during plant hypersensitive responses (Ali et al. 2007), suggesting that cytosolic Ca^{2+} signals and NO are also shared components in ABA and pathogen signaling.

4.6 G Protein Regulation of Whole-Leaf Water Status and Drought Response

Despite the numerous studies outlined above, which show that G proteins are involved in the regulation of stomatal movements, very little is known concerning G protein regulation of water status at the whole leaf or plant level. Excised leaf water loss assays (which should be interpreted with caution for reasons discussed above) show that *gpa1* mutants in the Ws background have increased the water loss; however, water loss assays of cotyledons of *gpa1* mutants in the Col background show reduced water loss whereas *agb1* mutant cotyledons show increased water loss (Wang et al. 2001; Mishra et al. 2006; Zhang et al. 2008a). In terms of drought tolerance, it has been reported that young *gpa1* seedlings (Col background) growing on agar media exposed to dry air exhibit drought tolerance while *agb1* mutants are drought sensitive (Zhang et al. 2008a). *gcr1* mutants show reduced water loss in excised leaf water loss assays and *gcr1* plants grown in soil exhibit improved survival following drought release (Pandey and Assmann 2004). Similar results were observed for transgenic plants overexpressing RGS1 (Chen et al. 2006b). The recent identification of GPA1 and AGB1 as regulators (positive and negative, respectively) of stomatal density in cotyledons (Zhang et al. 2008a) confounds any simple predictions of how G protein regulation of stomatal movements may contribute to whole-leaf and whole-plant water status and indicates the need for further experimentation.

5 Conclusions and Unanswered Questions

Pharmacological and molecular genetic approaches have identified heterotrimeric G proteins as regulators of ABA inhibition of stomatal opening, ABA promotion of stomatal closure, and pathogen-induced stomatal movements (summarized in Fig. 1 and Table 1). Electrophysiology experiments clearly indicate that ion channels are a target of G protein regulation. However, the mechanisms by which heterotrimeric G proteins regulate ion channels have yet to be elucidated and may involve both secondary messenger cascades and plasma membrane-based regulation. Further experimentation is needed to identify additional downstream effectors of G-protein signaling as well as G-protein-coupled receptor(s) that act directly upstream of the heterotrimeric G protein in the regulation of stomatal movements. Additional physiological studies of G protein mutants with the goal of examining stomatal conductance, stomatal density, and ultimately, plant fitness, under well-watered and tightly controlled drought stress conditions are warranted in order to obtain an integrated understanding of how heterotrimeric G proteins contribute to whole-plant water status and therefore of their potential utility as biotechnological targets for crop improvement.

Acknowledgments Research on G protein regulation of stomatal function in the authors' laboratory is supported by grants from the U.S. National Science Foundation and U.S. Department of Agriculture to Sarah M. Assmann. The authors also thank Sarah M. Assmann for her helpful comments and suggestions on this manuscript.

References

Adjobo-Hermans MJ, Goedhart J, Gadella TW (2006) Plant G protein heterotrimers require dual lipidation motifs of Gα and Gγ and do not dissociate upon activation. J Cell Sci 119:5087–5097

Ali R, Ma W, Lemtiri-Chlieh F, Tsaltas D, Leng Q, von Bodman S, Berkowitz GA (2007) Death don't have no mercy and neither does calcium: *Arabidopsis* CYCLIC NUCLEOTIDE GATED CHANNEL2 and innate immunity. Plant Cell 19:1081–1095

Allen GJ, Chu SP, Harrington CL, Schumacher K, Hoffmann T et al (2001) A defined range of guard cell calcium oscillation parameters encodes stomatal movements. Nature 411: 1053–1057

Anderson DJ, Botella JR (2007) Expression analysis and subcellular localization of the *Arabidopsis thaliana* G-protein beta-subunit AGB1. Plant Cell Rep 26:1469–1480

Armstrong F, Blatt MR (1995) Evidence for K^+ channel control in *Vicia* guard cells coupled by G-proteins to a 7TMS receptor mimetic. Plant J 8:187–198

Assmann SM (1993) Signal transduction in guard cells. Annu Rev Cell Biol 9:345–375

Blatt MR (1992) K^+ channels of stomatal guard cells. Characteristics of the inward rectifier and its control by pH. J Gen Physiol 99:615–644

Chen JG, Willard FS, Huang J, Liang J, Chasse SA, Jones AM, Siderovski DP (2003) A seven-transmembrane RGS protein that modulates plant cell proliferation. Science 301:1728–1731

Chen JG, Gao Y, Jones AM (2006a) Differential roles of *Arabidopsis* heterotrimeric G-protein subunits in modulating cell division in roots. Plant Physiol 141:887–897

Chen Y, Ji F, Xie H, Liang J (2006b) Overexpression of the regulator of G-protein signalling protein enhances ABA-mediated inhibition of root elongation and drought tolerance in *Arabidopsis*. J Exp Bot 57:2101–2110

Chen YL, Huang R, Xiao YM, Lu P, Chen J, Wang XC (2004) Extracellular calmodulin-induced stomatal closure is mediated by heterotrimeric G protein and H_2O_2. Plant Physiol 136: 4096–103

Coursol S, Fan LM, Le Stunff H, Spiegel S, Gilroy S, Assmann SM (2003) Sphingolipid signalling in *Arabidopsis* guard cells involves heterotrimeric G proteins. Nature 423:651–654

Coursol S, Le Stunff H, Lynch DV, Gilroy S, Assmann SM, Spiegel S (2005) *Arabidopsis* sphingosine kinase and the effects of phytosphingosine-1-phosphate on stomatal aperture. Plant Physiol 137:724–737

Cousson A, Vavasseur A (1998a) Putative involvement of cytosolic Ca^{2+} and GTP-binding proteins in cyclic-GMP-mediated induction of stomatal opening by auxin in *Commelina communis* L. Planta 206:308–314

Cousson A, Vavasseur A (1998b) Two potential Ca^{2+}-dependent transduction pathways in stomatal closing in response to abscisic acid. Plant Physiol Biochem 36:257–262

Davies WJ, Zhang JH (1991) Root signals and the regulation of growth and development of plants in drying soil. Annu Rev Plant Physiol Plant Mol Biol 42:55–76

Fairley-Grenot K, Assmann SM (1991) Evidence for G-protein regulation of inward K^+ channel current in guard cells of fava bean. Plant Cell 3:1037–1044

Fan LM, Zhang W, Chen JG, Taylor JP, Jones AM, Assmann SM (2008) Abscisic acid regulation of guard-cell K^+ and anion channels in Gβ- and RGS-deficient *Arabidopsis* lines. Proc Natl Acad Sci USA 105:8476–8481

Fan LM, Zhao Z, Assmann SM (2004) Guard cells: a dynamic signaling model. Curr Opin Plant Biol 7:537–546

Forestier C, Bouteau F, Leonhardt N, Vavasseur A (1998) Pharmacological properties of slow anion currents in intact guard cells of *Arabidopsis*. Application of the discontinuous single-electrode voltage-clamp to different species. Pflugers Arch 436:920–927

Gao Y, Zeng Q, Guo J, Cheng J, Ellis BE, Chen JG (2007) Genetic characterization reveals no role for the reported ABA receptor, GCR2, in ABA control of seed germination and early seedling development in *Arabidopsis*. Plant J 52:1001–1013

Gilman AG (1987) G Proteins: Transducers of receptor-generated signals. Annu Rev Biochem 56:615–649

Gilroy S, Read ND, Trewavas AJ (1990) Elevation of cytoplasmic calcium by caged calcium or caged inositol trisphosphate initiates stomatal closure. Nature 346:769–771

Gookin TE, Kim J, Assmann SM (2008) Whole proteome identification of plant candidate G-protein coupled receptors in *Arabidopsis*, rice, and poplar: computational prediction and in-vivo protein coupling. Genome Biol 9:R120

Guo J, Zeng Q, Emami M, Ellis BE, Chen J-G (2008) The GCR2 gene family is not required for ABA control of seed germination and early seedling development in *Arabidopsis*. PLoS ONE 3:e2982

Hedrich R, Busch H, Raschke K (1990) Ca^{2+} and nucleotide dependent regulation of voltage dependent anion channels in the plasma membrane of guard cells. EMBO J 9:3889–3892

Hetherington AM (1990) Abscisic acid-induced elevation of guard cell cytosolic Ca^{2+} precedes stomatal closure. Nature 343:186–188

Hille B (2001) Ion channels of excitable membranes. Sinauer Associates, Inc, Sunderland, MA, pp 441–470

Huang H, Weiss CA, Ma H (1994) Regulated expression of the *Arabidopsis* G protein α subunit gene GPA1. Int J Plant Sci 155:3–14

Humphrey TV, Botella JR (2001) Re-evaluation of the cytokinin receptor role of the *Arabidopsis* gene GCR1. J Plant Physiol 158:645–653

Illingworth CJR, Parkes KE, Snell CR, Mullineaux PM, Reynolds CA (2008) Criteria for confirming sequence periodicity identified by Fourier transform analysis: application to GCR2, a candidate plant GPCR? Biophys Chem 133:28–35

Jacob T, Ritchie S, Assmann SM, Gilroy S (1999) Abscisic acid signal transduction in guard cells is mediated by phospholipase D activity. Proc Natl Acad Sci USA 96:12192–12197

Josefsson L-G, Rask L (1997) Cloning of a putative G-protein-coupled receptor from *Arabidopsis thaliana*. Eur J Biochem 249:415–420

Kelly WB, Esser JE, Schroeder JI (1995) Effects of cytosolic calcium and limited, possible dual, effects of G protein modulators on guard cell inward potassium channels. Plant J 8:479–489

Kinoshita T, Nishimura M, Shimazaki K (1995) Cytosolic concentration of Ca^{2+} regulates the plasma membrane H^+-ATPase in guard cells of fava bean. Plant Cell 7:1333–1342

Kornreich BG (2007) The patch clamp technique: principles and technical considerations. J Vet Cardiol 9:25–37

Lee M, Choi Y, Burla B, Kim Y-Y, Jeon B, Maeshima M, Yoo J-Y, Martinoia E, Lee Y (2008) The ABC transporter AtABCB14 is a malate importer and modulates stomatal response to CO_2. Nat Cell Biol 10:1217–1223

Lee HJ, Tucker EB, Crain RC, Lee Y (1993) Stomatal opening is induced in epidermal peels of *Commelina communis* L. by GTP analogs or pertussis toxin. Plant Physiol 102:95–100

Lemtiri-Chlieh F, MacRobbie EAC (1994) Role of calcium in the modulation of *Vicia* guard cell potassium channels by abscisic acid: a patch-clamp study. J Membr Biol 137:99–107

Levchenko V, Konrad KR, Dietrich P, Roelfsema MRG, Hedrich R (2005) Cytosolic abscisic acid activates guard cell anion channels without preceding Ca^{2+} signals. Proc Natl Acad Sci USA 102:4203–4208

Liu X, Yue Y, Li B, Nie Y, Li W, Wu W-H, Ma L (2007) A G protein-coupled receptor is a plasma membrane receptor for the plant hormone abscisic acid. Science 315:1712–1716

Ma H, Yanofsky MF, Meyerowitz EM (1990) Molecular cloning and characterization of GPA1, a G protein alpha subunit gene from *Arabidopsis thaliana*. Proc Natl Acad Sci USA 87:3821–3825

MacRobbie EAC (1983) Effects of light/dark on cation fluxes in guard cells of *Commelina communis* L. J Exp Bot 34:1695–1710

MacRobbie EAC (1998) Signal transduction and ion channels in guard cells. Philos Trans R Soc Lond B Biol Sci 353:1475–1488

Marten H, Konrad KR, Dietrich P, Roelfsema MRG, Hedrich R (2007) Ca^{2+}-dependent and -independent abscisic acid activation of plasma membrane anion channels in guard cells of *Nicotiana tabacum*. Plant Physiol 143:28–37

Mason MG, Botella JR (2000) Completing the heterotrimer: isolation and characterization of an *Arabidopsis thaliana* G protein γ-subunit cDNA. Proc Natl Acad Sci USA 97:14784–1478

Mason MG, Botella JR (2001) Isolation of a novel G-protein γ-subunit from *Arabidopsis thaliana* and its interaction with Gβ. Biochim Biophys Acta 1520:147–153

McAinsh MR, Brownlee C, Hetherington AM (1990) Abscisic acid-induced elevation of guard cell cytosolic Ca^{2+} precedes stomatal closure. Nature 343:186–188

McAinsh MR, Brownlee C, Hetherington AM (1992) Visualizing changes in cytosolic-free Ca^{2+} during the response of stomatal guard cells to abscisic acid. Plant Cell 7:1113–1122

Melotto M, Underwood W, He SY (2008) Role of stomata in plant innate immunity and foliar bacterial diseases. Annu Rev Phytopathol 46:101–122

Melotto M, Underwood W, Koczan J, Nomura K, He SY (2006) Plant stomata function in innate immunity against bacterial invasion. Cell 126:969–980

Miles GP, Samuel MA, Jones AM, Ellis BE (2004) Mastoparan rapidly activates plant MAP kinase signaling independent of heterotrimeric G proteins. Plant Physiol 134:1332–1336

Mishra G, Zhang W, Deng F, Zhao J, Wang X (2006) A bifurcating pathway directs abscisic acid effects on stomatal closure and opening in *Arabidopsis*. Science 312:264–266

Moriyama E, Strope P, Opiyo S, Chen Z, Jones A (2006) Mining the *Arabidopsis thaliana* genome for highly-divergent seven transmembrane receptors. Genome Biol 7:R96

Neher E, Sakmann B, Steinbach JH (1978) The extracellular patch clamp: A method for resolving currents through individual open channels in biological membranes. Pflugers Arch 375: 219–228

Ng CK, Carr K, McAnish MR, Powell B, Hetherington AM (2001) Drought-induced guard cell signal transduction involves sphingosine-1-phosphate. Nature 410:596–599

Obrdlik P, Neuhaus G, Merkle T (2000) Plant heterotrimeric G protein beta subunit is associated with membranes via protein interactions involving coiled-coil formation. FEBS Lett 476: 208–212

Pandey S, Assmann SM (2004) The *Arabidopsis* putative G protein-coupled receptor GCR1 interacts with the G protein a subunit GPA1 and regulates abscisic acid signaling. Plant Cell 16:1616–1632

Pandey S, Chen JG, Jones AM, Assmann SM (2006) G-protein complex mutants are hypersensitive to abscisic acid regulation of germination and postgermination development. Plant Physiol 141:243–256

Pandey S, Nelson DC, Assmann SM (2009) Two novel GPCR-type G proteins are abscisic acid receptors in *Arabidopsis*. Cell 136:136–148

Pandey S, Zhang W, Assmann SM (2007) Roles of ion channels and transporters in guard cell signal transduction. FEBS Lett 581:2325–2336

Pei ZM, Kuchitsu K, Ward JM, Schwarz M, Schroeder JI (1997) Differential abscisic acid regulation of guard cell slow anion channels in *Arabidopsis* wild-type and *abi1* and *abi2* mutants. Plant Cell 9:409–423

Pei ZM, Murata Y, Benning G, Thomine S, Klusener B, Allen GJ, Grill E, Schroeder JI (2000) Calcium channels activated by hydrogen peroxide mediate abscisic acid signalling in guard cells. Nature 406:731–734

Peškan T, Oelmüller R (2000) Heterotrimeric G-protein β-subunit is localized in the plasma membrane and nuclei of tobacco leaves. Plant MolBiol 42:915–922

Riven I, Iwanir S, Reuveny E (2006) GIRK channel activation involves a local rearrangement of a preformed G protein channel complex. Neuron 51:561–573

Roelfsema MR, Steinmeyer R, Staal M, Hedrich R (2001) Single guard cell recordings in intact plants: light-induced hyperpolarization of the plasma membrane. Plant J 26:1–13

Schroeder JI, Allen GJ, Hugouvieux V, Kwak JM, Waner D (2001) Guard cell signal transduction. Annu Rev Plant Physiol Plant Mol Biol 52:627–658

Schroeder JI, Hagiwara S (1989) Cytosolic calcium regulates ion channels in the plasma membrane of *Vicia faba* guard cells. Nature 338:427–430

Schroeder JI, Hagiwara S (1990) Repetitive increases in cytosolic Ca^{2+} of guard cells by abscisic acid activation of nonselective Ca^{2+} permeable channels. Proc Natl Acad Sci USA 87: 9305–9309

Schroeder JI, Raschke K, Neher E (1987) Voltage dependence of K channels in guard-cell protoplasts. Proc Natl Acad Sci USA 84:4108–4112

Shimazaki K-i, Doi M, Assmann SM, Kinoshita T (2007) Light regulation of stomatal movement. Annu Rev Plant Biol 58:219–247

Staxen I, Pical C, Montgomery LT, Gray JE, Hetherington AM, McAinsh MR (1999) Abscisic acid induces oscillations in guard-cell cytosolic free calcium that involve phosphoinositide-specific phospholipase C. Proc Natl Acad Sci USA 96:1779–1784

Talbott L, Zeiger E (1998) The role of sucrose in guard cell osmoregulation. J Exp Bot 49:329–337

Trusov Y, Rookes JE, Tilbrook K, Chakravorty D, Mason MG, Anderson D, Chen J-G, Jones AM, Botella JR (2007) Heterotrimeric G protein γ subunits provide functional selectivity in G$\beta\gamma$ dimer signaling in *Arabidopsis*. Plant Cell 19:1235–1250

Trusov Y, Zhang W, Assmann SM, Botella JR (2008) Gγ1 + Gγ2 not equal to Gβ: heterotrimeric G protein Gg-deficient mutants do not recapitulate all phenotypes of Gb-deficient mutants. Plant Physiol 147:636–649

Underwood W, Melotto M, He SY (2007) Role of plant stomata in bacterial invasion. Cell Microbiol 9:1621–1629

Wang XQ, Ullah H, Jones AM, Assmann SM (2001) G protein regulation of ion channels and abscisic acid signaling in *Arabidopsis* guard cells. Science 292:2070–2072

Wang SY, Narendra S, Fedoroff N (2007) Heterotrimeric G protein signaling in the *Arabidopsis* unfolded protein response. Proc Natl Acad Sci USA 104:3817–3822

Wang SY, Assmann SM, Fedoroff NV (2008) Characterization of the *Arabidopsis* heterotrimeric G protein. J Biol Chem 283:13913–13922

Weiss CA, Garnaat CW, Mukai K, Hu Y, Ma H (1994) Isolation of cDNAs encoding guanine nucleotide-binding protein beta-subunit homologues from maize (ZGB1) and *Arabidopsis* (AGB1). Proc Natl Acad Sci USA 91:9554–9558

Weiss CA, Huang H, Ma H (1993) Immunolocalization of the G protein α subunit encoded by the GPA1 gene in *Arabidopsis*. Plant Cell 5:1513–1528

Weiss CA, White E, Huang H, Ma H (1997) The G protein α subunit (GPa1) is associated with the ER and the plasma membrane in meristematic cells of *Arabidopsis* and cauliflower. FEBS Lett 407:361–367

Wilkinson S, Davies WJ (2002) ABA-based chemical signalling: the co-ordination of responses to stress in plants. Plant Cell Environm 25:195–210

Willmer CM, Sexton R (1979) Stomata and plasmodesmata. Protoplasma 100:113–124

Worrall D, Liang YK, Alvarez S, Holroyd GH, Spiegel S, Panagopulos M, Gray JE, Hetherington AM (2008) Involvement of sphingosine kinase in plant cell signalling. Plant J 56:64–72

Wu WH, Assmann SM (1994) A membrane-delimited pathway of G-protein regulation of the guard-cell inward K^+ channel. Proc Natl Acad Sci USA 91:6310–6314

Yang Y, Costa A, Leonhardt N, Siegel RS, Schroeder JI (2008) Isolation of a strong guard cell promoter and its potential as a research tool. Plant Methods 4:6

Zeng Q, Wang X, Running MP (2007) Dual lipid modification of *Arabidopsis* Gγ-subunits is required for efficient plasma membrane targeting. Plant Physiol 143:1119–1131

Zhang L, Hu G, Cheng Y, Huang J (2008a) Heterotrimeric G protein α and β subunits antagonistically modulate stomatal density in *Arabidopsis thaliana*. Dev Biol 324:68–75

Zhang W, He SY, Assmann SM (2008b) The plant innate immunity response in stomatal guard cells invokes G-protein-dependent ion channel regulation. Plant J 56:984–996

Zhang W, Nilson SE, Assmann SM (2008c) Isolation and whole-cell patch clamping of *Arabidopsis* guard cell protoplasts. CSH Protocol 2008: pdb.prot5014.

Zhang W, Qin C, Zhao J, Wang X (2004) Phospholipase Dα1-derived phosphatidic acid interacts with ABI1 phosphatase 2C and regulates abscisic acid signaling. Proc Natl Acad Sci USA 101:9508–9513

Zhao J, Wang X (2004) *Arabidopsis* phospholipase Dα1 interacts with the heterotrimeric G-protein α-subunit through a motif analogous to the DRY motif in G-protein-coupled receptors. J Biol Chem 279:1794–1800

Zhao Z, Zhang W, Stanley BA, Assmann SM (2008) Functional proteomics of *Arabidopsis thaliana* guard cells uncovers new stomatal signaling pathways. Plant Cell 20:3210–3226

The Role of Seven-Transmembrane Domain MLO Proteins, Heterotrimeric G-Proteins, and Monomeric RAC/ROPs in Plant Defense

Justine Lorek, Ralph Panstruga, and Ralph Hückelhoven

Abstract MLO proteins are structurally reminiscent of G-protein-coupled receptors but act independently of heterotrimeric G-proteins as major susceptibility factors to powdery mildew fungi. In barley, monomeric RAC/ROPs, instead of heterotrimeric G-proteins, MLO-dependently modulate susceptibility to powdery mildew, which may involve functions in cytoskeleton remodeling. In contrast to the role of RAC/ROPs in barley susceptibility to powdery mildew, rice OsRAC1 exerts a central function in basal and effector-triggered immunity. In this context, a functional cooperation with the heterotrimeric G-protein subunit, Gα, and additional protein complexes with functions in plant immunity has been discovered. These polypeptides together modulate the oxidative burst and regulate the abundance of defense-associated messenger RNAs and defense proteins. This chapter highlights the interconnection of MLO, RAC/ROP, and heterotrimeric G-proteins in plant immunity.

1 Plant Defense Mechanisms

Plants are continuously exposed to a large range of pathogens with diverse life styles, but unlike animals, they are neither able to escape their enemies nor do they possess an adaptive immune system to protect themselves. Given that plants are resistant to the majority of ambient microbes, they obviously have evolved effective weapons to defeat their foes. Early defense responses in the battle against

J. Lorek and R. Panstruga (✉)
Department of Plant-Microbe Interactions, Max-Planck Institute for Plant Breeding Research, Carl von Linné Weg 10, Köln 50829, Germany
e-mail: panstrug@mpiz-koeln.mpg.de

R. Hückelhoven
Technische Universität München, Lehrstuhl für Phytopathologie, Am Hochanger 2, Freising-Weihenstephan, 85350, Germany
e-mail: hueckelhoven@wzw.tum.de

S. Yalovsky et al. (eds.), *Integrated G Proteins Signaling in Plants*, Signaling and Communication in Plants,
DOI 10.1007/978-3-642-03524-1_11, © Springer-Verlag Berlin Heidelberg 2010

pathogens are triggered immediately after the first contact with a potential intruder. They rely on the recognition of pathogen-derived molecules, the so-called pathogen-associated molecular patterns (PAMPs), which are perceived via plasma membrane-localized pattern recognition receptors (PRRs). This type of pathogen resistance is referred to as PAMP-triggered immunity (PTI) (Chisholm et al. 2006; Jones and Dangl 2006).

PAMPs are highly conserved essential microbial molecules, including bacterial flagellin, lipopolysaccharides (LPS), and elongation factor Tu (EF-Tu), as well as chitin and β-glucan, which are cell wall components of fungi and oomycetes (Schwessinger and Zipfel 2008). A highly conserved flagellin-derived amino acid epitope, flg22, or in case of EF-Tu an 18 amino acid peptide, elf18, are sufficient to trigger PTI responses (Felix et al. 1999; Kunze et al. 2004). PRRs can be broadly grouped into two families, the receptor-like kinases (RLKs) and receptor-like proteins (RLPs), which lack a cytoplasmic kinase domain (Zipfel 2008). The best-studied plant PRRs are the *Arabidopsis thaliana* RLKs FLS2 (for bacterial flagellin or flg22) (Gómez-Gómez and Boller 2000), EFR (for bacterial EF-Tu or elf18) (Zipfel et al. 2006), CERK1 (Miya et al. 2007) and CEBiP (Kaku et al. 2006) (for fungal chitin), and the RLPs LeEix (for fungal xylanase EIX) (Ron and Avni 2004).

Following PAMP recognition, a plethora of defense responses is triggered to defeat the pathogen(s). Seconds to minutes after PAMP treatment, extracellular alkalinization, and ROS (reactive oxygen species) production occur. Intracellular signaling cascades involving Ca^{2+} fluxes and mitogen-activated protein kinases (MAPKs) lead to biosynthesis and extrusion of antimicrobial products such as PR (pathogenesis-related) proteins and low-molecular weight compounds (phytoalexins). Furthermore, the (1,3)-β-D polyglucan callose is locally deposited at the cell wall. The significance of many of these stereotypical stress responses to pathogen defense remains, however, largely elusive.

Some microbes have evolved strategies to overcome the PAMP-based defense system. Successful pathogens deliver a range of effector molecules that suppress PTI, thereby enabling host colonization (Chisholm et al. 2006; Jones and Dangl 2006; Bent and Mackey 2007; da Cunha et al. 2007). Bacteria transfer proteinaceous effectors via a dedicated delivery apparatus, the type III secretion system, which penetrates through the host cell wall and plasma membrane (Block et al. 2008). The molecular mechanisms of the delivery of fungal effectors into plant cells are still poorly understood (Ellis et al. 2006). Many fungi and oomycetes penetrate the plant cuticle and cell wall through mechanical and/or enzymatic means. Subsequently, intracellular infection structures (haustoria or infection hyphae) are formed, which are thought to serve as feeding organs for nutrient uptake as well as for effector delivery.

In response to the subversion of PTI, plants evolved a further layer of defense to recognize effectors either directly or indirectly by special immune receptors referred to as resistance (R) proteins. Typically, R proteins possess a characteristic nucleotide-binding site (NB) and a leucine-rich repeat (LRR) domain (Bent and Mackey 2007). The so-called effector-triggered immunity (ETI) conferred by R proteins is race-specific and historically known as gene-for-gene resistance

(Flor 1942). ETI shares common signaling and execution pathways with PTI, but generally the effector-based response is faster and usually results in localized programmed death of the attacked cell, which is also known as hypersensitive response (HR).

Recently, a number of defense execution components have been identified that contribute to the ability of the dicotyledonous reference plant *A thaliana* to resist penetration by the nonadapted powdery mildew fungus *Blumeria graminis* f. sp. *hordei* (*Bgh*), which is a natural pathogen of barley (Collins et al. 2003; Lipka et al. 2005; Stein et al. 2006; Kwon et al. 2008; Underwood and Somerville 2008). Two genetically separable pathways have been proposed to play a major role in pre-invasion resistance against powdery mildew fungi: one pathway comprises targeted vesicle-mediated and PEN1-dependent exocytosis. PEN1 is a syntaxin, also known as t-SNARE (*SOLUBLE N*-E-SENSITIVE FACTOR ATTACHMENT PROTEIN RECEPTOR), which participates in vesicle fusion events through formation of ternary SNARE complexes. In *A. thaliana*, PEN1, SNAP33 (SYNAPTOSOMAL-ASSOCIATED PROTEIN OF 33 kDa), and VAMP (VESICLE-ASSOCIATED MEMBRANE PROTEIN) 721/722 assemble into a ternary SNARE complex during antifungal defense. This SNARE complex is thought to mediate exocytotic delivery of toxic and/or cell wall-related cargo to the plant apoplast (Kwon et al. 2008). A second antimicrobial delivery system implicates the activity of the plasma membrane ABC transporter PEN3, which is proposed to export PEN2-generated toxic compounds contributing to penetration resistance (Stein et al. 2006). PEN2 is an unconventional myrosinase, associated with the surface of peroxisomes, presumably catalyzing the formation of toxic indole glucosinolate hydrolysis products (Lipka et al. 2005; Bednarek et al. 2009).

2 MLO: A Negative Modulator of Defense Against Powdery Mildew Fungi

Powdery mildew is a common fungal disease of many plant species. The disease has economical significance causing great yield losses in agriculture. In barley (*Hordeum vulgare*), an important crop plant, recessive mutations in the *MLO* (*MILDEW RESISTANCE LOCUS O*) gene confer durable broad-spectrum resistance to all known isolates of the barley powdery mildew fungus *B. graminis* f. sp. *hordei* (*Bgh*). For this reason, natural and induced *mlo* mutant alleles have been widely adopted in barley breeding programs (Büschges et al. 1997; Jørgensen 1992; Lyngkjaer et al. 2000). Naturally occurring broad-spectrum resistance to *Bgh* was first observed, in 1937, in Ethiopian barley landraces, which were later found to carry a mutation at the *MLO* locus (Jørgensen 1992; Piffanelli et al. 2004). For more than 60 years, *mlo*-based resistance was considered a barley-specific phenomenon. Recently, however, a requirement for MLO proteins in powdery mildew pathogenesis in the dicotyledonous plants *A. thaliana* (Consonni et al. 2006) and tomato (*Solanum lycopersicum*) was reported (Bai et al. 2008).

The *A. thaliana* genome encodes 15 proteins with extensive sequence similarity to barley MLO, which according to phylogenetic analysis can be grouped into four clades (Devoto et al. 2003; Chen et al. 2006b). Mutation of *AtMLO2* was found to confer only partial resistance to the adapted powdery mildew pathogen *Golovinomyces orontii*, since fungal invasion and subsequent conidiation in *Atmlo2* mutant plants were diminished but not completely eliminated as in case of barley *mlo* mutants (Consonni et al. 2006). *AtMLO2* belongs to a phylogenetic clade comprising two additional *MLO* genes, *AtMLO6* and *AtMLO12* (Chen et al. 2006b). Reminiscent of barley *mlo* mutants, a respective *Atmlo2 Atmlo6 Atmlo12* triple mutant was fully resistant to *G. orontii*. This finding indicates an unequal genetic redundancy among *AtMLO2*, *AtMLO6*, and *AtMLO12* regarding susceptibility against *G. orontii*, with a predominant role for *AtMLO2* in this context (Consonni et al. 2006).

Accumulating data indicate that MLO negatively affects PEN1- and PEN2/PEN3-dependent defense pathways during penetration resistance to powdery mildew fungi. In both barley and *Arabidopsis*, syntaxins (PEN1 or the barley ortholog ROR2) are required for *mlo*-based resistance, as *pen1* or *ror2* mutations in a *mlo*-resistant background restore wild-type-like entry rates of the respective powdery mildew pathogen (Freialdenhoven et al. 1996; Collins et al. 2003; Consonni et al. 2006). Moreover, also *Atmlo2 pen2* and *Atmlo2 pen3* double mutants exhibit wild-type levels of powdery mildew invasion, indicative of MLO acting as a negative modulator of the PEN2/PEN3-associated defense pathway (Consonni et al. 2006). Unlike *Atmlo2 pen1* plants, these double mutants in addition display a significant increase in powdery mildew conidiation, suggesting a role for PEN2/PEN3 in both pre- and postpenetration defenses in the context of *Atmlo2*-conditioned resistance.

Devoto et al. (1999) experimentally uncovered MLO as an integral plasma membrane-resident protein with seven transmembrane (TM) helices, an extracellularly located *N*-terminus, and a cytoplasmic *C*-terminus. The latter was subsequently found to harbor a calmodulin-binding domain (CaMBD) (Kim et al. 2002a,b). The CaMBD is conserved throughout the MLO family, suggesting that CaM binding is a general feature of MLO proteins. Mutations in the MLO-CaMBD lowered MLO-mediated susceptibility by 50%, indicating that CaM is either an activator of MLO function or a factor involved in signaling downstream of MLO (Kim et al. 2002a,b).

MLO proteins constitute the largest 7TM domain protein family in *A thaliana*. The sequence diversity, subcellular localization, and topology of MLO proteins are reminiscent of the G-protein-coupled receptor (GPCR) superfamily in metazoans (Devoto et al. 1999). In animals and fungi, GPCRs relay extracellular stimuli into intracellular signaling events by the activation of heterotrimeric G-proteins (see also chapter "Bioinformatics of Seven-Transmembrane Receptors in Plant Genomes"). To date, only sparse knowledge about potential plant GPCRs is available and although no significant sequence similarity between mammalian GPCRs and MLO proteins exists, these plant-specific 7TM domain proteins remain obvious receptor candidates for G-protein binding and signaling. Moreover, it is known that several human pathogens exploit host GPCRs for successful infection. Prominent examples include the human immunodeficiency virus type 1 (HIV-1) and the bacterium *Stereptococcus pneumoniae*, which target GPCRs for host cell entry (Pease and Murphy 1998). Together, these facts raise the question whether MLO proteins might play a similar role during plant colonization by powdery mildew fungi. This topic as well as a putative involvement of MLO proteins in plant heterotrimeric G-protein signaling will be discussed in the present chapter. Furthermore, since the barley RAC/ROP protein HvRACB, a monomeric GTPase, operates in conferring susceptibility to *Bgh* in an MLO-dependent manner, the role of small GTPases during powdery mildew pathogenesis as well as in other defense-associated processes will be highlighted.

3 Plant Heterotrimeric G-Protein Signaling and Plant Defense

The canonical heterotrimeric G-protein signaling cascade is initiated upon cell surface perception of a ligand by the corresponding GPCR (Temple and Jones 2007). Like MLO proteins, GPCRs harbor 7TM domains and possess an extracellular amino- and intracellular carboxy-terminus. At the cytosolic face, GPCRs are associated with the G-protein, which consists of three distinct subunits, Gα, Gβ, and Gγ. The Gα subunit binds the guanine nucleotides GDP and GTP. In its GDP-bound state, the three subunits assemble to a heterotrimeric complex, which is associated to the GPCR. Extracellular binding of a cognate ligand to the receptor induces the exchange of GDP for GTP at the Gα subunit. In consequence, the heterotrimeric G-protein complex dissociates and Gα-GTP separates from the Gβγ dimer. Both, Gα-GTP and the Gβγ dimer detach from the receptor and can activate or inactivate downstream effectors. The intrinsic hydrolytic GTPase activity of Gα recovers the GDP-bound state, which promotes reassociation of the complex into its inactive form. Regulator of G-protein Signaling (RGS) proteins accelerate the GTPase activity of Gα to reinstate the inactive heterotrimeric complex (see also chapter "Plant G alpha Structure and Properties").

Based on the analysis of the complete genome sequences of the mono- and dicotyledonous reference plants rice (*Oryza sativa*) and *A thaliana*, there exist single copy genes for each of the Gα and Gβ subunits (*RGA1* and *RGB1* or *GPA1* and *AGB1*, respectively) and two genes encoding Gγ subunits (*RGG1* and *RGG2* or *AGG1* and *AGG2*, respectively). Thus, higher plants encode a much simpler repertoire of heterotrimeric G-protein components than other eukaryotes. However, plant G-protein subunits are involved in a wide range of processes including developmental events as well as responses to abiotic and biotic stresses (Perfus-Barbeoch et al. 2004; see also chapter "G proteins and plant innate immunity"). On the basis of pharmacological studies in cell cultures, a role for the heterotrimeric G-protein in plant defense has been originally proposed more than a decade ago (Legendre et al. 1992; Gelli et al. 1997; Beffa et al. 1995). Meanwhile, the involvement of the heterotrimeric G-protein complex in plant defense has been tested directly by the use of mutants, and the results of these studies are summarized and discussed in the following sections.

3.1 *Heterotrimeric G-Protein Signaling in Rice Defense Responses*

The first genetic evidence for an involvement of heterotrimeric G-proteins in defense mechanisms stems from research with the rice *dwarf1* (*d1*) mutant, lacking a functional Gα-encoding gene, *RGA1*, and its interaction with the rice blast fungus, *Magnaporthe oryzae* (previously *M. grisea*) (Suharsono et al. 2002; Lieberherr et al. 2005). Inoculation of *d1* mutants with an avirulent race of *M. oryzae* or treatment with a sphingolipid elicitor (SE) resulted in highly reduced defense responses, including diminished ROS production, lower accumulation of defense gene transcripts (*PR1* and *PBZ1*), as well as less HR-mediated cell death (Suharsono et al. 2002). Furthermore, expression of *RGA1* was induced by infection with the avirulent *M. oryzae* strain or upon treatment with SE (Suharsono et al. 2002). Likewise, in response to virulent rice blight bacteria, *Xanthomonas oryzae* pv. *oryzae* (*Xoo*), *d1* plants developed earlier and more severe disease symptoms and showed delayed accumulation of defense proteins, suggesting an involvement of the heterotrimeric G-protein also in defense responses to virulent pathogens (Komatsu et al. 2004). These results implicate the heterotrimeric G-protein α subunit as an important player in rice resistance to bacterial and fungal pathogens. However, inoculation of *d1* mutants with a virulent strain of *M. oryzae* caused disease symptoms that were indistinguishable from wild type (Suharsono et al. 2002). Furthermore, in *d1* mutant suspension-cultured cells treated with *N*-acetyl-chitooligosaccharide, the oligosaccharide elicitor chitin, the stimulation of defense responses such as extracellular alkalinization, ROS generation, phytoalexin accumulation, and induction of defense genes did not differ from that of wild-type cells (Tsukada et al. 2002). Taken together, these data indicate that the contribution of the heterotrimeric G-protein α subunit to defense signaling is pathogen- and elicitor-specific.

3.2 *Heterotrimeric G-Protein Signaling in Arabidopsis* Defense Responses

The involvement of heterotrimeric G-proteins in *Arabidopsis* defense responses has been documented mainly for necrotrophic pathogens. Mutants lacking a functional Gβ subunit, AGB1, showed increased susceptibility against the necrotrophic fungi *Plectosphaerella cucumerina*, *Alternaria brassicicola*, and *Fusarium oxysporum*, while Gα-deficient plants (*gpa1*) exhibited slightly enhanced resistance to these pathogens (Llorente et al. 2005; Trusov et al. 2006). The infection phenotype of double-knockout mutants lacking both subunits, Gα and Gβ, were indistinguishable from that of the single Gβ mutant (Trusov et al. 2006). This data strongly suggests that rather the Gβγ dimer, and not Gα, is the predominant factor involved in the defense signaling pathway that is active against necrotrophic fungi in *Arabidopsis*.

A potential participation of Gγ1, but not Gγ2, along with Gβ in defense mechanisms was initially indicated by gene expression studies in *Arabidopsis* β-glucuronidase (GUS)-reporter lines infected with *A. brassicicola* and *F. oxysporum* (Trusov et al. 2007). These observations were corroborated by the infection phenotypes of the corresponding knockout mutants with these pathogens. While Gβ- and all tested Gγ1-deficient mutants showed increased susceptibility to these fungi and also exhibited reduced defense gene (*PDF1.2*) induction, plants lacking the Gγ2 subunit, AGG2, resembled the wild type (Trusov et al. 2007). Furthermore, Gβ- and Gγ1-deficient mutants showed reduced responses to methyl jasmonate, a signaling compound mainly involved in plant defense against necrotrophic pathogens, supporting the hypothesis that heterotrimeric G-proteins could play a role in jasmonate-mediated defense signaling (Trusov et al. 2006, 2007). In summary, these findings emphasize the requirement and importance of the Gβγ1 dimer for defense against necrotrophic fungi and preclude any significant role of the Gβγ2 dimer in this context. The slight increase in resistance observed for Gα-deficient mutants suggests that, with respect to plant defense, Gα acts by keeping the Gβγ1 attached to the inactive heterotrimeric complex (Llorente et al. 2005; Trusov et al. 2006, 2007). So far the *agb1 agg1* double-knockout mutant has not been tested regarding its infection phenotype to any pathogen, which would be an interesting addition to the present set of experiments.

Recent infection studies performed in our (J.L. and R.P.) laboratory implicate the Gβγ1 dimer also in defense against biotrophic powdery mildew fungi. Both *Gβ* and *Gγ1* knockout mutants exhibited slightly increased susceptibility to adapted as well as nonadapted powdery mildew fungi (*Golovinomyces orontii* and *Erysiphe pisi*, respectively; unpublished data). Surprisingly, the adapted pathogen *G. orontii* showed highly enhanced sporulation upon infection of knockout mutants lacking either the Gα, Gβ, Gγ1, or Gγ2 subunit or the RGS1-protein. This finding indicates a putative role of all heterotrimeric G-protein components in basal defense mechanisms that act following successful invasion by the fungus (unpublished data).

To investigate the involvement of heterotrimeric G-proteins in defense responses to bacteria, Trusov and colleagues challenged Gα and Gβ null mutants with compatible and incompatible strains of *Pseudomonas syringae* pv *tomato* (Trusov et al. 2006). In both cases, no difference between mutant and wild-type lines were observed, neither phenotypically nor with respect to the expression levels of the defense gene *PR1*, indicating that responses to *P. syringae* appear to be independent of heterotrimeric G-protein subunits. However, other data connect heterotrimeric G-protein signaling with bacterial PAMP perception and PTI. For instance, inhibition of stomatal opening by flg22 as part of PTI seems to implicate the Gα subunit, as Gα mutants showed impaired flg22-mediated stomatal closure (Zhang et al. 2008). Furthermore, it was recently postulated that the Gβ subunit is involved in ROS production triggered by the bacterial PAMPs flg22 and elf18. Additionally, Gβ seems to be required for elf18-mediated restriction of plant transformation via *Agrobacterium tumefaciens* (Ishikawa 2009).

In summary, a range of studies indicate a role of heterotrimeric G-proteins in plant defense in both monocotyledonous as well as dicotyledonous plants. Interestingly, rice and *Arabidopsis* Gα-deficient mutants displayed different pathogen responses. While in *A thaliana* the lack of the Gα subunit caused rather increased resistance to fungal pathogens, rice mutants exhibited reduced defense responses. Moreover, mutations in the Gα subunit induced different morphological phenotypes in both plant species, leading to dwarfism in rice, while in *A. thaliana* the mutation produced rather the opposite effect, with mutants being slightly larger than the wild type (Fujisawa et al. 1999; Ullah et al. 2003). These differences suggest that the G-protein subunits could have functionally diverged during evolution in monocots and dicots. The studies also indicate that in both plant clades the extent of heterotrimeric G-protein signaling in response to avirulent and virulent pathogens is pathogen- and/or elicitor-dependent.

4 MLO: A Putative Plant GPCR?

As outlined above, GPCRs are 7TM domain-containing proteins with an extracellularly localized *N*-terminus and a cytosolic *C*-terminus (Temple and Jones 2007). Approximately 1,000 GPCRs have been estimated to be encoded by mammalian genomes, whereas in plants only a few candidates for GPCRs have been identified. Given that even human GPCRs do not show extensive sequence conservation between subfamilies, searches on the basis of sequence homology alone could fail to detect plant GPCRs. Owing to this constraint, Moriyama et al. (2006) developed biocomputational tools by combining multiple protein classification methods, including alignment-free approaches, to identify the highly divergent GPCR candidates in plants (Kim et al. 2000; Moriyama et al. 2006; see also chapter "Bioinformatics of Seven-Transmembrane Receptors in Plant Genomes"). In a related approach, computational analysis of the entire virtual proteomes of the three model plant species, *A. thaliana*, *O. sativa*, and *Populus trichocarpa* were performed to identify plant protein sequences that most likely represent GPCRs (Gookin et al. 2008). Although there was some overlap between both studies, there were also considerable differences, emphasizing the importance of experimental verification of GPCR candidates based on functional studies.

At present, there are few proteins/protein families annotated as putative GPCRs in *Arabidopsis*. Among these candidates, only GCR1 shares extended (approximately 20%) sequence identity with known GPCRs, the cyclic AMP receptor, CAR1, found in *Dictyostelium discoideum* (slime mold), and the Class B Secretin family GPCRs (Josefsson and Rask 1997; Plakidou-Dymock et al. 1998). GCR1 physically interacts with the *Arabidopsis* Gα subunit, GPA1, but a ligand for GCR1 has not been identified (Pandey and Assmann 2004). Likewise, the *Arabidopsis* RGS1-protein is also predicted to represent a 7TM domain protein and has been shown to physically interact with the Gα subunit (Chen et al. 2006a). Lately, it has been proposed that RGS1 acts together with the heterotrimeric G-protein complex as an extracellular glucose receptor (Jeffrey et al. 2008). Owing to seemingly erroneous biocomputational predictions, it has been proposed that the GCR2 protein functions as a GPCR for the phytohormone abscisic acid (ABA) (Liu et al. 2007), which lately has been challenged by several independent studies (Johnston et al. 2007; Gao et al. 2007; Guo et al. 2008). Recently, two novel GPCR-type G-proteins, named GTG1 and GTG2, were proposed as ABA receptors in *Arabidopsis* (Pandey et al. 2009). These newly discovered proteins combine dual functions, seemingly representing a new type of G-protein with classic GTP-binding and GTPase activity as well as operating as GPCRs that interact with the Gα subunit, GPA1, and specifically bind ABA.

The remaining GPCR candidates are represented by the plant-unique MLO proteins, which have a predicted 7TM domain topology that has been confirmed experimentally for barley MLO (Devoto et al. 1999). Loss-of-function mutations of the *MLO* gene confer resistance to pathogenic powdery mildew fungi in barley, *Arabidopsis* and tomato (Büschges et al. 1997; Consonni et al. 2006; Bai et al. 2008) (see *above*). A combined pharmacological and genetic study indicated, however, that powdery mildew susceptibility/disease resistance in barley is independent of heterotrimeric G-protein function (Kim et al. 2002b). In these experiments, the contribution of the HvGα subunit on susceptibility to the powdery mildew fungus *Bgh* was tested by transient expression of constitutive active and dominant negative HvGα variants in single barley leaf epidermal cells. None of these Gα variants did alter fungal entry rates in either susceptible wild-type *MLO* or resistant mutant *mlo* genotypes. Similarly, application of pharmacological G-protein activators did not change infection phenotypes. Taken together, these data provided first evidence that MLO proteins function independently of the heterotrimeric G-protein. Given that these results were based on transient expression of Gα variants and G-protein activators derived and known from studies in the animal but not the plant field, the findings of Kim et al. (2002b) have to be regarded with caution. Recently, our group (J.L. and R.P.), thus, chose a genetic approach using stable *Arabidopsis* knockout mutants lacking either the Gα, Gβ, Gγ1, or Gγ2 subunit, or the RGS1 protein to address the same question. These mutants exhibited susceptibility to *G. orontii* that was indistinguishable from wild-type, except for Gβ- and Gγ1-deficient mutants, which showed increased susceptibility to the powdery mildew pathogen, independently of the presence or absence of MLO. The findings suggest a role for these heterotrimeric G-protein components in antifungal defense mechanisms that are separate from MLO functions (unpublished data). Taken together, our data support the previous results from Kim et al. (2002b), indicating that susceptibility conferred by presence of MLO does not implicate heterotrimeric G-protein signaling, precluding a role of MLO as a GPCR in this context. However, since the biochemical core function of MLO proteins is still unknown, the possibility remains that members of the MLO family may turn out to operate as GPCRs in processes distinct from pathogen defense. Alternatively, MLO proteins may function as cell surface receptors via a signaling cascade that does not involve the heterotrimeric G-protein complex. It, nevertheless, remains also possible that the function of MLO proteins is entirely unrelated to ligand binding and signal transduction.

5 Plant Rho-Like Proteins

In plants, small monomeric GTPases of the Rho-superfamily regulate the production of reactive oxygen species (ROS), Ca^{2+} fluxes, and cytoskeleton organization throughout plant development and during interactions with the environment. These processes are considered as key events in elicitor-triggered signal transduction and in the context of cell wall-associated defense mechanisms (Garcia-Brugger et al. 2006; Hückelhoven 2007). Intriguingly, in barley, MLO modulates local ROS production at the plant–pathogen interface, interacts with the cytoplasmic calcium sensor calmodulin in a Ca^{2+}-dependent manner, and affects actin cytoskeleton polarization during barley-powdery mildew interactions. These findings, thus, point to a possible link between Rho and MLO functions in powdery mildew susceptibility (Kim et al. 2002a,b; Hückelhoven and Kogel 2003; Opalski et al. 2005). The subclass of plant-specific Rho GTPases is called RAC or ROP (ROP: Rho of Plants) and constitutes a comparatively small protein family (Fu and Yang 2001). Winge et al. (2000) subdivided the 11 *Arabidopsis* RAC/ROP proteins into two major subgroups that can be distinguished by length due to an additional exon in group II. In contrast to *Arabidopsis*, grasses seem to express only six to nine RAC/ROP genes (Fu and Yang 2001; Christensen et al. 2003; Schultheiss et al. 2003, see chapter “ROP Evolution and ROPs in Grasses”).

5.1 RAC/ROPs in Disease Resistance and Susceptibility

RAC/ROP proteins have been implicated in defense-related signal transduction, thus modulating the outcome of plant–pathogen interactions. Expression of a gene encoding a ROP-binding kinase that interacts with ROPs in vivo is locally activated when adapted or nonadapted pathogens attack *Arabidopsis* (Molendijk et al. 2008). It was also shown that a soybean RAC-like GTPase integrates into the microsomal membrane fraction following elicitation of the oxidative burst, suggesting that membrane localization of this RAC/ROP requires a biotic stress stimulus. Heterologous expression of constitutively activated GTP-bound (CA) or dominant negative (DN, GDP-bound or nucleotide-free) mutants of human HsRAC1 in soybean cells boosted or reduced, respectively, the oxidative burst in response to different elicitor preparations (Park et al. 2000). Vice versa, CA maize RAC proteins provoked ROS production in mammalian NIH 3 T3 cells (Hassanain et al. 2000). Additionally, DN OsRAC1 and antisense-mediated gene silencing of tobacco *NtRAC1* were able to suppress HR-mediated cell death in response to tobacco mosaic virus infection or to treatment with elicitor preparations in tobacco (Schiene et al. 2000; Moeder et al.

2005). However, a more detailed insight into the role of RAC/ROPs in interactions with microbes is only available for rice OsRAC1 and some barley HvRAC/ROPs as outlined below in detail.

5.1.1 RAC1 in Rice Disease Resistance

In the context of disease resistance, rice RAC1 is the best-characterized RAC/ROP protein (Table 1). Kawasaki et al. (1999) have shown that pathogen-triggered cell death in the *sl* lesion mimic mutant of rice could be modulated by the expression of CA or DN OsRAC1 in opposite directions: CA OsRAC1 supported cell death whereas DN OsRAC1 reduced cell death. CA OsRAC1 provoked the generation of ROS via a flavin-dependent oxidase, which was suggested to be a respiratory burst oxidase homolog (RBOH, see *below*). Subsequently, it was shown that expression of CA OsRAC1 was sufficient to confer resistance to virulent *M. oryzae*. Fungal invasion into transgenic CA OsRAC1-expressing rice plants was stopped coincident with the execution of an HR, which included the local generation of ROS. In contrast, DN OsRAC1 strongly suppressed race-specific resistance to avirulent *M. oryzae* but did not limit basal resistance to a virulent race (Ono et al. 2001). OsRAC1, thus, appeared to be a regulator of race-specific resistance to *M. oryzae*. Besides this, CA OsRAC1 supported basal resistance to virulent *X oryzae* pv. *oryzae* (Table 1). CA OsRAC1 could further complement loss of basal resistance to *M. oryzae* and *Xoo* in *OsRAR1* (*REQUIRED FOR MLA12-MEDIATED RESISTANCE*)-silenced rice RNA interference (RNAi) plants. Since the RAR1 zinc finger protein is considered to function as a cochaperone in race-specific immune complexes, the data support a function of OsRAC1 in ETI.

The biological effects of OsRAC1 in rice disease resistance described above are reminiscent of functions of Gα. Accordingly, expression of CA OsRAC1 in the rice *d1* mutant restored resistance to avirulent *M. oryzae*, execution of HR, defense gene expression, and ROS formation. This places OsRAC1 parallel to or downstream of Gα in resistance to *M. oryzae* (see also"Introduction" of the Chapter "Structure and function of ROPs and their GEFs"; Table 1) (Suharsono et al. 2002). Coimmunoprecipitation experiments showed association of OsRac1 with OsMAPK6, a mitogen-activated protein kinase activated during responses to pathogens or pathogen-derived elicitors (Lieberherr et al. 2005). In either *d1* or *OsRac1*-silenced cell lines treated with sphingolipid elicitor, OsMAPK6 protein levels and activity were reduced but mRNA accumulation was unaltered, suggesting posttranslational regulation of OsMPKA6 accumulation levels by Gα and OsRac1 (Table 1) (Lieberherr et al. 2005). Together, these results support a defense signaling cascade from the heterotrimeric G-protein α subunit via the small GTPase OsRac1 to OsMAPK6.

Consistent with a more general function of OsRAC1 in modulating defense-associated protein abundance, expression of CA OsRAC1 induced changes in the proteome of cultured rice cells that were similar to those induced by the sphingolipid elicitor. Among the upregulated proteins were many defense-related proteins, chaperones, proteases and protease inhibitors, phenylpropanoid biosynthesis

Table 1 Rice RAC/ROPs and interactors that operate in disease resistance

RAC/ROP Protein	RAC/ROP interactors[a] and function		Reference
	RAC/ROP interactor	Function	
OsRAC1		Signaling in cell death, reactive oxygen production, and race-specific resistance to *Magnaporthe oryzae*	Kawasaki et al. (1999); Ono et al. (2001)
		Signaling in basal resistance to *Xanthomonas oryzae* pv *oryzae*	Ono et al. (2001)
		Inhibits expression of the antioxidant metallothionein MT2b	Wong et al. (2004)
		Modulates host defense proteome similar to a sphingolipid elicitor	Fujiwara et al. (2006)
	Heterotrimeric G-protein Gα subunit	Acts upstream of RAC1 in race-specific resistance and elicitor signaling	Suharsono et al. (2002)
	MPK6	Is regulated by Gα and RAC1, complexes with active RAC1 and is involved in elicitor-activated gene expression	Lieberherr et al. (2005)
	RBOH-type NADPH oxidase	Is activated by active RAC1 via protein–protein interaction in planta and produces reactive oxygen species	Wong et al. (2004)
	Cinnamoyl-CoA reductase 1	Interacts with activated RAC1 and functions in biosynthesis of lignin	Kawasaki et al. (2006)
	RAR1	Is found in an immune complex with HSP90 and RAC1 and functions in PAMP-triggered immunity of rice. Additionally, OsRAC1 regulates RAR1 expression	Thao et al. (2007)
	HSP90	Is found in an immune complex with RAR1 and RAC1 and stabilizes the complex	Thao et al. (2007)
	RACK1	Complexes active RAC1, RAR1, and SGT1 and supports resistance to *M. oryzae* and reactive oxygen production	Nakashima et al. (2008)

[a] An interactor either interacts functionally or physically where indicated

enzymes, polyamine and ethylene-related proteins, redox proteins, and enzymes of the alcoholic fermentation pathway (Fujiwara et al. 2006). This strongly supports the view that OsRAC1 is a central node for the regulation of protein abundance in several pathways that are crucial for pathogen defense (Table 1).

However, what may actually stop invading pathogens from growth in CA OsRAC1-expressing rice plants? A truncated variant of the monolignol biosynthesis pathway enzyme cinnamoyl-CoA reductase was identified in a yeast two-hybrid assay to interact with CA OsRAC1 but not with DN OsRAC1. GTP-bound OsRAC1 interacted with cinnamoyl-CoA reductase in vitro and stimulated its enzymatic activity. Expression of CA OsRAC1 also elevated lignin contents in transgenic rice cell cultures and enhanced the activity of cinnamoyl-CoA reductase (Kawasaki et al. 2006). Together with the fact that CA OsRAC1 promotes ROS production, the data suggest that OsRAC1 orchestrates lignification of the plant cell wall (Table 1), which may be crucial for arresting invasive growth of *M. oryzae* in resistant plants (Schaffrath et al. 1995).

The role of OsRAC1 and related RAC/ROPs in ROS production was recently elucidated in more detail (Wong et al. 2004). OsRAC1 was found to interact with the *N*-terminal cytoplasmic extension of the plasma membrane-localized RBOH NADPH oxidase, which carries two potential calcium-binding EF-hand motifs. This interaction was demonstrated by in vitro pull-down experiments, yeast two-hybrid studies, and in vivo fluorescence resonance energy transfer (FRET) experiments. Depending on the presence of intact EF-hands, FRET efficiency dropped under high calcium concentrations, which indicates a role of calcium in controlling the OsRAC1-RBOH interaction. Transient coexpression of CA OsRAC1 and OsRBOHB in leaves of *Nicotiana benthamiana* resulted in enhanced ROS production when compared with the expression of each single protein, suggesting that both proteins synergistically contribute to ROS production. In contrast to OsRAC1, tobacco NtRAC5 attenuated an elicitor-activated burst and negatively regulated abundance of NtRBOHD (Morel et al. 2004). In humans, HsRAC is crucial for the activation of at least three types of RBOH-like NADPH oxidases that partially contribute to innate immunity and programmed cell death (Bedard and Krause 2007). Hence, NADPH oxidase activation by Rho-like GTPases is a conserved phenomenon in mammals and plants, although the structural basis for protein complex formation may differ in the two kingdoms (Table 1) (Kao et al. 2008).

OsRAC1 also directly interacts with OsRAR1 and the heat shock protein HSP90. Both are important components of *R* gene-mediated disease resistance (ETI). Interaction in vivo was supported by coimmunoprecipitation of OsRAC1 with RAR1, HSP90, and HSP70. CA OsRAC1-mediated boosting of elicitor responses was dependent on RAR1 and HSP90. OsRAC1 also regulates RAR1 expression at both the mRNA and the protein level (Table 1) (Thao et al. 2007). Most recently, RACK1 (RECEPTOR FOR ACTIVATED C-KINASE 1) was isolated via affinity chromatography using glutathione-*S* transferase (GST) epitope-tagged CA OsRAC1. RACK1 appears to form a protein complex by linking RBOH and OsRAC1 to RAR1 and SGT1 and, when overexpressed, it was sufficient to enhance resistance to virulent *M. oryzae* (Nakashima et al. 2008). The authors, thus, suggested that RACK1 acts as a scaffolding protein in rice immune protein complexes (Table 1). In summary, OsRAC1 appears to be a key player in the rice PTI and ETI in the context of different plant–pathogen interactions.

5.1.2 RAC/ROPs in Barley Disease Resistance and Susceptibility

In contrast to the role of OsRAC1 in disease resistance of rice, the barley RAC/ROP protein HvRACB operates in susceptibility to the biotrophic barley powdery mildew fungus *B. graminis* f. sp. *hordei* (*Bgh*) (Table 2) (Schultheiss et al. 2002, 2003). Knockdown of HvRACB by RNAi in single epidermal cells, transformed via microprojectile-mediated gene delivery, rendered cells more resistant to fungal penetration. RNAi-mediated penetration resistance was not efficient in *ror1*-mutants, which are impaired in basal and nonspecific *mlo*-mediated resistance (Table 2). In contrast, expression of CA HvRACB supported fungal penetration success, whereas nonactivated wild-type HvRACB or closely related CA HvRACD had no effect. However, CA HvRACB did not break the highly effective *mlo*-mediated resistance. Together, this suggests that HvRACB modulates basal

Table 2 Barley RAC/ROPs and interactors that operate in disease resistance or susceptibility

RAC/ROP protein	RAC/ROP interactors[a] and function		Reference
	RAC/ROP interactor	Function	
HvRACB		Is required for full susceptibility to *Bgh*	Schultheiss et al. (2002, 2003)
		Supports entry by *Bgh*	Schultheiss et al. (2003, 2005)
		Functions in cell polarity and organization of actin microfilaments	Opalski et al. (2005)
	MLO	Is required for RACB function in susceptibility	Schultheiss et al. (2003)
	ROR1	Is required for RACB function in susceptibility and for resistance mediated by RACB RNAi	Schultheiss et al. (2002, 2003)
	RIC171	Interacts with RACB in planta and supports entry by *B. graminis*	Schultheiss et al. (2008)
HvRAC3		Supports entry by *B. graminis*	Schultheiss et al. (2003); Pathuri et al. (2008)
		Functions in cell polarity	Pathuri et al. (2008), Pathuri et al. unpublished
		CA Overexpression enhances cell size and susceptibility to *P.s.tabaci* in tobacco	Pathuri et al. unpublished
HvRAC1		Supports entry by *B. graminis*	Pathuri et al. (2008)
		Functions in cell polarity	Pathuri et al. (2008)
		Supports callose deposition and H_2O_2 production	Pathuri et al. (2008)
		Supports basal resistance to *M. oryzae*	Pathuri et al. (2008)

[a]An interactor either interacts functionally or physically where indicated

susceptibility of barley to *Bgh* in an MLO- and ROR1-dependent manner (Table 2) (Schultheiss et al. 2002, 2003).

The role of RAC/ROPs in dicot–microbe interactions is not yet understood. However, the ectopic expression of barley CA HvRACB or CA HvRAC3 in tobacco enhanced susceptibility to powdery mildew, and tobacco plants expressing CA HvRAC3 showed additional super-susceptibility to the bacterial pathogen *P syringae* pv. *tabaci* (Indira Pathuri and R.H. unpublished). Furthermore, an Arabidopsis *Rho-GTPase ACTIVATING PROTEIN* (*GAP*) T-DNA insertion allowed for accelerated fungal development and enhanced sporulation of powdery mildew (Christina Huesmann and R.H. unpublished). Together, this suggests an involvement of RAC/ROPs also in dicot susceptibility to various phytopathogens.

Since RAC/ROPs are key regulators of the cytoskeleton (see also chapter "ROP GTPases and the Cytoskeleton"), the role of HvRACB in filamentous F-actin organization under attack from *Bgh* was analyzed (Table 2) (Opalski et al. 2005). Knockdown of *HvRACB* led to more polarization of F-actin to the site of attempted penetration, which was correlated with enhanced resistance. In contrast, expression of CA HvRACB induced actin filament depolarization, supporting susceptibility. Together with the observation that virulent *Bgh* seemed to inhibit polarization of attacked cells in an MLO-dependent manner, this suggests that *Bgh* might target HvRACB to suppress polar plant defense, or to support haustorial establishment. CA HvRACB also partially inhibited polarization of *mlo* barley cells, however, without inducing susceptibility. Hence, HvRACB requires functional MLO in susceptibility, but can affect F-actin organization independently from MLO (Opalski et al. 2005).

Transgenic barley plants stably expressing CA HvRACB displayed enhanced susceptibility to powdery mildew. Additionally, CA HvRACB-expressing plants showed pleiotropic effects in root and shoot development as well as in water retention capacity, when cut off from water supply or when treated with abscisic acid. This suggests that HvRACB might have a physiological role in plant development and in biotic as well as abiotic stress responses (Schultheiss et al. 2005). In transient expression experiments, it was shown that other barley RAC/ROPs might fulfill HvRACB-redundant functions in susceptibility to *Bgh*. When stably expressed in barley, CA HvRAC1 and CA HvRAC3 exhibited similar effects on plant development as expression of CA HvRACB (Table 2). In particular, all three CA HvRAC/POPs abolished polarity in tip-growing root hairs (Pathuri et al. 2008). Additionally, transgenic barley plants expressing CA HvRACB or CA HvRAC1 showed significantly longer epidermal cells and aberrant development of stomata (Pathuri et al. 2008, 2009). Together, the data suggest that similar to what is known from *Arabidopsis* (Yalovsky et al. 2008), monocot RAC/ROPs have conserved functions in cell expansion and polarized tip growth. This supports the idea that virulent *Bgh* corrupts a plant tip growth program (see also chapter RAC/ROP GTPases in the "Regulation of Polarity and Polar Cell Growth") to establish a rapidly growing haustorium surrounded by a host derived extrahaustorial mem brane in intact epidermal cells of barley (Schultheiss et al. 2003; Opalski et al. 2005). This assumption was further corroborated by the observation of host-derived actin rings, which can also be observed below the apical dome of tip-growing plant cells (Yalovsky et al. 2008), around the tip of emerging haustoria (Opalski et al. 2005).

Interestingly, similar to transgenic CA HvRACB barley lines, CA HvRAC1-expressing barley plants were super-susceptible to *Bgh*. This could be explained by enhanced success of fungal penetration. However, CA HvRAC1 barley plants displayed significantly more cells with whole cell hydrogen peroxide (H_2O_2) accumulation as visualized by 3,3′-diaminobenzidine (DAB) staining. This phenomenon was restricted to cells where *Bgh* failed to penetrate and can, thus, be considered as part of a secondary defense reaction. The same plants also reacted more frequently with localized callose deposition to attack by *Bgh* in cells that did not support resistance to fungal penetration (Table 2) (Pathuri et al. 2008). Hence, although barley RAC/ROPs function in conferring susceptibility, they might have an additional role in positively modulating cellular defense reactions, which is similar to the situation of OsRAC1 in rice. In accordance with this, CA HvRAC1-expressing barley plants showed enhanced basal resistance to *M. oryzae*, which could be explained by enhanced resistance to fungal penetration in the first attacked epidermal cell (Pathuri et al. 2008). Hence, both OsRAC1 and its closest relative in barley, HvRAC1, can support resistance to *M. oryzae* (Tables 1 and 2). However, CA OsRAC1 mediates fungus-induced HR whereas the CA HvRAC1 supports penetration resistance of living cells, which form localized cell wall appositions. This situation in barley is reminiscent of the role of MLO, which is required for penetration by *Bgh* but limits the penetration success of *M. oryzae* (Jarosch et al. 1999). These findings additionally support the above-mentioned functional link or partial redundancy of MLO and RAC/ROPs in barley. However, direct evidence for a cooperative function of MLO and RAC/ROPs is currently missing.

A possible link between RAC/ROPs and MLO might be the actin cytoskeleton (Opalski et al. 2005; Miklis et al. 2007). RAC/ROPs are well known as regulators of actin nucleation and dynamics. For instance, downstream of AtROP2, the Arabidopsis RIC proteins (RAC/ROP Interactive Cdc42/Rac Interactive Binding (CRIB)-Motif Containing Proteins), AtRIC1 and AtRIC4, regulate the establishment of spatial arrays of F-actin and microtubules during lobe and neck formation of interlocked epidermal pavement cells (Fu et al. 2005). A role of RAC/ROPs in actin nucleation is supported because RAC/ROPs interact with components of the actin-polymerizing WAVE complex, which is involved in epidermis development. Yeast two-hybrid experiments showed that AtROP2 interacts with PIR121/SRA1 subunits of this complex, suggesting that WAVE activity in plants may be regulated by RAC/ROPs (Basu et al. 2004). Recently, AtROP2 activation by the DOCK family protein SPIKE1, which has RAC/ROP-stimulating guanidine nucleotide exchange factor activity, has been evidenced. Hence, SPIKE1–ROP2–SRA1 signaling appears to operate during establishment of actin nucleation complexes (Basu et al. 2008). *Arabidopsis* AtICR1 (INTERACTOR OF CONSTITUTIVE ACTIVE ROPs 1) has been found to interact with both active RAC/ROPs and SEC3, which is associated with Rho in the exocyst complex in mammals (Lavy et al. 2007; Berken and Wittinghofer 2008). It also has been suggested that tobacco NtRAC1 controls the activity of ACTIN DEPOLYMERIZING FACTOR NtADF1 during pollen tube growth. In analogy to mammalian systems, this might be facilitated via a RAC/ROP-activated kinase that phosphorylates NtADF1, leading to protein inactivation and subsequent actin polymerization. Because RAC/ROP activity is spatially and temporarily fine-tuned during pollen tube growth, this may contribute to the dynamics of F-actin throughout this morphogenetic process (Chen et al. 2003). Since *mlo*-resistance is partially compromised by overexpression of HvADF3, and because barley RAC/ROPs presumably inhibit the activity of HvADF3, it has been suggested that functional MLO in concert with RAC/ROPs inhibits F-actin reorganization for polar defense reactions or orchestrates actin dynamics during fungal entry (Opalski et al. 2005; Miklis et al. 2007).

In a targeted yeast two-hybrid assay, HvRACB was shown to interact with a 171 amino acid CRIB-motif-containing protein of barley designated HvRIC171. Interaction of HvRACB and HvRIC171 proteins was supported by bimolecular fluorescence complementation (BiFC), which indicated that HvRIC171 interacts with CA HvRACB but not with DN HvRACB in planta, and thus is likely involved in downstream effects of HvRACB-GTP (Schultheiss et al. 2008). Accordingly, similar to CA HvRACB, overexpression of HvRIC171 supported susceptibility to *Bgh*. In contrast, a presumably nonfunctional CRIB-containing HvRIC171 fragment of 46 amino acids bound CA HvRACB in planta but had a dominant negative effect on fungal penetration success when transiently expressed in barley epidermal cells (Table 2). A red fluorescing HvRIC171–DsRED fusion protein was recruited to the cell periphery by membrane-associated CA HvRACB, but not by DN HvRACB, and accumulated at sites of fungal penetration attempts. This suggests focal HvRACB activity at sites of attempted fungal penetration (Schultheiss et al. 2008). Further investigations have to show whether HvRIC171 interferes with F-actin organization or whether other barley RAC/ROP-interacting proteins could explain how *Bgh* corrupts RAC/ROPs for compatibility. Interestingly, type III effectors of bacterial pathogens target Rho family proteins of mammals. *Yersinia* outer protein effectors (YOPs) have GAP or guanine nucleotide dissociation inhibitor (GDI) functions (see also chapter "Regulatory and Cellular Functions of Plant RhoGAPs and RhoGDIs") or are Rho-cleaving cysteine proteases involved in actin reorganization for invasion of nonphagocytic cells (Gruenheid and Finlay 2003; Aepfelbacher et al. 2007). It remains to be seen whether in analogy, *Bgh* effectors target barley RAC/ROPs during powdery mildew pathogenesis.

5.2 ROPs and Lipid Rafts

Recently, it has been shown that a type I *Arabidopsis* RAC/ROP in an activity-dependent manner inserts into detergent-resistant membrane fractions, and that this recruitment is mediated via reversible *S*-acylation of a conserved cysteine residue (e.g., C156 in AtROP6) (Sorek et al. 2007). Together with earlier findings that further carboxy-terminal cysteine residues can be prenylated in type I RAC/ROPs and/or acylated in type II RAC/ROPs (Lavy et al. 2002; Yalovsky et al. 2008; see also chapter "ROPs, Vesicle Trafficking and Lipid Modifications"), these data support that signaling downstream of RAC/ROPs may operate from specific lipid domains, which have been found to be enriched with other signaling proteins such as RLKs, NADPH oxidases, and syntaxins (Mongrand et al. 2004; Morel et al. 2004; Bhat and Panstruga 2005). A GFP-tagged version of CA AtROP6 was recently imaged at sites of attack from virulent powdery mildew on *Arabidopsis* supporting recruitment of RAC/ROPs into specialized membrane domains at intimate sites of fungal contact (Hoefle and Hückelhoven 2008). In this context, it is also noteworthy that truncated CA type I HvRACB or CA type II HvRAC3, in which presumably lipid-modified cysteine residues were removed, were dislocated from the plasma membrane and could no longer support fungal entry by *Bgh* (Schultheiss et al. 2003). This suggests that membrane or lipid raft association could be crucial for RAC/ROP function in susceptibility to *Bgh*.

6 Perspectives

OsRAC1 is linked to Gα functions and both are important in resistance to avirulent *M. oryzae* and virulent *X. oryzae*. OsRAC1 and Gα, thus, likely represent common elements of PTI and ETI in rice. In barley, RAC/ROPs rather than heterotrimeric G-proteins are modulators of MLO-mediated susceptibility to powdery mildew and of basal resistance to *M. oryzae,* which is also dependent on MLO. It remains, however, elusive how the pathogen recognition machinery connects to G-protein signaling. Despite the well-documented involvement of these proteins in interactions of grasses with pathogenic microbes and the conserved function of MLO in dicots, little is understood about the role of RAC/ROPs and heterotrimeric G-proteins and their interplay with MLO proteins in disease resistance of dicot plant species. Additional studies are, thus, required to shed light on the contribution of G-proteins in interactions of dicots with pathogenic organisms and on the potential role of heterotrimeric G-proteins in physiological functions of MLO. Additionally, the important question whether MLO and G-proteins might be direct or indirect targets of microbial effector molecules needs future clarification.

References

Aepfelbacher M, Trasak C, Ruckdeschel K (2007) Effector functions of pathogenic Yersinia species. Thromb Haemost 98:521–529

Bai Y, Pavan S, Zheng Z, Zappel NF, Reinstädler A, Lotti C, De Giovanni C, Ricciardi L, Lindhout P, Visser R, Theres K, Panstruga R (2008) Naturally occurring broad-spectrum powdery mildew resistance in a central American tomato accession is caused by loss of Mlo function. Mol Plant Microbe Interact 21:30–39

Basu D, El-Din El-Assal D, Le J, Mallery EL, Szymanski DB (2004) Interchangeable functions of *Arabidopsis* PIROGI and the human WAVE complex subunit SRA1 during leaf epidermal development. Development 131:4345–4355

Basu D, Le J, Zakharova T, Mallery EL, Szymanski DB (2008) A SPIKE1 signaling complex controls actin-dependent cell morphogenesis through the heteromeric WAVE and ARP2/3 complexes. Proc Natl Acad Sci USA 105:4044–4049

Bedard K, Krause KH (2007) The NOX family of ROS-generating NADPH oxidases: physiology and pathophysiology. Physiol Rev 87:245–313

Bednarek P, Pislewska-Bednarek M, Svatos A, Schneider B, Doubsky J, Mansurova M, Humphry M, Consonni C, Panstruga R, Sanchez-Vallet A, Molina A, Schulze-Lefert P (2009) A glucosinolate metabolism pathway in living plant cells mediates broad-spectrum antifungal defense. Science 323:101–106

Beffa R, Szell M, Meuwly P, Pay A, Vögeli-Lange R, Métraux JP, Neuhaus G, Meins F, Nagy JR (1995) Cholera toxin elevates pathogen resistance and induces pathogenesis-related gene expression in tobacco. EMBO J 14:5753–5761

Bent AF, Mackey D (2007) Elicitors, effectors, and R genes: the new paradigm and a lifetime supply of questions. Ann Rev Phytopathol 45:399–436

Berken A, Wittinghofer A (2008) Structure and function of Rho-type molecular switches in plants. Plant Physiol Biochem 46:380–393

Bhat RA, Panstruga R (2005) Lipid rafts in plants. Planta 223:5–19

Block A, Li G, Fu ZQ, Alfano JR (2008) Phytopathogen type III effector weaponry and their plant targets. Curr Opin Plant Biol 11:396–403

Büschges R, Hollricher K, Panstruga R, Simons G, Wolter M, Frijters A, van Daelen R, van der Lee T, Diergaarde P, Groenendijk J, Töpsch S, Vos P, Salamini F, Schulze-Lefert P (1997) The rarley Mlo gene: a novel control element of plant pathogen resistance. Cell 88:695–705

Chen CY, Cheung AY, Wu HM (2003) Actin-depolymerizing factor mediates Rac/Rop GTPase-regulated pollen tube growth. Plant Cell 15:237–49

Chen Y, Ji F, Xie H, Liang J, Zhang J (2006a) The regulator of G-protein signaling proteins involved in sugar and abscisic acid signaling in *Arabidopsis* seed germination. Plant Physiol 140:302–310

Chen Z, Hartmann HA, Wu MJ, Friedman EJ, Chen JG, Pulley M, Schulze Lefert P, Panstruga R, Jones AM (2006b) Expression analysis of the AtMLO gene family encoding plant-specific seven-transmembrane domain proteins. Plant Mol Biol 60:583–597

Chisholm ST, Coaker G, Day B, Staskawicz BJ (2006) Host-microbe interactions: shaping the evolution of the plant immune response. Cell 124:803–814

Christensen TM, Vejlupkova Z, Sharma YK, Arthur KM, Spatafora JW, Albright CA, Meeley RB, Duvick JP, Quatrano RS, Fowler JE (2003) Conserved subgroups and developmental regulation in the monocot RAC/ROP gene family. Plant Physiol 133:1791–1808

Collins NC, Thordal-Christensen H, Lipka V, Bau S, Kombrink E, Qiu J-L, Hückelhoven R, Stein M, Freialdenhoven A, Somerville SC, Schulze-Lefert P (2003) SNARE-protein-mediated disease resistance at the plant cell wall. Nature 425:973–977

Consonni C, Humphry ME, Hartmann HA, Livaja M, Durner J, Westphal L, Vogel J, Lipka V, Kemmerling B, Schulze-Lefert P, Somerville SC, Panstruga R (2006) Conserved requirement for a plant host cell protein in powdery mildew pathogenesis. Nat Genet 38:716–720

da Cunha L, Sreerekha M-V, Mackey D (2007) Defense suppression by virulence effectors of bacterial phytopathogens. Curr Opin Plant Biol 10:349–357

Devoto A, Piffanelli P, Nilsson I, Wallin E, Panstruga R, von Heijne G, Schulze-Lefert P (1999) Topology, subcellular localization, and sequence diversity of the Mlo family in plants. J Biol Chem 274:34993–35004

Devoto A, Hartmann HA, Piffanelli P, Elliott C, Simmons C, Taramino G, Goh CS, Cohen FE, Emerson BC, Schulze-Lefert P (2003) Molecular phylogeny and evolution of the plant-specific seven-transmembrane MLO family. J Mol Evol 56:77–88

Ellis J, Catanzariti A-M, Dodds P (2006) The problem of how fungal and oomycete avirulence proteins enter plant cells. Trends Plant Sci 11:61–63

Felix G, Duran JD, Volko S, Boller T (1999) Plants have a sensitive perception system for the most conserved domain of bacterial flagellin. Plant J 18:265–276

Flor HH (1942) Inheritance of pathogenicity in *Melampsora lini*. Phytopathol 32:653–669

Freialdenhoven A, Peterhansel C, Kurth J, Kreuzaler F, Schulze-Lefert P (1996) Identification of genes required for the function of non-race-specific mlo resistance to powdery mildew in barley. Plant Cell 8:5–14

Fu Y, Yang Z (2001) Rop GTPase: a master switch of cell polarity development in plants. Trends Plant Sci 6:545–547

Fu Y, Gu Y, Zheng Z, Wasteneys G, Yang Z (2005) *Arabidopsis* interdigitating cell growth requires two antagonistic pathways with opposing action on cell morphogenesis. Cell 120:687–700

Fujisawa Y, Kato T, Ohki S, Ishikawa A, Kitano H, Sasaki T, Asahi T, Iwasaki Y (1999) Suppression of the heterotrimeric G protein causes abnormal morphology, including dwarfism, in rice. Proc Natl Acad Sci USA 96:7575–7580

Fujiwara M, Umemura K, Kawasaki T, Shimamoto K (2006) Proteomics of Rac GTPase signalling reveals its predominant role in elicitor-induced defense response of cultured rice cells. Plant Physiol 140:734–745

Gao Y, Zeng Q, Guo J, Cheng J, Ellis BE, Chen JG (2007) Genetic characterization reveals no role for the reported ABA receptor, GCR2, in ABA control of seed germination and early seedling development in *Arabidopsis*. Plant J 52:1001–1013

Garcia-Brugger A, Lamotte O, Vandelle E, Bourque S, Lecourieux D, Poinssot B, Wendehenne D, Pugin A (2006) Early signaling events induced by elicitors of plant defenses. Mol Plant Microbe Interact 19:711–724

Gelli A, Higgins VJ, Blumwald E (1997) Activation of plant plasma membrane Ca2+-permeable channels by race-specific fungal elicitors. Plant Physiol 113:269–279

Gómez-Gómez L, Boller T (2000) FLS2: An LRR receptor-like kinase involved in the perception of the bacterial elicitor flagellin in *Arabidopsis*. Mol Cell 5:1003–1011

Gookin T, Kim J, Assmann S (2008) Whole proteome identification of plant candidate G-protein coupled receptors in *Arabidopsis*, rice, and poplar: computational prediction and in-vivo protein coupling. Genome Biol 9:R120

Guo J, Zeng Q, Emami M, Ellis BE, Chen J-G (2008) The GCR2 gene family is not required for ABA control of seed germination and early seedling development in *Arabidopsis*. PLoS ONE 3:e2982

Gruenheid S, Finlay BB (2003) Microbial pathogenesis and cytoskeletal function. Nature 422:775–781

Hassanain HH, Sharma YK, Moldovan L, Khramtsov V, Berliner LJ, Duvick JP, Goldschmidt-Clermont PJ (2000) Plant rac proteins induce superoxide production in mammalian cells. Biochem Biophys Res Commun 272:783–788

Hoefle C, Hückelhoven R (2008) Enemy at the gates – traffic at the plant cell pathogen interface. Cell Microbiol 10:2400–2407

Hückelhoven R (2007) Cell wall-associated mechanisms of disease resistance and susceptibility. Ann Rev Phytopathol 45:101–127

Hückelhoven R, Kogel K-H (2003) Reactive oxygen intermediates in plant-microbe interactions: who is who in powdery mildew resistance? Planta 216:891–902

Ishikawa A (2009) The *Arabidopsis* G-protein beta;-subunit is required for defense response against *Agrobacterium tumefaciens*. Biosci Biotech and Biochem 73:47–52

Jarosch B, Kogel K-H, Schaffrath U (1999) The ambivalence of the barley *Mlo* Locus: mutations conferring resistance against powdery mildew (*Blumeria graminis* f.sp. *hordei*) enhance susceptibility to the rice blast fungus *Magnaporte grisea*. Mol Plant Microbe Interact 12:508–514

Jeffrey CG, Daniel O, Wolf-Rüdiger S, Chenggang L, Mark S, Alan MJ (2008) d-Glucose sensing by a plasma membrane regulator of G signaling protein, AtRGS1. FEBS Lett 582:3577–3584

Johnston CA, Taylor JP, Gao Y, Kimple AJ, Grigston JC, Chen J-G, Siderovski DP, Jones AM, Willard FS (2007) GTPase acceleration as the rate-limiting step in *Arabidopsis* G protein-coupled sugar signaling. Proc Natl Acad Sci 104:17317–17322

Jones JDG, Dangl JL (2006) The plant immune system. Nature 444:323–329

Jørgensen IH (1992) Discovery, characterization and exploitation of Mlo powdery mildew resistance in barley. Euphytica 63:141–152

Josefsson LG, Rask L (1997) Cloning of a putative G-protein-coupled receptor from *Arabidopsis thaliana*. Eur J Biochem 249:415–420

Kaku H, Nishizawa Y, Ishii-Minami N, Akimoto-Tomiyama C, Dohmae N, Takio K, Minami E, Shibuya N (2006) Plant cells recognize chitin fragments for defense signaling through a plasma membrane receptor. Proc Natl Aca Sci USA 103:11086–11091

Kao YY, Gianni D, Bohl B, Taylor RM, Bokoch GM (2008) Identification of a conserved Rac-binding site on NADPH oxidases supports a direct GTPase regulatory mechanism. J Biol Chem 283:12736–12746

Kawasaki T, Henmi K, Ono E, Hatakeyama S, Iwano M, Satoh H, Shimamoto K (1999) The small GTP-binding protein Rac is a regulator of cell death in plants. Proc Natl Acad Sci USA 96:10922–10926

Kawasaki T, Koita H, Nakatsubo T, Hasegawa K, Wakabayashi K, Takahashi H, Umemura K, Umezawa T, Shimamoto K (2006) Cinnamoyl-CoA reductase, a key enzyme in lignin biosynthesis, is an effector of small GTPase Rac in defense signaling in rice. Proc Natl Acad Sci USA 103:230–235

Kim J, Moriyama EN, Warr CG, Clyne PJ, Carlson JR (2000) Identification of novel multi-transmembrane proteins from genomic databases using quasi-periodic structural properties. Bioinformatics 16:767–775

Kim MC, Lee SH, Kim JK, Chun HJ, Choi MS, Chung WS, Moon BC, Kang CH, Park CY, Yoo JH, Kang YH, Koo SC, Koo YD, Jung JC, Kim ST, Schulze-Lefert P, Lee SY, Cho MJ (2002a) Mlo, a modulator of plant defense and cell death, is a novel calmodulin-binding protein. Isolation and characterization of a rice Mlo homologue. J Biol Chem 277:19304–19314

Kim MC, Panstruga R, Elliott C, Muller J, Devoto A, Yoon HW, Park HC, Cho MJ, Schulze-Lefert P (2002b) Calmodulin interacts with MLO protein to regulate defence against mildew in barley. Nature 416:447–451

Komatsu S, Yang G, Hayashi N, Kaku H, Umemura K, Iwasaki I (2004) Alterations by a defect in a rice G protein alpha subunit in probenazole and pathogen-induced responses. Plant Cell Environ 27:947–957

Kunze G, Zipfel C, Robatzek S, Niehaus K, Boller T, Felix G (2004) The *N* terminus of bacterial elongation factor Tu elicits innate immunity in *Arabidopsis* plants. Plant Cell 16:3496–3507

Kwon C, Neu C, Pajonk S, Yun HS, Lipka U, Humphry M, Bau S, Straus M, Kwaaitaal M, Rampelt H, Kasmi FE, Jurgens G, Parker J, Panstruga R, Lipka V, Schulze-Lefert P (2008) Co-option of a default secretory pathway for plant immune responses. Nature 451:835–840

Lavy M, Bracha-Drori K, Sternberg H, Yalovsky S (2002) A cell-specific, prenylation-independent mechanism regulates targeting of type II RACs. Plant Cell 14:2431–2450

Lavy M, Bloch D, Hazak O, Gutman I, Poraty L, Sorek N, Sternberg H, Yalovsky S (2007) A Novel RAC/ROP/RAC effector links cell polarity, root-meristem maintenance, and vesicle trafficking. Curr Biol 17:947–952

Legendre L, Heinstein PF, Low PS (1992) Evidence for participation of GTP-binding proteins in elicitation of the rapid oxidative burst in cultured soybean cells. J Biol Chem 267:20140–20147

Lieberherr D, Thao NP, Nakashima A, Umemura K, Kawasaki T, Shimamoto K (2005) A sphingolipid elicitor-inducible mitogen-activated protein kinase is regulated by the small GTPase OsRac1 and heterotrimeric G-protein in rice. Plant Physiol 138:1644–1652

Lipka V, Dittgen J, Bednarek P, Bhat R, Wiermer M, Stein M, Landtag J, Brandt W, Rosahl S, Scheel D, Llorente F, Molina A, Parker J, Somerville S, Schulze-Lefert P (2005) Pre- and postinvasion defenses both contribute to nonhost resistance in *Arabidopsis*. Science 310:1180–1183

Liu X, Yue Y, Li B, Nie Y, Li W, Wu W-H, Ma L (2007) A G protein-coupled receptor is a plasma membrane receptor for the plant hormone abscisic acid. Science 315:1712–1716

Llorente F, Alonso-Blanco C, Sánchez-Rodriquez C, Jorda L, Molina A (2005) ERECTA receptor-like kinase and heterotrimeric G protein from *Arabidopsis* are required for resistance to the necrotrophic fungus *Plectosphaerella cucumerina*. Plant J 43:165–180

Lyngkjaer MF, Newton AC, Atzema JL, Baker SJ (2000) The barley *mlo*-gene: an important powdery mildew resistance source. Agronomie 20:745–756

Miklis M, Consonni C, Bhat RA, Lipka V, Schulze-Lefert P, Panstruga R (2007) Barley MLO modulates actin-dependent and actin-independent antifungal defense pathways at the cell periphery. Plant Physiol 144:1132–1143

Miya A, Albert P, Shinya T, Desaki Y, Ichimura K, Shirasu K, Narusaka Y, Kawakami N, Kaku H, Shibuya N (2007) CERK1, a LysM receptor kinase, is essential for chitin elicitor signaling in *Arabidopsis*. Proc Natl Aca Sci USA 104:19613–19618

Moeder W, Yoshioka K, Klessig DF (2005) Involvement of the small GTPase Rac in the defense responses of tobacco to pathogens. Mol Plant Microbe Interact 18:116–124

Molendijk AJ, Ruperti B, Singh MK, Dovzhenko A, Ditengou FA, Milia M, Westphal L, Rosahl S, Soellick TR, Uhrig J, Weingarten L, Huber M, Palme K (2008) A cysteine-rich receptor-like kinase NCRK and a pathogen-induced protein kinase RBK1 are RAC/ROP GTPase interactors. Plant J 53:909–923

Mongrand S, Morel J, Laroche J, Claverol S, Carde JP, Hartmann MA, Bonneu M, Simon-Plas F, Lessire R, Bessoule JJ (2004) Lipid rafts in higher plant cells: purification and characterization of Triton X-100-insoluble microdomains from tobacco plasma membrane. J Biol Chem 279:36277–36286

Morel J, Fromentin J, Blein JP, Simon-Plas F, Elmayan T (2004) Rac regulation of NtrbohD, the oxidase responsible for the oxidative burst in elicited tobacco cell. Plant J 37:282–293

Moriyama E, Strope P, Opiyo S, Chen Z, Jones A (2006) Mining the *Arabidopsis thaliana* genome for highly-divergent seven transmembrane receptors. Genome Biol 7:R96

Nakashima A, Chen L, Thao NP, Fujiwara M, Wong HL, Kuwano M, Umemura K, Shirasu K, Kawasaki T, Shimamoto K (2008) RACK1 functions in rice innate immunity by interacting with the Rac1 immune complex. Plant Cell 20:2265–2279

Ono E, Wong HL, Kawasaki T, Hasegawa M, Kodama O, Shimamoto K (2001) Essential role of the small GTPase Rac in disease resistance of rice. Proc Natl Acad Sci USA 98:759–764

Opalski KS, Schultheiss H, Kogel K-H, Hückelhoven R (2005) The receptor-like MLO protein and the RAC/ROP family G-protein HvRACB modulate actin reorganization in barley attacked by the biotrophic powdery mildew fungus *Blumeria graminis* f.sp. *hordei*. Plant J 41:291–303

Pandey S, Assmann SM (2004) The *Arabidopsis* putative G protein-coupled receptor GCR1 interacts with the G protein alpha-subunit GPA1 and regulates abscisic acid signaling. Plant Cell 16:1616–1632

Pandey S, Nelson DC, Assmann SM (2009) Two novel GPCR-type G proteins are abscisic acid receptors in *Arabidopsis*. Cell 136:136–148

Park J, Choi HJ, Lee S, Lee T, Yang Z, Lee Y (2000) Rac-related GTP-binding protein in elicitor-induced reactive oxygen generation by suspension-cultured soybean cells. Plant Physiol 124:725–732

Pathuri PI, Hensel G, Kumlehn J, Eichmann R, Hückelhoven R (2008) Constitutively activated barley ROPs modulate epidermal cell size, defense reactions and interactions with fungal leaf pathogens. Plant Cell Rep 27:1877–1887

Pathuri PI, Eichmann R, Hückelhoven R (2009) Plant small monomeric G-proteins (RAC/ROPs) of barley are common elements of susceptibility to fungal leaf pathogens, cell expansion and stomata development. Plant Signal Behav 4:109–110

Pease JE, Murphy PM (1998) Microbial corruption of the chemokine system: an expanding paradigm. Sem Immunol 10:169–178

Perfus-Barbeoch L, Jones AM, Assmann SM (2004) Plant heterotrimeric G protein function: insights from *Arabidopsis* and rice mutants. Curr Opin Plant Biol 7:719–731

Piffanelli P, Ramsay L, Waugh R, Benabdelmouna A, D'Hont A, Hollricher K, Jorgensen JH, Schulze-Lefert P, Panstruga R (2004) A barley cultivation-associated polymorphism conveys resistance to powdery mildew. Nature 430:887–891

Plakidou-Dymock S, Dymock D, Hooley R (1998) A higher plant seven-transmembrane receptor that influences sensitivity to cytokinins. Curr Biol 8:315–324

Ron M, Avni A (2004) The receptor for the fungal elicitor ethylene-inducing xylanase is a member of a resistance-like gene family in tomato. Plant Cell 16:1604–1615

Schaffrath U, Scheinpflug H, Reisener HJ (1995) An elicitor from *Pyricularia oryzae* induces resistance responses in rice: isolation, characterization and physiological properties. Physiol Mol Plant Pathol 46:293–307

Schiene K, Pühler A, Niehaus K (2000) Transgenic tobacco plants that express an antisense construct derived from a *Medicago sativa* cDNA encoding a Rac-related small GTP-binding protein fail to develop necrotic lesions upon elicitor infiltration. Mol Genet 263:761–770

Schultheiss H, Dechert C, Kogel K-H, Hückelhoven R (2002) A Small GTP-binding host protein is required for entry of powdery mildew fungus into epidermal cells of barley. Plant Physiol 128:1447–1454

Schultheiss H, Dechert C, Kogel K-H, Hückelhoven R (2003) Functional analysis of barley RAC/ ROP G-protein family members in susceptibility to the powdery mildew fungus. Plant J 36:589–601

Schultheiss H, Hensel G, Imani J, Broeders S, Kumlehn J, Kogel K-H, Sonnewald U, Hückelhoven R (2005) Ectopic expression of constitutively activated HvRACB in barley enhances susceptibility to powdery mildew and abiotic stress. Plant Physiol 139:353–362

Schultheiss H, Preuss J, Pircher T, Eichmann R, Hückelhoven R (2008) Barley HvRIC171 interacts with HvRACB in planta and supports entry of the powdery mildew fungus. Cell Microbiol 10:1815–1826

Schwessinger B, Zipfel C (2008) News from the frontline: recent insights into PAMP-triggered immunity in plants. Curr Opin Plant Biol 11:389–395

Sorek N, Poraty L, Sternberg H, Bar E, Lewinsohn E, Yalovsky S (2007) Activation status-coupled transient *S* acylation determines membrane partitioning of a plant Rho-related GTPase. Mol Cell Biol 27:2144–2154

Stein M, Dittgen J, Sanchez-Rodriguez C, Hou B-H, Molina A, Schulze-Lefert P, Lipka V, Somerville S (2006) *Arabidopsis* PEN3/PDR8, an ATP binding cassette transporter, contributes to nonhost resistance to inappropriate pathogens that enter by direct penetration. Plant Cell 18:731–746

Suharsono U, Fujisawa Y, Kawasaki T, Iwasaki Y, Satoh H, Shimamoto K (2002) The heterotrimeric G protein alpha subunit acts upstream of the small GTPase Rac in disease resistance of rice. Proc Natl Acad Sci USA 99:13307–13312

Temple BRS, Jones AM (2007) The plant heterotrimeric G-protein complex. Ann Rev Plant Biol 58:249–266

Thao NP, Chen L, Nakashima A, Hara S, Umemura K, Takahashi A, Shirasu K, Kawasaki T, Shimamoto K (2007) RAR1 and HSP90 form a complex with RAC/ROP GTPase and function in innate-immune responses in rice. Plant Cell 19:4035–4045

Trusov Y, Rookes JE, Chakravorty D, Armour D, Schenk PM, Botella JR (2006) Heterotrimeric G proteins facilitate *Arabidopsis* resistance to necrotrophic pathogens and are involved in jasmonate signaling. Plant Physiol 140:210–220

Trusov Y, Rookes JE, Tilbrook K, Chakravorty D, Mason MG, Anderson D, Chen J-G, Jones AM, Botella JR (2007) Heterotrimeric G protein γ-subunits provide functional selectivity in G$\beta\gamma$ dimer signaling in *Arabidopsis*. Plant Cell 19:1235–1250

Tsukada K, Ishizaka M, Fujisawa Y, Iwasaki Y, Yamaguchi T, Minami E, Shibuya N (2002) Rice receptor for chitin oligosaccharide elicitor does not couple to heterotrimeric G-protein: elicitor responses of suspension cultured rice cells from Daikoku dwarf (d1) mutants lacking a functional G-protein alpha-subunit. Physiol Plant 116:373–382

Ullah H, Chen J-G, Temple B, Boyes DC, Alonso JM, Davis KR, Ecker JR, Jones AM (2003) The beta-subunit of the *Arabidopsis* G protein negatively regulates auxin-induced cell division and affects multiple developmental processes. Plant Cell 15:393–409

Underwood W, Somerville SC (2008) Focal accumulation of defences at sites of fungal pathogen attack. J Exp Bot 59:3501–3508

Winge P, Brembu T, Kristensen R, Bones AM (2000) Genetic structure and evolution of RAC-GTPases in *Arabidopsis thaliana*. Genetics 156:1959–1971

Wong HL, Sakamoto T, Kawasaki T, Umemura K, Shimamoto K (2004) Down-regulation of metallothionein, a reactive oxygen scavenger, by the small GTPase OsRac1 in rice. Plant Physiol 135:1447–1456

Yalovsky S, Bloch D, Sorek N, Kost B (2008) Regulation of membrane trafficking, cytoskeleton dynamics, and cell polarity by ROP/RAC GTPases. Plant Physiol 147:1527–1543

Zhang W, He SY, Assmann SM (2008) The plant innate immunity response in stomatal guard cells invokes G-protein-dependent ion channel regulation. Plant J 56:984–996

Zipfel C (2008) Pattern-recognition receptors in plant innate immunity. Curr Opin Immunol 20:10–16

Zipfel C, Kunze G, Chinchilla D, Caniard A, Jones JDG, Boller T, Felix G (2006) Perception of the bacterial PAMP EF-Tu by the receptor EFR restricts Agrobacterium-mediated transformation. Cell 125:749–760

G Proteins and Plant Innate Immunity

Yuri Trusov, Lucía Jordá, Antonio Molina, and Jose Ramon Botella

Abstract Under the peaceful appearance of lovely green meadow, the different plant communities are engaged in a continuous struggle for life. Plants use every imaginable mechanism to enhance their defenses in order to survive attacks from an enormous number of pathogens. Plant innate immunity strongly relies on signal transduction pathways connecting pathogen recognition with the establishment of specific defense responses. Heterotrimeric and small GTP-binding proteins provide such signaling between plasma membrane receptors and cytoplasm localized effector molecules. Recent studies, mostly in *Arabidopsis* and rice, have revealed very important roles for G proteins in plant resistance to fungal pathogens. Experimental evidence implicating G proteins in plant innate immunity and putative signaling mechanisms is presented and discussed in this chapter.

1 Introduction

Members of the superfamily of GTP hydrolyzing proteins (G proteins) are present in most living organisms. Their functions are extremely divergent as are the cellular processes in which they are involved. In this chapter, we review the involvement of G proteins in plant innate immunity.

The connection between G proteins and disease resistance was first established in medical research. Defects in G proteins or their associated receptors leading to dysfunctional signal transduction pathways result in an impressive variety of

Y. Trusov and J. Ramon Botella (✉)
Plant Genetic Engineering Laboratory, School of Biological Sciences, University of Queensland, Brisbane, 4072, Australia

L. Jordá and A. Molina
Centro de Biotecnología y Genómica de Plantas (UPM-INIA), E.T.S. Ingenieros Agronomos, Universidad Politécnica de Madrid, Campus Montegancedo, E-28223-Pozuelo de Alarcón (Madrid), Spain

S. Yalovsky et al. (eds.), *Integrated G Proteins Signaling in Plants*, Signaling and Communication in Plants,
DOI 10.1007/978-3-642-03524-1_12,

diseases in humans (Farfel et al. 1999; Spiegel 1996). Specific mutations in Gα_s, an alpha subunit of the heterotrimeric G proteins, are established molecular basis for Albright hereditary osteodystrophy, acromegaly, and hyperfunctional thyroid nodules (Spiegel 1996). Cholera is a devastating disease caused by the bacterium *Vibrio cholera*. An enterotoxin secreted by the harmful strains of the bacteria, known as cholera toxin (CTX), is responsible for the infectious gastroenteritis characteristic of cholera infection (Sharp and Hynie 1971). CTX exerts its action by ribosylating Gα subunits and hence locking the G proteins in a permanently active state. Other medical conditions caused by the malfunction of G-protein-mediated pathways are color blindness, mental retardation, cancer, familial male precocious puberty, Nephrogenic diabetes insipidus, and congenital bleeding (Bos 1989, Farfel et al. 1999; Ropers and Hamel 2005; Spiegel 1996).

2 Heterotrimeric G Proteins and Plant Innate Immunity

Plants have faced their arch nemeses – plant pathogens for millions of years. It is impossible to trace when this war begun, but it is certainly far from over. Plants constantly develop new defense mechanisms while pathogens continuously search for ways to overcome the plant defense. As a result, innate immunity of modern plants possesses an enormous arsenal of defense pathways including multiple receptors recognizing bacterial and fungal elicitors, signal transduction pathways conveying information to transcription machinery, and a multitude of protein and metabolic compounds providing an appropriate response to the attacker. The quicker the response, the more chances to survive. Heterotrimeric GTP-binding proteins are well known in animal systems as one of the fastest signal transducing elements and soon after their discovery in plants (Ma et al. 1990) their involvement in plant innate immunity was also studied (Legendre et al. 1992).

2.1 Pharmacological Studies

Initially, the absence of G proteins mutants prevented the use of genetic approaches to study heterotrimeric G proteins in plants. However, a broad assortment of compounds able to modulate heterotrimeric G protein activity was well known from animal and medical studies. Pharmacological approaches, modulating heterotrimeric G protein activity with different chemical and biochemical agents had been widely used in animal systems to study their function. The most common G protein modulators used in such experiments were mastoporan, cholera toxin, pertussis toxin, GTPγS, and suramin. Mastoparan is a 14-amino acid peptide present in wasp venom. When bound to the phospholipid membrane, mastoparan mimics two basic intracellular loops of the G protein-coupled receptors; it interacts with and activates the Gα subunit (Higashijima et al. 1988). Cholera toxin is a protein complex secreted by the bacterium *V. cholerae*. It modifies the Gα subunit by disabling its intrinsic GTPase activity, and thus, locks Gα in a constant active state. Pertussis toxin is a protein produced by the human pathogenic bacterium *Bordetella pertussis*. It inactivates G proteins by catalyzing the ADP-ribosylation of the Gα subunit (Burns 1988). GTPγS is a nonhydrolysable GTP analog, once bound to Gα GTPγS retains it in the active state (Seifert et al. 1986). Suramin, polysulfonated naphthylurea, prevents binding between Gα and GTP, thus keeping the heterotrimer in the inactive form (Beindl et al. 1996).

2.1.1 Modulation of G Proteins Activity Results in Altered Resistance to Various Pathogens and/or Deregulation of Defense Pathways

The first report describing the possible involvement of G proteins in plant defense dates back to 1992 (Legendre et al. 1992). The authors studied a specific defense related process – the oxidative burst. In general, the defense- or pathogenesis-related oxidative burst is a part of the plant response to a pathogen invasion manifested by the rapid release of reactive oxygen species such as peroxides, superoxide, and free radicals. Upon infection, bacterial and fungal pathogens produce cell wall degrading enzymes to gain access inside the host cells. Degradation of cell wall pectin with pectinases results in the production of polygalacturonic acid which acts as a signal of pathogen intrusion. It has been shown that application of polygalacturonic acid almost linearly correlates with production of hydrogen peroxide molecules in cultured soybean cells. Using purified polygalacturonic acid and elicitors extracted directly from the pathogenic fungus *Verticillium dahliae* (Legendre et al. 1992) demonstrated that antigen-Gα-binding fragments were able to facilitate hydrogen peroxide production. In concert with this observation, the G protein activator mastoparan induced defense-related oxidative burst even without elicitor stimulation. On the contrary, other G protein modulators such as pertussis and cholera toxins had very small or no effect on the induced hydrogen peroxide production (Legendre et al. 1992). Follow-up studies by the same group established that phosphaditylinositol-directed phospholipase C (PI-PLC) might be involved in the defense pathway connecting G proteins and peroxide production (Legendre et al. 1993a, b). Importantly, most members of the PI–PLC family were shown to be direct downstream effectors of G proteins in animals (Suh et al. 2008).

A number of pharmacological studies by different research groups have repeatedly confirmed the involvement of G-proteins in plant defense (Beffa et al. 1995; Beindl et al. 1996; Han and Yuan 2004; Higashijima et al. 1988; Mahady et al. 1998; Perekhod et al. 1998; Rajasekhar et al. 1999; Roos et al. 1999; Vera-Estrella et al. 1994b). Collectively, most of those experiments were organized as follows: a pathogen or an elicitor/s was introduced to a plant or cultured plant cells either pretreated or not pretreated with a particular pharmacological agent; thereafter, the effect of the agent on the specific process was estimated and interpreted. These experiments encompass a wide range of plant species, numerous pathogens and elicitors, and a number of important biological processes such as activation/

inactivation of specific elements in signaling and biosynthetic pathways, production of reactive oxygen species and phytoalexins, and direct disease progress evaluation. It has been suggested that G proteins could be involved in the activation of plasma membrane-localized H^+-ATPase, Ca^{2+}-permeable channels, and redox reactions in tomato plants treated with elicitors from *Cladosporium fulvum* (Gelli et al. 1997; Vera-Estrella et al. 1994a, b). Transgenic tobacco plants expressing cholera toxin show accumulation of salicylic acid resulting in upregulation of PR genes and increased resistance to *Pseudomonas tabaci* (Beffa et al. 1995). Using pharmacological agents, G proteins have been linked to resistance to *Phytophthora infestans* in potato (Perekhod et al. 1998), *Pseudomonas syringae*-induced oxidative burst in soybean cell cultures (Rajasekhar et al. 1999), stimulation of phospholipase A (PLA) activity in response to yeast elicitors in Californian poppy (Roos et al. 1999), production of phytoalexin 6-methoxymellein in carrot cell cultures treated with oligogalacturonide elicitor (Kurosaki et al. 2001), generation of active oxidative species induced by shear stress in suspension cultures of *Taxus cuspidata* (Han and Yuan 2004), fungal-induced benzophenanthridine alkaloid biosynthesis in *Sanguinaria canadensis* suspension cell cultures (Mahady et al. 1998), production of phytoalexin scoparone as part of the hypersensitive response against *Alternaria alternata* in lemon (Ortega et al. 2002), and mediation of jasmonic acid pathway leading to biosynthesis of the phytoalexin, β-thujaplicin, in *Cupressus lusitanica* cell cultures (Zhao and Sakai 2003).

2.1.2 Limits of the Pharmacological Approach in Plants

The unquestionable advantage of the pharmacological approach in heterotrimeric G protein research is that a modulating agent can be easily applied to any plant species of interest and the effect can be readily analyzed. However, it also has substantial pitfalls. First, although most of the G protein modulators are directed to the Gα subunit, they inescapably affect the Gβγ dimer. In plants, just like in animals, Gα and Gβγ act as two functional subunits transmitting signals to their distinct pathways (Chen et al. 2006; Joo et al. 2005; Trusov et al. 2008; Ullah et al. 2003). As a consequence, it is hard, if not impossible, to interpret the specific involvement of each G protein subunit in an especific process. Second, application of a modulator to an entire plant cannot be uniform and even for all cells, for that reason most of the pharmacological studies were carried out on suspension cell cultures. Therefore, the role of G protein in processes such as plant development could not be studied. Finally, despite the wide use and acceptance of the pharmacological agents, their specificity for modulation of canonical heterotrimeric G-proteins in plants has been questioned (Fujisawa et al. 2001; Miles et al. 2004). A demonstrative example is the amphiphilic tetradecapeptide mastoparan, used as a G-protein activator in many of the studies mentioned above. In animals, mastoparan is capable of direct stimulation of the Gα-subunit, which in turn activates multiple downstream effectors, including mitogen-activated protein kinases (MAPK). However, in plants, it has been shown that mastoparan-mediated induction of MAPK signaling does not require the participation of either the Gα or Gβ subunits of the plant heterotrimeric G-proteins (Miles et al. 2004). Thus, final conclusions about the involvement of heterotrimeric G-proteins in specific process should be entirely based on pharmacological studies, but need to be supported by comprehensive genetic analyzes.

2.2 *Genetic Studies*

The production and discovery of mutants lacking G protein subunits in rice and *Arabidopsis* have provided an invaluable tool for plant heterotrimeric G protein research. The study of G protein mutants has provided conclusive evidence of the involvement of heterotrimeric G proteins in plant development, physiology, and defense.

2.2.1 The Gα Subunit Facilitates Resistance to Pathogens in Rice

The first heterotrimeric G protein mutant was found in rice. The Daikoku *d1* mutant is a dwarf that produces small seeds (Ashikari et al. 1999). Several *d1* alleles were shown to be defective in the only canonical Gα subunit present in rice (*RGA1*) (Ashikari et al. 1999). The involvement of *RGA1* in plant defense system and the establishment of the corresponding signaling pathway was elegantly documented by the Shimamoto's group (Lieberherr et al. 2005; Suharsono et al. 2002). As a plant–pathogen model, the authors used the rice–*Magnaporthe grisea* interaction. This fungal pathogen is responsible for rice blast – one of the most important rice diseases worldwide. The authors found that *M. grisea* infection as well as treatment with sphingolipid elicitors extracted from *M. grisea* membranes caused alterations in *RGA1* transcription. At an early stage, both avirulent and virulent races of the fungus suppressed *RGA1* expression. Later, however, from day 1 to day 3 after infection, *RGA1* transcript levels increased when using the avirulent race, while the virulent race kept *RGA1* expression suppressed. When transgenic plants expressing the β galacturonidase (GUS) gene under the control of the native *RGA1* promoter were subjected to the same treatments, the increased expression was localized at the infection spots.

To test if Gα has a role in resistance against avirulent races of this pathogen, four Daikoku *d1* alleles were exposed to avirulent *M. grisea* and the intensity of hypersensitive response quantified. All the four *d1* mutants showed greatly reduced hypersensitive response, indicating reduced resistance to the avirulent race of rice blast. In agreement with these observations, the induction of pathogenesis related genes *PR1* and *PBZ1* in response to *M. grisea* infection was significantly delayed in all *d1* mutants compared with wild-type control (Suharsono et al. 2002).

Further analysis of the rice Gα subunit-mediated defense signaling was performed in suspension cell cultures generated from embryo-derived calli from four *d1* mutant alleles (Suharsono et al. 2002). It was shown that production of hydrogen peroxide in response to *M. grisea* sphingolipid elicitors was dramatically decreased or even abolished in *d1* cell cultures, depending on the allele tested. Moreover, induction of the *PBZ1* gene in response to the elicitors did not occur in any of the *d1* cell cultures.

Once Gα's involvement in plant defense and peroxide production was established, it was important to investigate the signal transduction pathway used. OsRac1, a rice homolog of the small GTPase Rac, plays an important regulatory role in the production reactive oxygen species and disease resistance in rice (Kawasaki et al. 1999). To test if heterotrimeric G proteins and OsRac1 operate in the same pathway, three independent *d1* mutants were transformed with a constitutively active form of OsRac1 (Suharsono et al. 2002). Cell cultures were subsequently generated and hydrogen peroxide production as well as *PBZ1* expression in response to elicitors treatment were analyzed. Interestingly, sphingolipid elicitors-induced peroxide production was restored in all three *d1* mutants expressing active OsRac1 to the same levels than the WT plants expressing active OsRac1, suggesting that OsRac1 acts downstream of Gα in the peroxide production pathway (Suharsono et al. 2002). Very different observations were reported when *PBZ1* expression was studied. While this gene was completely suppressed in *d1* mutants regardless of induction, in transgenic *d1* expressing active OsRac1 the *PBZ1* mRNA was clearly detectable even without induction. However, upon sphingolipid elicitors treatment, *PBZ1* was induced in WT and in WT expressing active OsRac1, but remained in steady state level in *d1* mutants expressing active OsRac1. The authors, thus, hypothesized that Gα could control *PBZ1* induction independently from OsRac1 (Suharsono et al. 2002). Importantly, the transgenic *d1* plants expressing active OsRac1 were fully resistant to avirulent races of *M. grisea*, but they were still dwarf and set small round seeds. This observation provides further evidence that Gα is involved in different pathways and OsRac1 does not operates in all of them.

At least three other downstream elements playing a part in defense-related Gα–OsRac1 signaling in rice have been reported: a mitogen-activated protein kinase, OsMAPK6 (Lieberherr et al. 2005); cinnamoyl-CoA reductase 1 (OsCCR1), an enzyme involved in lignin biosynthesis (Kawasaki et al. 2006); and metallothionein (OsMT2b) acting as ROS scavenger (Wong et al. 2004). Homologs of OsMAPK6 in *Arabidopsis*, tomato, and tobacco have a well-recorded involvement in plant innate immunity (Mayrose et al. 2004; Menke et al. 2004; Yang et al. 2001; Zhang and Liu 2001). In WT plants, OsMAPK6 is posttranslationally activated in response to pathogen attack or elicitor application (Lieberherr et al. 2005). In *d1* mutants and in transgenic rice plants with OsRac1 silenced by RNAi, OsMAPK6 protein levels were substantially reduced. Moreover, coimmunoprecipitation experiments showed that OsMAPK6 assembles a complex with active OsRac1, but not with the inactive form of OsRac1. Nevertheless, yeast two-hybrid assays failed to detect direct interaction between these proteins, suggesting the involvement of other proteins in the complex (Lieberherr et al. 2005). Direct interaction between OsRac1 and OsCCR1 has been demonstrated. This interaction led to the enzymatic activation of OsCCR1, which resulted in stimulation of lignin biosynthesis. The increase in lignin production strengthens cell walls, providing a mechanical barrier for the fungus (Kawasaki et al. 2006). OsRac1 suppresses the expression of metallothionein (OsMT2b), a potent ROS scavenger, thus increasing the ROS pool in the infected cells (Wong et al. 2004).

The above mentioned studies have demonstrated the complexity of the Gα-mediated defense signaling with bifurcating pathways. However, since there are no Gβ or Gγ mutants available in rice, a complete picture of the role of heterotrimeric G proteins in rice innate immunity is still missing.

2.2.2 Gβγ1 Signaling is Involved in *Arabidopsis* Defense Against Necrotrophic Fungi

Arabidopsis boasts with a complete set of mutants lacking each of the heterotrimeric G protein subunits, allowing a comprehensive analysis of their role in a particular biological process. Similarly to rice, *Arabidopsis* has a single canonical Gα subunit gene, GPA1 (Ma 2001; Ma et al. 1990). The amino acid sequences of the rice RGA1 and the *Arabidopsis* GPA1 are approximately 80% identical. Such high level of similarity would usually imply functional resemblance of the proteins. However, a mere look at rice *d1* (RGA1 deficient) "dwarves" and *Arabidopsis gpa1* (GPA1 deficient) "giants" with statistically significantly enlarged rosettes contradict such hypothesis. Furthermore, *gpa1* mutants displayed increased resistance to fungal pathogens compared with WT controls (Llorente et al. 2005; Trusov et al. 2006; Trusov et al. 2007). Importantly, mutants lacking the single canonical Gβ subunit (*agb1* mutants) showed dramatically increased susceptibility to these pathogens. Analysis of double mutants lacking both Gα and Gβ subunits revealed that Gβ-mediated signaling plays the leading role in *Arabidopsis* defense, while the observed Gα effect on resistance is probably due to its ability to impound Gβ into the inactive heterotrimer (Trusov et al. 2006). This does not necessarily preclude participation of the Gα subunit in defense role in *Arabidopsis*, especially considering that Gα is required for the late, cell death-associated oxidative burst (Joo et al. 2005).

The *Arabidopsis* genome contains only one gene encoding a canonical Gβ subunit, *AGB1* (Weiss et al. 1994). Two Gβ-deficient mutants have been identified (Lease et al. 2001; Ullah et al. 2003). Genetic studies implicating Gβ signaling in plant innate immunity were recently reported by two independent groups (Llorente et al. 2005; Trusov et al. 2006). Both reports described straightforward experiments in which disease progression was quantified in WT control plants and *agb1* mutants after pathogen infection. Collectively, five fungal species *Botrytis cinerea, Fusarium oxysporum, Plectosphaerella cucumerina, Alternaria brassicicola*, and *Peronospora parasitica* and one bacterium *P. syringae* were tested. Disease symptoms were significantly more severe in *agb1* mutants infected with the necrotrophic *B. cinerea, P. cucumerina, A. brassicicola* and the vascular fungus *F. oxysporum*. In contrast, no differences between WT and mutants were observed for the oomycete *P. parasitica* and the bacterium *P. syringae*, even though in both cases virulent and avirulent races were tested. Expression analysis of pathogenesis-related genes revealed that induction of *PDF1.2*, *OPR3*, and *PAD3* in response to *A. brassicicola* and methyl jasmonate (MeJA) treatments was significantly delayed (Trusov et al. 2006). At the same time, expression of the salicylic acid dependent gene *PR1* was increased upon *P. cucumerina* infection (Llorente et al. 2005). These observations, together with the necrotrophic nature of the fungi studied, suggest a possible interaction between G proteins and the JA/ethylene pathways. Indeed, MeJA-dependent inhibition of seed germination, postgermination development and root elongation were, to some extent, weakened in the *agb1* mutants compared with WT, suggesting a role for Gβ signaling promoting several JA-mediated processes (Trusov et al. 2006). On the other hand, it was found that *agb1* mutants were incapable of accumulating callose – a high-molecular weight β-1,3-glucan depositing in plant tissues challenged with a pathogen. Interestingly, the callose deposition was only arrested in the *agb1* mutants in response to necrotrophic pathogen *P. cucumerina*, while the oomycete *P. parasitica* induced callose deposition similarly in WT and *agb1* mutants. This indicates that the *agb1* mutation affected a rather specific signaling pathway triggering callose deposition, but not its biosynthesis pathway (Llorente et al. 2005).

The *Arabidopsis* genome contains two genes encoding Gγ subunits, *AGG1* and *AGG2* (Mason and Botella 2000, 2001). Gβ and Gγ subunits form a tightly bound dimer that acts as a single functional unit and can only be dissociated using denaturing conditions (Gautam et al. 1998). Despite possessing a similar domain structure, AGG1 and AGG2 are quite different in primary sequence (only about 55% identity), tissue-specific expression patterns (Trusov et al. 2007, 2008), and in their ability to target the plasma membrane (Adjobo-Hermans et al. 2006; Zeng et al. 2007), and consequently they mediate different cellular processes (Trusov et al. 2007, 2008). Infection with *F. oxysporum* or *A. brassicicola* induced *AGG1*, but not *AGG2* gene expression (Trusov et al. 2007). Consequently, when mutants lacking Gγ1 (*agg1*), Gγ2 (*agg2*), or both proteins simultaneously (*agg1 agg2*) were infected with either *F. oxysporum* or *A. brassicicola*, they showed very different behaviors. When plant decay and rosette growth inhibition for *F. oxysporum* inoculation and lesion expansion for *A. brassicicola* infection were monitored, *agg1* and double *agg1 agg2* mutants were more susceptible to pathogen attack than WT plants, while *agg2* mutants showed WT behavior. These results suggested that Gβγ1 is the specific subunit involved in plant defense (Trusov et al. 2007). However, subsequent large-scale analysis using a more reliable quantitative Fusarium wilt monitoring technique – counts of yellow leaves per plant, revealed that both Gγ subunits additively contribute to the resistance. Interestingly, at an early stage of symptom development (5–9 days) both *agg1* and *agg2* mutants displayed similar levels of susceptibility – somewhat intermediate between levels of *agb1* or double *agg1 agg2* mutants and WT. In contrast, at a later infection stage (10 days to plant decay), *agg1* mutants accumulated yellow leaves significantly faster than *agg2* mutants, resulting in similar values to *agb1* and double *agg1*/agg2 mutants at the end of the experiment, while *agg2* mutants ended up with values similar to WT (Trusov and Botella unpublished data).

The presented data, from both pharmacological and genetic studies, provide unequivocal evidence of heterotrimeric G proteins involvement in plant innate immunity. It is also clear that distinct roles of the subunits can vary depending on the plant species and the pathogen studied. It is, therefore, not possible to predict with certainty the role which G proteins will play in a particular plant–pathogen interaction.

3 Heterotrimeric G Protein and Plant Cell Death

In multicellular organisms, programmed cell death (PCD) plays an essential role in the control of several biological processes, including immunity, development, and senescence (Patel et al. 2006). In plants, the most studied form of PCD is the localized cell death associated with the hypersensitive response (HR), which is a constituent of a significant number of immune responses against different type of pathogens. Although the biological relevance of HR-PCD is not fully understood, its timely activation is thought to be essential for containment of the pathogen at the penetration site (Patel et al. 2006). Several studies have implicated a number of signaling molecules in HR-PCD. Among these molecules are nitric oxide (NO), salicylic acid (SA), and reactive oxygen species (ROS) (Torres and Dangl 2005).

The production of ROS is one of the earliest events in the plant immune response. It occurs shortly after pathogen recognition and involves the transient increase of, predominantly, superoxide (O_2^-) and hydrogen peroxide (H_2O_2) in the apoplast leading to initiation of the oxidative burst. ROS were initially thought to function just as executioners of PCD. However, a broader role of ROS in cell signalling and regulation of gene expression in response to biotic and abiotic stresses and to developmental cues has emerged recently. ROS production can be also elicited by phytohormones such as auxin or ABA (Joo et al. 2001; Pei et al. 2000; Schopfer et al. 2002; Zhang et al. 2001), and by a large variety of abiotic stresses, including drought, high temperatures, high light intensity and ozone (Joo et al. 2005; Overmyer et al. 2003). Moreover, ROS can be produced inside the plant cell, in chloroplasts, mitochondria, and peroxisomes as by-products of processes such as photosynthesis and respiration (Apel and Hirt 2004). The molecular basis of these novel ROS functions is still poorly understood.

Heterotrimeric G proteins have been involved in the regulation of cell death and ROS generation during defense responses in rice (Suharsono et al. 2002). In *Arabidopsis*, G proteins were shown to control ROS production stimulated by extracellular calmodulin in guard cells (Chen et al. 2004) and upon plant exposure to ozone (Joo et al. 2005). Together, these studies suggest that heterotrimeric G proteins might be a common component of ROS and PCD signaling in response to different stresses or developmental cues.

3.1 Function of G Protein in Activation of Oxidative Stress

ROS were initially proposed to orchestrate the establishment of plant defense response and the HR-PCD associated to pathogen recognition, and latter on have been implicated in the regulation of molecular responses associated with some abiotic stresses. The production of ROS in these processes has been explained by several biochemical mechanisms. In apoplast, H_2O_2 can be produced by the coordinated action of cell-wall peroxidases, amino oxidases, or germin-like oxalate oxidases (Apel and Hirt 2004). Also, ROS can be generated by plant NADPH oxidases upon pathogen challenge or ABA signaling during stomata closure. Plant NADPH oxidases are similar to those producing superoxide during the respiratory burst in mammalian phagocytes (Groom et al. 1996). The mammalian NADPH oxidases are also known as respiratory burst oxidases (RBOs), a protein complex comprised of a membrane-bound NADPH-binding flavocytochrome b558 and cytosolic regulatory proteins. The activity of mammalian NADPH oxidases is regulated by additional proteins, such as RAC GTPases. Plant NADPH oxidase homologs were first identified in rice and named respiratory burst oxidase homologs (Rboh) (Groom et al. 1996). In *Arabidopsis*, there are ten *AtRboh* genes, and among them *AtRbohC, D*, and *F* are the best characterized (Torres and Dangl 2005).

The first evidence of the putative involvement of heterotrimeric G proteins in the regulation of ROS production and cell death came from the demonstration that the Dikoku *dwarf1* (*d1*) rice mutant, lacking a functional Gα subunit, showed a significant reduction in HR response upon infection with avirulent races of the blast fungus *Magnaporthe grisea* (Suharsono et al. 2002). This attenuation of HR in *d1* was accompained by 24-h delayed induction of *PR* genes expression and compromised resistance to blast fungus compared to wild-type plants. The H_2O_2 production was also compromised in *d1* cell cultures upon treatment with elicitors, such as sphingolipid derivatives. This phenotype was restored to wild-type levels by expression of the constitutively active OsRAC1 (CA-OsRAC1), a ROP protein (see Chapter "ROP GTPases and the Cytoskeleton" and Section "Small G proteins in plant innate immunity"), which can induce H_2O_2 production in the absence of pathogen infection or an elicitor treatment. These results indicated that *d1*-enhanced susceptibility was due to a H_2O_2 deficient production and that OsRAC1 function downstream from Gα in the rice heterotrimeric G-protein signaling pathway (Suharsono et al. 2002). In *Arabidopsis*, production of ROS upon pathogen infection does not seem to be altered in mutants defective in heterotrimeric G-protein subunits. For example, the *Arabidopsis* Gβ-deficient mutant *agb1* shows enhanced susceptibility to necrotrophic fungi but the production of ROS upon inoculation with *P. cucumerina* does not differ from that of wild-type plants (Llorente et al. 2005). However, a detailed characterization of ROS production in heterotrimeric G protein mutants upon infection with different types of pathogens is required to determine the function of the GPA1, AGB1, and AGG1/AGG2 subunits in the regulation of ROS production and signaling in plant innate immunity.

Ozone (O_3) causes oxidative stress in living organisms and is a major pollutant affecting plants. Plant exposure to ozone (O_3) induces a bimodal oxidative burst that has some similarities to the pathogen-induced oxidative burst during plant immune response (Overmyer et al. 2003). In *Arabidopsis*, ozone exposure leads to production of ROS and activation of ROS-mediated signaling. Interestingly, heterotrimeric G proteins control this ROS production and signaling, as well as the sensitivity to ozone. Gα- and Gβ-deficient mutants are, respectively, less and more sensitive to O_3 damage than wild-type plants. In *gpa1* and *agb1* mutants, the first peak of the oxidative burst elicited by O_3 is missing, whereas the late peak is missing in *gpa1*, but not in *agb1* mutants. ROS production is primarily detectable in chloroplast of leaf epidermal guard cells, and expands to adjacent cells by the action of the membrane-bound AtRbohD and AtRbohF NADPH oxidases. The ROS produced by the action of these NADPH oxidases act on guard cell plasma membrane calcium channels, stimulate ROS production in adjacent cells, and influence intercellular ROS signaling (Kwak et al. 2003). ROS signaling from the chloroplast is central to the O_3-induced oxidative stress response, as it has been shown to mediate light and signaling (Chang et al. 2004; Fryer et al. 2003, 2002). AtRbohD and AtRbohF also participate in the O_3-induced oxidative stress response that leads to intercellular signaling and ultimate to PCD. Interestingly, the observed cell death- associated with ozone exposure requires only the Gα protein (Joo et al. 2005). These results indicate that *Arabidopsis* heterotrimeric G proteins act both synergistically and separately in activating different intracellular ROS-generating systems.

Heterotrimeric G proteins have also been implicated in the regulation of ROS production through activation of extracellular calmodulin (ExtCaM) during stomatal closure (Chen et al. 2004). ExtCaM activation triggers a significant increase in NO levels associated with stomatal closure in wild-type plants. This effect is abolished in the *Arabidopsis atnoa1* (nitric oxide associated 1) mutant, which is defective in NO production. Interestingly, ExtCAM-mediated production of NO is regulated by GPA1, as it is blocked in *gpa1* mutants and enhanced by overexpression of a constitutive active form of GPA1 (cGα) (Li et al. 2009). In line with these results, the defect in NO production in *gpa1* was rescued by overexpression of *AtNOA1*. G protein-mediated activation of NO production depends on H_2O_2 accumulation, which is impaired in the *atrbohD atrbohF* double mutant. Previous results demonstrated that stomatal closure and plasma membrane calcium channel activation are reduced in *atrbohD atrbohF* double mutants, but can be restored by treatment with H_2O_2, indicating that ROS serve as second messengers in ABA signaling (Kwak et al. 2003). The data discussed above suggest the existence of a signaling pathway leading to ExtCaM-induced stomatal closure, which involves GPA1-dependent activation of H_2O_2 production and subsequent AtNOA1-dependent NO accumulation (Li et al. 2009). This signaling pathway provides an additional link between G proteins and Rboh NADPH oxidases that deserves an additional investigation.

3.2 Regulation of Cell Death by G Proteins

After O_3 stress, the PCD signaling system is intact in Gβ-deficient mutants. The increased resistance to O_3 observed in the *gpa1-4* mutant is consistent with the suggested role for the Gα subunit in activating the membrane-bound NADPH oxidases to produce damaging levels of ROS (Suharsono et al. 2002). The association of cell death with ROS production by membrane-bound NADPH oxidases is further supported by the observation that mutants lacking either the AtrbohD, or the AtrbohF, or both NADPH oxidase proteins were deficient in the late oxidative burst and were more resistant to O_3 than wild-type plants (Joo et al. 2005).

Another oxidative stress that leads to cell death is produced when hypocotyls of far red (FR) grown seedlings are exposed to white light. ROS are again critical mediators of the cell death. This type of cell death is phytochrome A-mediated but heterotrimeric G proteins are also involved (Wei et al. 2008). The *gpa1* mutant plants exhibit reduced cell death, while the *agb1* mutant shows enhanced death, indicating that GPA1 and AGB1 play antagonistic roles in the phyA-mediated cell death pathway.

Heterotrimeric G proteins have been recently involved in the cell death process associated with the unfolded protein response (UPR) in *Arabidopsis* (Wang et al. 2007). The UPR is a protective response that takes place in the cells whose protein folding and modification machineries at the endoplasmic reticulum (ER) are disrupted. This complex response is quite well characterized in yeast and mammalian cells, where transcriptional and translational changes occur to promote protein folding and disposal of misfolded proteins. When these mechanisms fail, the UPR activates apoptotic cell death (Schroder and Kaufman 2005). In plants, mutations that affected ER chaperone and secretory protein genes also compromised defense responses to pathogen attack, suggesting that the folding and secretory machinery are critical for plant immunity (Bilgin et al. 2003; Jelitto-Van Dooren et al. 1999; Wang et al. 2005). Interestingly, Gβ-deficient mutants are more resistant than wild-type plants to the cell death induced by the antibiotic tunicamycin (Tm), an inhibitor of the *N*-linked protein glycosylation (Wang et al. 2007). Moreover, the majority of Gβ protein localizes to the ER, cofractionates with ER lumenal chaperone protein and is degraded during the UPR, whereas Gα is not (Wang et al. 2007). It can be concluded that the Gβγ dimer mediates cell death signaling in the UPR and that ER stress-related proteins are probable downstream targets of Gβγ signaling. These observations have lead to the hypothesis that heterotrimeric G signaling is not exclusively confined to the plasma membrane but can also occur in cellular compartments, where hormone receptors have been identified (Wang et al. 2007). Further research is needed to establish the role of ER in heterotrimeric G protein signaling.

A putative connection between the plant heterotrimeric G protein and PCD through the regulation of Ca^{2+} ion channel remains to be elucidated. In mammals, Gβγ complex is known to interact with calcium channels, and calcium signaling plays a pivotal role in plant stress responses (Klusener et al. 2002). Interestingly, it has been suggested that the chloroplast-localized Ca^{2+} ion channel PPF1 functions as an inhibitor of programmed cell death in apical meristems (Li et al. 2004). Moreover, as indicated above, plasma membrane calcium channel activation are reduced in *atrbohD atrbohF* double mutants, but can be restored by treatment with H_2O_2 (Kwak et al. 2003). Together these data suggest a putative connection between calcium channels, ROS production, and PCD that could be regulated by G proteins through Rboh oxidases, but this model needs further characterization.

4 Innate Immunity Networks Regulated by Heterotrimeric G Proteins

Plant innate immunity encompasses an enormous variety of molecules conferring resistance to a not lesser variety of plant pathogens. These molecules are organized in an intricate network of pathways starting from pathogen recognition, following activation of signal transduction, leading to transcription induction, protein synthesis, and metabolite production, and eventually resulting in formation of a protective response. Heterotrimeric G proteins are apparently involved in the signal transduction part of the response. It is generally accepted that most of the plant defense responses are governed by several parallel signaling pathways. The most important of them are regulated by low-molecular weight signaling molecules, such as SA, jasmonic acid (JA), and ethylene (ET) (Ausubel 2005; Kachroo and Kachroo 2007). Brassinosteroids, and more recently abscisic acid (ABA) and auxin, have also been added to this list of immune signaling molecules (Adie et al. 2007; Asselbergh et al. 2008; Hernandez-Blanco et al. 2007; Llorente et al. 2008; Navarro et al. 2006; Robert-Seilaniantz et al. 2007; Wang et al. 2007). These distinct parallel pathways nevertheless interact or intersect at certain crucial regulatory steps (Kunkel and Brooks 2002; Pieterse and Van Loon 2004). Importantly, either acting independently or as a part of the network, these signaling pathways contribute differently to the resistance against a particular pathogen. It has been hypothesized that fungal life style could be used as a predictor of the induced plant defense response (Dangl and Jones 2001; Mcdowell and Dangl 2000). For instance, plants usually activate JA/ET-mediated signaling pathways in response to necrotrophic fungi, while biotrophs and bacteria generally provoke SA signaling (Glazebrook 2005; Schenk et al. 2000; Thomma et al. 1998, 1999, 2001b; Van Wees et al. 2003). The activation of an appropriate signaling pathway with external application of the hormones prior to an infection is an effective method to protect plants from pathogens. On the other hand, mutations disrupting a particular pathway can compromise resistance to the corresponding group of pathogens, while they may have no effect or even a beneficial effect on resistance against a different group of pathogens (Grant and Lamb 2006; Kunkel and Brooks 2002; Maleck and Dietrich 1999; Schenk et al. 2003; Takahashi et al. 2004; Thomma et al. 2001a). However, exceptions to this general rule have been increasingly reported over the last few years, indicating the complexity of the innate immunity network controlling plant resistance to a particular type of pathogen (Berrocal-Lobo and Molina 2008; Edgar et al. 2006; Glazebrook 2005; Thatcher et al. 2009; Van Wees et al. 2003). The effectiveness of a signaling pathway for plant protection also depends on the especific growth conditions and inoculation technique used (Trusov and Botella, unpublished). Therefore, it is important to keep in mind that the specific inputs from the pathways involved in resistance for a specific plant–pathogen system may vary between reported experimental data.

Heterotrimeric G proteins have been implicated in plant resistance to necrotrophic and vascular fungi (Llorente et al. 2005; Trusov et al. 2006), but their place and interactions in the innate immunity network remains largely unknown. Recently, an attempt has been made to establish relations between Gβ defense signaling and SA-, JA-, ET-, and ABA-mediated pathways (Trusov et al. 2009). It has been shown that mutants lacking the Gβ subunits, *agb1-2*, were extremely susceptible to *F. oxysporum* compared with wild type, but displayed only mildly decreased resistance to *A. brassicicola*. On the other hand, when a number of mutants defective in SA-, JA-, ET-, and ABA-mediated signaling pathways were tested against these pathogens in the same conditions, a range of responses was found as follows. Mutants deficient in SA biosynthesis (*NahG* expressing transgenic plants and *eds5-1* mutant) were susceptible to *F. oxysporum* at levels similar to *agb1-2* mutants, while *npr1-2* and *eds1-22* mutants impaired in SA signaling were similar to wild-type control. Neither of these SA-related mutations had a measurable effect on resistance against *A. brassicicola*. On the contrary, mutants deficient in ABA (*aba1-6*), insensitive to ethylene (*ein2-1*) or JA (*coi1-21* and *jin1-9*) were more resistant to *F. oxysporum* compared with wild-type plants; at the same time, when challenged with *A. brassicicola aba1-6* and *coi1-21*, but not *ein2-1* or *jin1-9*, displayed high levels of susceptibility. Interestingly, another mutant impaired in JA response, *jar1-1*, was to some extent more susceptible to *F. oxysporum* than wild type and fully resistant to *A. brassicicola*, showing an opposite behavior compared to *coi1-21* and suggesting that JA mediate at least two different pathways involved in plant defense (Trusov et al. 2009). Pretreatment of wild-type *Arabidopsis* with SA, ethylene, methyl jasmonate, or ABA and subsequent resistance analysis confirmed the genetic results obtained for *F. oxysporum*. The effect of hormonal pretreatment on *A. brassicicola* infection was not tested. To establish whether G proteins act as a part of these pathways or interact with their components, the *agb1-2* mutant was crossed with all mutants mentioned above. The study of double mutants is proven to be useful for establishing genetic interactions between two elements in a particular response. If two elements are agonistically involved in the same pathway, the double mutant should behave like one of the single mutants, while if they control different mechanisms, the double mutant should display an enhanced phenotype compared with the single mutants, if such an increase is possible. This last scenario was observed for *agb1NahG*, *agb1eds5*, and *agb1jar1* when infected with *F. oxysporum*. Similarly, the double mutants *agb1coi1* and *agb1aba1* showed additive effect after infection with

A. brassicicola, indicating that Gβ subunit signaling is independent of the tested pathways. The double mutant approach is also useful to analyze the interaction between elements involved in the same process in antagonistic manner. In this case one mutation should suppress the effect of the other. Since *agb1-2* influenced resistance to *F. oxysporum* in an opposite way than *coi1-21, jin1-9, ein2-1*, and *aba1-6*, the suppression effect was analyzed on the corresponding double mutants infected with *F. oxysporum.* It was found that the *agb1coi1, agb1aba1, agb1jin1* and *agb1ein2* double mutants had intermediate resistance levels (quantified as number of yellow leaves) between the *agb1-2* and all the single mutants, suggesting that Gβ acts independently from the pathways in which these proteins are involved. On the other hand, when percentage of decayed plants was counted, it was observed that *agb1coi1* and *agb1jin1* plants eventually survived the infection, while *agb1aba1* and *agb1ein2* perished. Development of yellow leaves' counts and percentage of decayed plants are two methods to evaluate disease progression in plants infected with *F. oxysporum.* Both parameters usually show a good correlation, but it should be noted that yellow leaves' counts reflect early stages of infection, while plant decay indicates the final outcome of the disease. The described results, therefore, suggested that Gβ might act upstream of COI1 and ATMYC2 at the later stages of infection. It has been recently established that COI1 is a receptor for several JA derivatives (Katsir et al. 2008), which activate the transcription factor ATMYC2 which in turn regulates expression of several pathogenesis-related genes. ATMYC2 plays a central role in switching between drought- and pathogenesis-related signals (Anderson et al. 2004; Lorenzo et al. 2004). *ATMYC2* expression is originally increased in response to *F. oxysporum* intrusion, but then quickly returns to the steady-state level (Trusov et al. 2009) or could even be suppressed (Anderson et al. 2004). In the *agb1-2* mutant, *ATMYC2* expression was significantly higher compared with that of wild-type levels during infection, but not in uninfected plants, suggesting that Gβ signaling suppresses *ATMYC2* during *F. oxysporum* attack. Expression analysis of the pathogenesis-related genes *PDF1.2, CHITINASE*, and *PR4* in single *agb1-2, jin1-9*, and double *agb1jin1* mutants revealed a complex interaction between Gβ and ATMYC2 during *F. oxysporum* infection. It was observed that Gβ and ATMYC2 suppressed *CHITINASE* independently, while the positive effect of Gβ signaling on *PDF1.2* and *PR4* is due to Gβ suppression of ATMYC2 (Trusov et al. 2009). Taken together, these observations point out that Gβ signaling apparently intersect with one of the JA-mediated pathways at ATMYC2; it is possible that this interaction accounts for part of the Gβ immune response. However, the nature of the other part of the Gβ-facilitated defense remains unknown.

The next logical question to ask is how plasma membrane-localized heterotrimeric G proteins regulate gene expression in the nucleus? One candidate to fill this gap is the oxidative burst-induced signaling. Involvement of oxidative burst in plant innate immunity has been established, although it is not fully understood (Berrocal-Lobo and Molina 2008; Galletti et al. 2008; Lamb and Dixon 1997; Torres and Dangl 2005). It has been demonstrated that Gβ signaling is required for an early component of the oxidative burst triggered by ozone (Joo et al. 2005).

On the other hand, accumulation of ROS caused by *P. cucumerina* inoculation was independent of Gβ signaling (Llorente et al. 2005). To reconcile these contradicting facts, one may assume that during pathogen attack, as opposite to ozone treatment, Gβ signaling and oxidative burst interact downstream of ROS production. Such scenario converges a possible G protein signaling pathways with pathogen-induced ROS at oxidative signal inducible kinase OXI1 has been proposed (Anthony et al. 2004, 2006). The protein kinase OXI1 (also known as AGC2-1) was independently discovered and studied by two groups (Anthony et al. 2004; Rentel et al. 2004). It was shown that expression of *OXI1* gene is strongly induced by hydrogen peroxide, cellulose, the fungus *P. parasitica*, and wounding (Rentel et al. 2004). At the same time, activation of OXI1 kinase activity depends on binding to activated 3-phosphoinositide-dependent protein kinase PDK1 (Anthony et al. 2004). Activation of PDK1 occurs upon binding to phosphatidic acid (PA), which relies on phospholipases C (PLC) and D (PLD) for its production. Importantly, there is pharmacological evidence implicating G proteins in PA production (Munnik et al. 1995). Moreover, it was observed that the heterotrimeric G proteins agonist mastoparan activates OXI1 in a PDK1-dependent manner (Anthony et al. 2004). Further analysis revealed that activated OXI1 directly binds and phosphorylates serine/threonine protein kinase PTI1 (Anthony et al. 2006). PTI1 is a homolog of a tomato PTI1 kinase implicated in hypersensitive response and resistance to bacterial speck disease (Sessa et al. 1998; Zhou et al. 1995). On the other hand, OXI1 is required for full activation of the MAPK MPK3 and MPK6 (Rentel et al. 2004). Involvement of MPK3 and MPK6 in plant innate immunity is well documented (Asai et al. 2002; Menke et al. 2004; Ren et al. 2008). Finally, the potato ortholog of *Arabidopsis* MPK6, StMPK1, has been shown to phosphorylate a number of nuclear proteins (Katou et al. 2005), including two jasmonate ZIM-domain (JAZ) proteins with closest homology to tomato and *Arabidopsis* JAZ3 and tobacco JAZ2 (Katsir et al. 2008; Shoji et al. 2008) and a transcription factor highly similar to *Arabidopsis* WRKY33. Importantly, JAZ proteins, and JAZ3 in particular, were shown to be a missing link between COI1 and ATMYC2 (Chini et al. 2007). During JA signaling, the activated receptor COI1 as a part of SCF^{COI1} complex targets JAZ3 for degradation; this in turn releases ATMYC2 from the JAZ3–ATMYC2 complex allowing transcription of the corresponding genes (Chini et al. 2007). Phosphorylation of the JAZ proteins has not yet been studied, but it was shown for other proteins targeted by cullin-containing E3 ubiquitin ligases similar to SCF^{COI1} that phosphorylation effectively prevented their degradation (Schwechheimer and Calderon Villalobos 2004).

The following scenario connecting all the steps could be envisioned. Pathogen intrusion activates yet-to-discover G protein coupled receptors, resulting in release of the Gβγ1 dimer, which induces PLC/PLD dependent PA production and activation of PDK1. Simultaneous and independent ROS accumulation also takes place. PDK1 and ROS synergistically activate OXI1. Activated OXI1 binds to and phosporylates PTI1 and through its activation or by yet another way activates MPK6. Upon activation, MPK6 moves into the nucleus and phosphorilates JAZ proteins, protecting them from degradation and hence preventing ATMYC2 transcriptional activity. Bear in mind that ATMYC2 plays a negative role in resistance to *F. oxysporum*, and covers only a part of the Gβ defense response (Trusov et al. 2009). On the other hand, MPK6 could possibly phosphorylate WRKY33, as well stabilizing it. WRKY33 plays a positive role in resistance to necrotrophic fungi (Zheng et al. 2006), including *F. oxysporum* (Trusov and Botella, unpublished data) and could be accountable for the second part of the Gβ-mediated defense response. Despite the existence of indirect evidence supporting this hypothesis, the direct test is yet to be done for all components in a single plant–pathogen model. Also, pharmacological data involving Gβ signaling in PA production and subsequent activation of the serine/threonine kinases should be confirmed by genetic analysis.

A potential interconnection between heterotrimeric G proteins and the ERECTA receptor-like kinase (RLK) has been also described (Lease et al. 2001; Llorente et al. 2005). RLKs are the predominant cell surface receptors in plants (Shiu et al. 2004) and are involved in perception of pathogen-associated molecular patterns (PAMPs) and activation of innate immunity signaling cascades (Chinchilla et al. 2007; Kemmerling et al. 2007). *Arabidopsis* mutants lacking the Gβ subunit of the heterotrimeric G protein exhibit an erecta-like developmental phenotype similar to plants impaired in the ERECTA RLK (*er*), suggesting that they may form a part of the same pathway (Lease et al. 2001; Llorente et al. 2005). In line with this hypothesis, *er* and *agb1* mutants showed an enhanced susceptibility to the necrotrophic fungus *P. cucumerina* (Llorente et al. 2005). In contrast, no enhanced susceptibility to *F. oxysporum* and *B. cinerea* was observed in *er* mutants, whereas *agb1* mutants showed an enhanced susceptibility to these fungal pathogens (Llorente et al. 2005). Interestingly, ROP GTPases have been shown to interact with several RLKs (see "Small G proteins in plant innate immunity"). Based on these published data, potential direct or indirect (through ROP proteins) interactions of G proteins with different RLKs can be hypothesized (Fig. 1).

5 Small G Proteins in Plant Innate Immunity

ROP (from Rho-related GTPases from plants) proteins are plant-specific small GTPases that act as signaling transducers connecting extracellular inputs/stimuli to the cell (see Chapter "ROP GTPases and the Cytoskeleton"). These monomers are activated by a wide range of upstream signals that trigger the conversion of the guanosine diphosphate (GDP)-bound inactive form of ROP to the active guanosine triphosphate (GTP)-bound form. This active form is able to interact with one or more specific downstream effector proteins and therefore exert its biological function. Activation can be catalyzed by guanine nucleotide exchange factors (GEFs) (see chapter "Structure and Function of ROPs and Their GEFs"), while GTPase activating proteins (GAPs) promote GTP cleavage to return the switch back to the inactive GDP state and guanine nucleotide dissociation inhibitors (GDIs) prevent the activation process (see chapter "Regulatory and Cellular

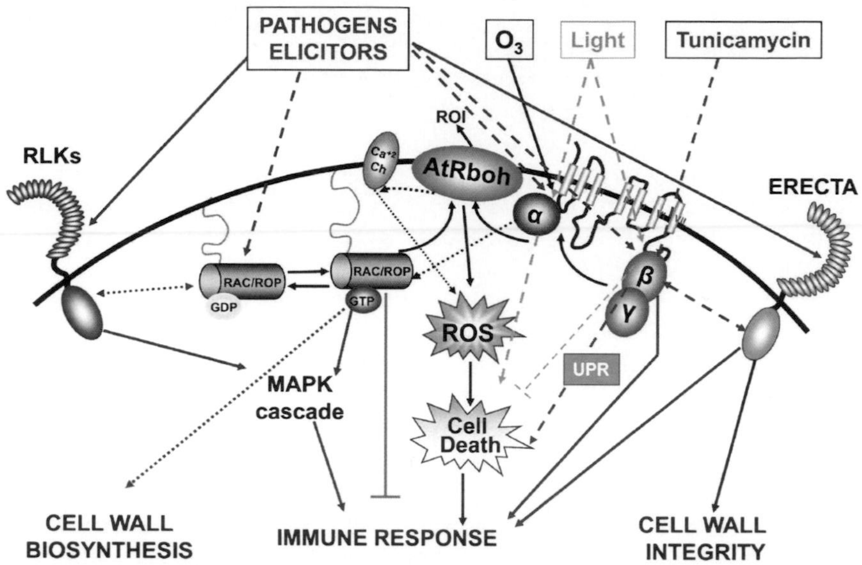

Fig. 1 *G and ROP proteins regulatory network controlling the production of ROS and the activation of plant cell death in response to different external treatments.* The different external stimuli/treatments that have been described to induce the production of ROS or/and the development of cell death through the activation/inactivation of G and ROP proteins are indicated in different colors. In some cases, immune responses have been associated to ROS production and cell death activation. *Dotted arrows* indicate biochemical/genetic interactions or regulatory effects not fully demonstrated or just suggested by the published data. *Ca^{+2} Ch* calcium channel. See the text for more details.

Functions of Plant RhoGAPs and RhoGDIs"). Since the first plant cDNA encoding a Rho-related GTPase (Rho1Ps) from pea was identified in 1993 (Yang and Watson 1993), a large number of Rho GTPases have been identified in mosses and higher plants (Delmer et al. 1995; Li et al. 1998; Winge et al. 1997; Yang and Watson 1993). The *Arabidopsis* genome encodes a gene family of 11 ROP Small GTPases, while rice comprise seven members and barley appears to express six ROP genes (Nibau et al. 2006; Schultheiss et al. 2008).

ROPs are the key switches that regulate many pathways important for development, such as cell polarity establishment and cell growth and differentiation, but also some are required for plant immunity (Yang and Fu 2007; Zheng and Yang 2000). The involvement of ROPs in plant defense responses was first demonstrated by the extensive work done on rice OsRAC1 (Kawasaki et al. 1999; Ono et al. 2001; Suharsono et al. 2002). Latter on, the studies performed on species as diverse as soybean, tobacco, maize, barley, and *Arabidopsis* have corroborated the function of ROPs in plant innate immunity (Agrawal et al. 2003).

In rice, constitutively active forms of OsRac1 (CA-OsRac1) activate H_2O_2 production and induces spontaneous PCD in leaves. The HR-like responses triggered by the overexpression of CA-OsRac1 also induced an enhanced resistance against a virulent race of rice blast fungus (*Magnaporthe grisea*, race 007) and the bacterial blight (*Xanthomonas oryzae* pv. *oryzae*, race 1). This resistance was also associated to the higher accumulation of the phytoalexin momilactone A, and altered gene expression of some defense/stress-related genes, such as the *D9* and *POX22.3* encoding, respectively, a terpenoid cyclase and a peroxidase. By contrast, the dominant negative form of OsRac1 (DN-OsRac1) inhibited H_2O_2 production and lesion formation induced by the avirulent race of blast fungus (*M. Grisea*, race 031). (Kawasaki et al. 1999). CA-OsRac1-induced H_2O_2 production was inhibited by the NADPH oxidase inhibitor diphenylene iodonium (DPI). Similar results were seen when mutant forms of cotton ROP (GhRac13) and human HsRac1 were manipulated, respectively, in the heterologous *Arabidopsis* and soybean cultured cells (Agrawal et al. 2003). OsRAC1 also suppresses expression of a metallothionein, a reactive oxygen scavenger, further enhancing the ROS-signaled defense response (Wong et al. 2004). In mammals, Rac2 GTPase interacts with the p67phox regulatory subunit of NADPH oxidase; however, plants, apparently, do not contain p67phox and p47phox regulatory subunits, and just the p91phox homologues are present in plants (Sagi and Fluhr 2001). Interestingly, it has been suggested that OsRAC1 activates the NADPH-mediated production of H_2O_2 by directly binding to the N-terminus of Rboh protein (the catalytic subunit of NADPH oxidase) (Wong et al. 2007). From the large series of studies performed in rice, it can be concluded that OsRAC1 activates defense responses in rice acting as a positive regulator of the NADPH oxidase complex, transiently stimulating ROS production, and enhancing pathogen resistance (Kawasaki et al. 1999; Ono et al. 2001).

Recent studies support the proposed function of plant ROPs in mediating plant immunity, mainly through modulating ROS production and accumulation, PCD, and defense signalling cascades (Jung et al. 2006; Molendijk et al. 2008; Opalski et al. 2005; Pathuri et al. 2008; Schultheiss et al. 2008; Thao et al. 2007). Tobacco NtRAC5 downregulates NADPH oxidase in tobacco cells in response to elicitors (Morel et al. 2004). Also, tobacco transgenic plants expressing heterologous *Medicago sativa*, *MsRac1* gene, results in development of necrotic lesions and cell death, while antisense expression have no effect or any other visible defense reaction (Schiene et al. 2000). In tomato, a Rac2-specific antibody detects a 21-kDa tomato protein that could be translocated to microsomal membranes in response to elicitor treatments (Agrawal et al. 2003).

A connection between ROPs and G proteins in the regulation of pant innate immunity has been established. As explained before, the rice *d1* mutant impaired in the Gα subunits of the heterotrimeric G protein exhibited reduced resistance to *M. grisea* and H_2O_2 production, and delayed *PR* gene expression, that were restored to wild-type phenotype by the constitutive overexpression in *d1* plants and cell cultures of CA-OsRac1 (Suharsono et al. 2002). All these data indicated that OsRac1 operates downstream of Gα as a positive regulator for defense pathway (Suharsono et al. 2002). It is believed that pathogen/elicitor-derived signals are likely to be received by as yet unknown receptor(s), and transmitted to OsRac1 through Gα. OsRac1 could also regulate defense response by additional molecular mechanisms. Thus, a recent study has revealed that OsRac1 forms a complex with RAR1, HSP90, and HSP70 in vivo, and coordinates the activity of these key components of the plant innate immunity (Thao et al. 2007). The putative connection of these immunity regulators with Gα subunit is currently unknown.

In contrast to rice OsRac1 that is functioning in the defensive responses to the hemibiotrophic fungus *Magnaporthe grisea*, barley HvRacB appears to be involved in susceptibility to the biotrophic fungi *Blumeria graminis* f. sp. *hordei* (*Bgh*). Transient knocking-down of barley *HvRacB* gene renders epidermal cells more resistant to penetration by *Bgh*, indicating that HvRacB negatively regulates the defense pathway (Schultheiss et al. 2008). Barley susceptibility to *Bgh* is mediated by the host seven transmembrane domain receptor MLO protein, as mutants that do not express functional MLO are completely resistant to penetration by *Bgh,* and by *ROR1* and *ROR2* genes (Assaad et al. 2004; Bhat et al. 2005; Collins et al. 2003). The role of HvRacB on susceptibility is dependent on functional *MLO* and *ROR1* genes. The function of HvRacB appears to be linked to the actin cytoskeleton, as overactivation of HvRacB in single-cell hampers actin-filament reorganization in *Bgh* attacked cells, while lack of HvRacB promotes polarization (Schultheiss et al. 2008). HvRacB interacts *in planta* with RIC171, a protein whose overexpression leads to more susceptible epidermal cells. Additionally, RIC171 accumulates at sites of fungal attack, suggesting enhanced ROP activity at sites of attempted fungal penetration (Schultheiss et al. 2008). The accumulation of ROP at the infection sites has also been observed in parsley cells infected with the oomycete *P. infestans,* exactly to the membrane subtending the oomycetous appressorium, where the actin cables focus (Schutz et al. 2006). The accumulation of ROP small GTPases at the infection sites can be explained because, some pathogens direct the plant to follow a tip growth mimicking program by hijacking host signalling via ROP and RIC proteins for accommodation of the fungal haustorium (Schultheiss et al. 2008). Recent studies have showed that constitutively activated expression of other barley ROP members of the family, such as HvRac1 and HvRac3, also lead to enhanced susceptibility to *Bgh*. However, from these barley ROPs, only HvRac1 overexpression promoted H_2O_2 accumulation. Interestingly, HvRac1 is also involved in callose deposition and resistance to *Magnaporthe oryzae,* hence, barley ROPs might have overlapping and specific roles in plant–microbe interactions (Pathuri et al. 2008). The isolation of rice closest ortholog to HvRacB and OsRacB confirmed the existence of specific functions of the different members of the small GTPases family, since OsRacB functions as a negative regulator for basal disease resistance (Jung et al. 2006).

The use of yeast two-hybrid strategies has allowed the identification of potential ROP interactors involved in plant defense signaling and responses. These assays have confirmed the existence of cross-talk between ROP and specific receptor-like kinases (RLKs) through direct molecular interaction (Molendijk et al. 2008). Coimmunoprecipitation of ROPs with RLKs, such as the *Arabidopsis* CLAVATA1 (Trotochaud et al. 1999) or the tomato LePRKs (Wengier et al. 2003), have been described. Also, in vitro and in vivo interactions of *Arabidopsis* NRCK, a receptor kinase belonging to the RLCK Class VIII, with AtROP4 and AtROP11 GTPases have been found (Molendijk et al. 2008). RLKs are the predominant cell surface receptors in plants. Receptor-like serine/threonine kinase family encompasses more than 400 members in *Arabidopsis* (Shiu et al. 2004). The role of RLKs in plant immune responses is an exciting field that has attracted attention of many research groups in recent years, as RLKs are involved in perception of pathogen-associated molecular patterns (PAMPs), regulation of PCD upon pathogen infection, and activation of innate immunity signaling cascades (Chinchilla et al. 2007; Kemmerling et al. 2007; Llorente et al. 2005). The interaction of ROPs with RLKs suggests that ROPs could mediate cell surface-associated signal perception and transductions. All ROPs contain two putative serine/threonine phosphorylation sites, SYR and SSK, which might be the targets of the RLKs identified. A connection between RLK ERECTA and heterotrimeric G proteins has been also hypothesized (Lease et al. 2001; Llorente et al. 2005). *Arabidopsis* mutant plants lacking Gβ subunit of the heterotrimeric G protein exhibit an erecta-like developmental phenotype similar to plants impaired in ERECTA RLK, suggesting that they may form a part of the same pathway (Lease et al. 2001; Llorente et al. 2005). In line with this hypothesis, *er* and *agb1* mutants showed an enhanced susceptibility to the necrotrophic fungus *P. cucumerina* (Llorente et al. 2005).

Recent studies have revealed the interaction between the small GTPases and downstream kinase-mediated immune signaling pathways. Silencing of the small GTPase OsRac1 by RNA interference prevents the elicitor-induced accumulation and posttranslational activation of the rice kinase OsMAPK6, although the mechanism of OsRAC1-dependent OsMAPK6 activation is still not known (Lieberherr et al. 2005). As indicated above, AtROP4 interacts, in vitro and *in planta*, with several kinases, including RBK1 and RBK2 – RLCK VIb family members (Molendijk et al. 2008). The expression of *RBK1* gene is upregulated in leaves inoculated with *P. infestans* and *B. cinerea* (Molendijk et al. 2008). Two other plant RLCK VI family members, the *Medicago truncatula* MtRRK1 and MtRRK2, as well as the *Arabidopsis* homolog AtRLCK VI_A2 kinase (AtRRK1), can also directly interact with ROP GTPases in yeast two-hybrid. In vitro kinase activation experiments showed that their kinase activity is specifically and strongly increased in the presence of active GTP-bound, but not of inactive GDP-bound, ROP GTPases, supporting a hypothesis that these kinases are potential downstream ROP GTPase effectors (Dorjgotov et al. 2009). However, further *in planta* experimentation are necessary to confirm these interactions and to verify the possible role in the activation of plant defense responses.

Plant cell wall integrity has emerged in the last years as a key modulator of some defense responses in *Arabidopsis* (Hematy and Hofte 2008; Hernandez-Blanco et al. 2007). Currently, it is not well understood how wall integrity changes are perceived by the plant cells. Like it has been described in yeast, some receptor (e.g., RLKs) or cell wall sensors may perceive the change produced in the wall by environmental stresses, such as pathogen attack (Philip and Levin 2001). Several studies linked ROPs with the biosynthesis of secondary cell walls (Delmer et al. 1995), as well as with some modifications in the cell walls, such as callose and lignin depositions, that take place after pathogen attack. Thus, GTP–AtROP1, but not GDP–AtROP1, interacts with a putative subunit of *Arabidopsis* callose synthase, and OsRac1 stimulates cinnamoyl-CoA reductase 1 (CCR1), a key enzyme required for the biosynthesis of lignin polymers, which guard off pathogens attack (Hong et al. 2001, Kawasaki et al. 2006). Also, OsRac1 coordinately activates the production of two intermediates (monolignol and hydrogen peroxide) required for lignin biosynthesis. Therefore, it cannot be ruled out a possible new interconnection between ROPs function and some plant cell wall mediated-defense responses. Interestingly, a specific function of ERECTA RLK in regulating cell wall-mediated disease resistance has recently been proposed (Sánchez-Rodríguez et al. 2009). Mutations in *ERECTA* gene led to altered cell wall composition and increased disease susceptibility to the necrotrophic pathogen *P. cucumerina*. Mutations suppressing the *erecta* phenotype and thereby restoring the resistance to *P. cucumerina* up to wild-type levels were identified (Sánchez-Rodríguez et al. 2009). The potential interaction among ROPs, RLKs (e.g., ERECTA), and heterotrimeric G proteins in the regulation of cell wall integrity is a research area that deserves additional attention in the future.

Studies on auxin signaling have revealed a previously uncharacterized cellular role of ROPs positively modulating the ubiquitin/26S proteasome-regulated proteolysis (Tao et al. 2005). In addition, ROP GTPases are important negative regulators of abscisic acid (ABA) signaling, and ABA has been shown to inactivate RAC/ROPs interaction (Lemichez et al. 2001; Xin et al. 2005; Zheng et al. 2002). The signaling pathways regulated by these hormones have been recently involved in modulating some plant defense responses to particular pathogens (Hernandez-Blanco et al. 2007; Llorente et al. 2008; Navarro et al. 2006), thus, again it is mandatory to gain a deeper understanding of the functionality of these small GTPases, and to determine the interconnection between their function in hormone regulation and their putative function in resistance to pathogens.

In summary, we can conclude that plant ROPs small GTPases have emerged as relevant signaling modules in plant immunity, as they regulate plant response to cellular oxidative environments that could lead to host PCD. Our current data indicate that ROPs roles in plant immunity responses are specific and different depending on the plant–pathogen interaction analyzed, since ROPs may act as positive or negative regulators of plant immunity. Additionally, the interaction of ROPs with cell surface-associated signal perception apparatus (e.g., RLKs receptors) to perceive a broad range of extracellular stimuli, including pathogen insult, may directly regulate downstream kinase signaling and activation of immune responses. Further experiments are necessary to determine whether these promising interactions between some RLKs/kinases and ROP observed using the in vitro yeast two-hybrid system also occur in vivo, and also to clarify the genetic and biochemical interaction of ROPs with G proteins in plant innate immunity.

6 Future Directions

A plethora of data is now available linking G proteins, small and large, with plant defense. Our present knowledge emphasizes that, although there is bound to be common themes, the roles played by G proteins can vary depending on the specific plant–pathogen system studied. The next challenge will be to establish the signaling pathways used by G proteins to mediate plant innate immunity; some components of these pathways are now emerging but many more need to be identified. Interactions identified using the yeast two hybrid system need to be confirmed *in planta*. Functional studies need to be complemented with genetic studies to determine the position of each component in the overall signal transduction cascade. Arguably, the most important components of these cascades will be the receptors, either seven transmembrane G-protein-coupled receptors, RLKs, or other as yet unidentified receptors.

Acknowledgments We thank Dr. Miguel Angel Torres for helpful discussions. Heterotrimeric G protein work in A. Molina's laboratory is supported by Spanish MEC (Grant BIO2006-00488). G-protein research in J. Botella's laboratory is supported by the Australian Research Council (grant DP0772145).

References

Adjobo-Hermans MJW, Goedhart J, Gadella TWJ Jr (2006) Plant G protein heterotrimers require dual lipidation motifs of Ga and Gg and do not dissociate upon activation. J Cell Sci 119:5087–5097

Adie BAT, Perez-Perez J, Perez-Perez MM, Godoy M, Sanchez-Serrano JJ, Schmelz EA, Solano R (2007) ABA is an essential signal for plant resistance to pathogens affecting JA biosynthesis and the activation of defenses in Arabidopsis. Plant Cell 19:1665–1681

Agrawal GK, Iwahashi H, Rakwal R (2003) Small GTPase 'Rop': molecular switch for plant defense responses. FEBS Letts 546:173–180

Anderson JP, Badruzsaufari E, Schenk PM, Manners JM, Desmond OJ, Ehlert C, Maclean DJ, Ebert PR, Kazan K (2004) Antagonistic interaction between abscisic acid and jasmonate-ethylene signaling pathways modulates defense gene expression and disease resistance in *Arabidopsis*. Plant Cell 16:3460–3479

Anthony RG, Henriques R, Helfer A, Meszaros T, Rios G, Testerink C, Munnik T, Deak M, Koncz C, Bogre L (2004) A protein kinase target of a PDK1 signalling pathway is involved in root hair growth in *Arabidopsis*. EMBO J 23:572–581

Anthony RG, Khan S, Costa J, Pais MS, Bogre L (2006) The *Arabidopsis* protein kinase PTI1–2 is activated by convergent phosphatidic acid and oxidative stress signaling pathways downstream of PDK1 and OXI1. J Biol Chem 281:37536–37546

Apel K, Hirt H (2004) Reactive oxygen species: metabolism, oxidative stress, and signal transduction. Ann Rev Plant Biol 55:373–399

Asai T, Tena G, Plotnikova J, Willmann MR, Chiu WL, Gomez-Gomez L, Boller T, Ausubel FM, Sheen J (2002) MAP kinase signalling cascade in Arabidopsis innate immunity. Nature 415:977–83

Ashikari M, Wu JZ, Yano M, Sasaki T, Yoshimura A (1999) Rice gibberellin-insensitive dwarf mutant gene Dwarf 1 encodes the a-subunit of GTP-binding protein. Proc Nat Acad Sci USA 96:10284–10289

Assaad FF, Qiu JL, Youngs H, Ehrhardt D, Zimmerli L, Kalde M, Wanner G, Peck SC, Edwards H, Ramonell K, Somerville CR, Thordal-Christensen H (2004) The PEN1 syntaxin defines a novel cellular compartment upon fungal attack and is required for the timely assembly of papillae. Mol Biol Cell 15:5118–5129

Asselbergh B, De Vleesschauwer D, Hofte M (2008) Global switches and fine-tuning-ABA modulates plant pathogen defense. Molecular Plant-Microbe Interactions 21:709–719

Ausubel FM (2005) Are innate immune signaling pathways in plants and animals conserved? Nature Immunology 6:973–979

Beffa R, Szell M, Meuwly P, Pay A, Vogeli-Lange R, Metraux J-P, Neuhaus G, Meins F Jr, Nagy F (1995) Cholera toxin elevates pathogen resistance and induces pathogenesis-related gene expression in tobacco. EMBO J 14:5753–5761

Beindl W, Mitterauer T, Hohenegger M, Ijzerman AP, Nanoff C, Freissmuth M (1996) Inhibition of receptor/G protein coupling by suramin analogues. Mol Pharmacol 50:415–423

Berrocal-Lobo M, Molina A (2008) *Arabidopsis* defense response against *Fusarium oxysporum*. Trends Plant Sci 13:145–150

Bhat RA, Miklis M, Schmelzer E, Schulze-Lefert P, Panstruga R (2005) Recruitment and interaction dynamics of plant penetration resistance components in a plasma membrane microdomain. Proc Nat Acad Sci USA 102:3135–3140

Bilgin DD, Liu Y, Schiff M, Dinesh-Kumar SP (2003) P58(IPK), a plant ortholog of double-stranded RNA-dependent protein kinase PKR inhibitor, functions in viral pathogenesis. Dev Cell 4:651–661

Bos JL (1989) RAS oncogenes in human cancer. Canc Res 49:4682–4689

Burns DL (1988) Subunit structure and enzymic activity of pertussis toxin. Microbiol Sci 5:285–287

Chang CC, Ball L, Fryer MJ, Baker NR, Karpinski S, Mullineaux PM (2004) Induction of ASCORBATE PEROXIDASE 2 expression in wounded *Arabidopsis* leaves does not involve known wound-signalling pathways but is associated with changes in photosynthesis. Plant J 38:499–511

Chen J-G, Gao Y, Jones AM (2006) Differential roles of *Arabidopsis* heterotrimeric G-protein subunits in modulating cell division in roots. Plant Physiol 141:887–897

Chen YL, Huang R, Xiao YM, Lu P, Chen J, Wang XC (2004) Extracellular calmodulin-induced stomatal closure is mediated by heterotrimeric G protein and H_2O_2. Plant Physiol 136:4096–4103

Chinchilla D, Zipfel C, Robatzek S, Kemmerling B, Nurnberger T, Jones JD, Felix G, Boller T (2007) A flagellin-induced complex of the receptor FLS2 and BAK1 initiates plant defence. Nature 448:497–500

Chini A, Fonseca S, Fernandez G, Adie B, Chico JM, Lorenzo O, Garcia-Casado G, Lopez-Vidriero I, Lozano FM, Ponce MR, Micol JL, Solano R (2007) The JAZ family of repressors is the missing link in jasmonate signalling. Nature 448:666–671

Collins NC, Thordal-Christensen H, Lipka V, Bau S, Kombrink E, Qiu JL, Huckelhoven R, Stein M, Freialdenhoven A, Somerville SC, Schulze-Lefert P (2003) SNARE-protein-mediated disease resistance at the plant cell wall. Nature 42:973–977

Dangl JL, Jones JDG (2001) Plant pathogens and integrated defence responses to infection. Nature 411:826–833

Delmer DP, Pear JR, Andrawis A, Stalker DM (1995) Genes encoding small GTP-binding proteins analogous to mammalian rac are preferentially expressed in developing cotton fibers. Mol Gen Genet 248:43–51

Dorjgotov D, Jurca ME, Fodor-Dunai C, Szucs A, Otvos K, Klement E, Biro J, Feher A (2009) Plant Rho-type (Rop) GTPase-dependent activation of receptor-like cytoplasmic kinases in vitro. FEBS Letts 583:1175–1182

Edgar CI, Mcgrath KC, Dombrecht B, Manners JM, Maclean DC, Schenk PM, Kazan K (2006) Salicylic acid mediates resistance to the vascular wilt pathogen Fusarium oxysporum in the model host Arabidopsis thaliana. Australasian Plant Pathology 35:581–591

Farfel Z, Bourne HR, Iiri T (1999) The expanding spectrum of G protein diseases. New Engl J Med 340:1012–1020

Fryer MJ, Ball L, Oxborough K, Karpinski S, Mullineaux PM, Baker NR (2003) Control of Ascorbate Peroxidase 2 expression by hydrogen peroxide and leaf water status during excess light stress reveals a functional organisation of *Arabidopsis* leaves. Plant J 33:691–705

Fryer MJ, Oxborough K, Mullineaux PM, Baker NR (2002) Imaging of photo-oxidative stress responses in leaves. J Exp Bot 53:1249–1254

Fujisawa Y, Kato H, Iwasaki Y (2001) Structure and function of heterotrimeric G proteins in plants. Plant Cell Physiol 42:789–794

Galletti R, Denoux C, Gambetta S, Dewdney J, Ausubel FM, De Lorenzo G, Ferrari S (2008) The AtrbohD-mediated oxidative burst elicited by oligogalacturonides in *Arabidopsis* is dispensable for the activation of defense responses effective against *Botrytis cinerea*. Plant Physiol 148:1695–1706

Gautam N, Downes GB, Yan K, Kisselev O (1998) The G-protein bg complex. Cell Signal 10:447–455

Gelli A, Higgins VJ, Blumwald E (1997) Activation of plant plasma membrane Ca^{2+}-permeable channels by race-specific fungal elicitors. Plant Physiol 113:269–279

Glazebrook J (2005) Contrasting mechanisms of defense against biotrophic and necrotrophic pathogens. Annual Review of Phytopathology 43:205–227

Grant M, Lamb C (2006) Systemic immunity. Current Opinion in Plant Biology 9:414–420

Groom QJ, Torres MA, Fordham-Skelton AP, Hammond-Kosack KE, Robinson NJ, Jones JD (1996) rbohA, a rice homologue of the mammalian gp91phox respiratory burst oxidase gene. Plant J 10:515–522

Han RB, Yuan YJ (2004) Oxidative burst in suspension culture of *Taxus cuspidata* induced by a laminar shear stress in short-term. Biotech Prog 20:507–513

Hematy K, Hofte H (2008) Novel receptor kinases involved in growth regulation. Curr Opin Plant Biol 11:321–328

Hernandez-Blanco C, Feng DX, Hu J, Sanchez-Vallet A, Deslandes L, Llorente F, Berrocal-Lobo M, Keller H, Barlet X, Sanchez-Rodriguez C, Anderson LK, Somerville S, Marco Y, Molina A (2007) Impairment of cellulose synthases required for *Arabidopsis* secondary cell wall formation enhances disease resistance. Plant Cell 19:890–903

Higashijima T, Uzu S, Nakajima T, Ross EM (1988) Mastoparan, a peptide toxin from wasp venom, mimics receptors by activating GTP-binding regulatory proteins (G proteins). J Biol Chem 263:6491–6494

Hong Z, Zhang Z, Olson JM, Verma DP (2001) A novel UDP-glucose transferase is part of the callose synthase complex and interacts with phragmoplastin at the forming cell plate. Plant Cell 13:769–779

Jelitto-Van Dooren EP, Vidal S, Denecke J (1999) Anticipating endoplasmic reticulum stress. A novel early response before pathogenesis-related gene induction. Plant Cell 11:1935–1944

Joo J, Wang S, Chen J, Jones A, Fedoroff N (2005) Different signaling and cell death roles of heterotrimeric G protein a and b subunits in the *Arabidopsis* oxidative stress response to ozone. Plant Cell 17.957–970

Joo JH, Bae YS, Lee JS (2001) Role of auxin-induced reactive oxygen species in root gravitropism. Plant Physiol 126:1055–1060

Jung YH, Agrawal GK, Rakwal R, Kim JA, Lee MO, Choi PG, Kim YJ, Kim MJ, Shibato J, Kim SH, Iwahashi H, Jwa NS (2006) Functional characterization of OsRacB GTPase–a potentially negative regulator of basal disease resistance in rice. Plant Physiol Biochem 44:68–77

Kachroo A, Kachroo P (2007) Salicylic acid-, jasmonic acid- and ethylene-mediated regulation of plant defense signaling. Genetic engineering 28:55–83

Katou S, Yoshioka H, Kawakita K, Rowland O, Jones JD, Mori H, Doke N (2005) Involvement of PPS3 phosphorylated by elicitor-responsive mitogen-activated protein kinases in the regulation of plant cell death. Plant Physiology 139:1914–26

Katsir L, Schilmiller AL, Staswick PE, He SY, Howe GA (2008) COI1 is a critical component of a receptor for jasmonate and the bacterial virulence factor coronatine. Proc Natl Acad Sci USA 105:7100–7105

Kawasaki T, Henmi K, Ono E, Hatakeyama S, Iwano M, Satoh H, Shimamoto K (1999) The small GTP-binding protein Rac is a regulator of cell death in plants. Proc Natl Acad Sci USA 96:10922–10926

Kawasaki T, Koita H, Nakatsubo T, Hasegawa K, Wakabayashi K, Takahashi H, Umemura K, Umezawa T, Shimamoto K (2006) Cinnamoyl-CoA reductase, a key enzyme in lignin biosynthesis, is an effector of small GTPase Rac in defense signaling in rice. Proc Natl Acad Sci USA 103: 230–235

Kemmerling B, Schwedt A, Rodriguez P, Mazzotta S, Frank M, Qamar SA, Mengiste T, Betsuyaku S, Parker JE, Mussig C, Thomma BP, Albrecht C, De Vries SC, Hirt H, Nurnberger T (2007) The BRI1-associated kinase 1, BAK1, has a brassinolide-independent role in plant cell-death control. Curr Biol 17:1116–1122

Klusener B, Young JJ, Murata Y, Allen GJ, Mori IC, Hugouvieux V, Schroeder JI (2002) Convergence of calcium signaling pathways of pathogenic elicitors and abscisic acid in *Arabidopsis* guard cells. Plant Physiol 130:2152–2163

Kunkel BN, Brooks DM (2002) Cross talk between signaling pathways in pathogen defense. Curr Opin Plant Biol 5:325–331

Kurosaki F, Yamashita A, Arisawa M (2001) Involvement of GTP-binding protein in the induction of phytoalexin biosynthesis in cultured carrot cells. Plant Sci 161:273–278

Kwak JM, Mori IC, Pei ZM, Leonhardt N, Torres MA, Dangl JL, Bloom RE, Bodde S, Jones JD, Schroeder JI (2003) NADPH oxidase AtrbohD and AtrbohF genes function in ROS-dependent ABA signaling in *Arabidopsis*. EMBO J 22:2623–2633

Lamb C, Dixon RA (1997) The oxidative burst in plant disease resistance. Annu Rev Plant Physiol Plant Mol Biol 48:251–275

Lease KA, Wen J, Li J, Doke JT, Liscum E, Walker JC (2001) A mutant *Arabidopsis* heterotrimeric G-protein b subunit affects leaf, flower, and fruit development. Plant Cell 13:2631–2641

Legendre L, Heinstein P, Low P (1992) Evidence for participation of GTP-binding proteins in elicitation of the rapid oxidative burst in cultured soybean cells. J Biol Chem 267:20140–20147

Legendre L, Rueter S, Heinstein PF, Low PS (1993a) Characterization of the oligogalacturonide-induced oxidative burst in cultured soybean (Glycine max) cells. Plant Physiol 102:233–240

Legendre L, Yueh Y, Crain R, Haddock N, Heinstein P, Low P (1993b) Phospholipase C activation during elicitation of the oxidative burst in cultured plant cells. J Biol Chem 268:24559–24563

Lemichez E, Wu Y, Sanchez JP, Mettouchi A, Mathur J, Chua NH (2001) Inactivation of AtRac1 by abscisic acid is essential for stomatal closure. Genes Dev 15:1808–1816

Li H, Wu G, Ware D, Davis KR, Yang Z (1998) *Arabidopsis* Rho-related GTPases: differential gene expression in pollen and polar localization in fission yeast. Plant Physiol 118:407–417

Li J, Wang DY, Li Q, Xu YJ, Cui KM, Zhu YX (2004) PPF1 inhibits programmed cell death in apical meristems of both G2 pea and transgenic *Arabidopsis* plants possibly by delaying cytosolic Ca^{2+} elevation. Cell Calcium 35:71–77

Li JH, Liu YQ, Lu P, Lin HF, Bai Y, Wang XC, Chen YL (2009) A signaling pathway linking nitric oxide production to heterotrimeric G protein and H_2O_2 regulates extracellular calmodulin induction of stomatal closure in *Arabidopsis*. Plant Physiol 150(1):114–124

Lieberherr D, Thao NP, Nakashima A, Umemura K, Kawasaki T, Shimamoto K (2005) A sphingolipid elicitor-inducible mitogen-activated protein kinase is regulated by the small GTPase OsRac1 and heterotrimeric G-protein in rice. Plant Physiol 138:1644–1652

Llorente F, Alonso-Blanco C, Sanchez-Rodriguez C, Jorda L, Molina A (2005) ERECTA receptor-like kinase and heterotrimeric G protein from *Arabidopsis* are required for resistance to the necrotrophic fungus *Plectosphaerella cucumerina*. Plant J 43:165–180

Llorente F, Muskett P, Sanchez-Vallet A, Lopez G, Ramos B, Sanchez-Rodriguez C, Jorda L, Parker J, Molina A (2008) Repression of the auxin response pathway increases *Arabidopsis* susceptibility to necrotrophic fungi. Mol Plant 1:496–509

Lorenzo O, Chico JM, Sanchez-Serrano JJ, Solano R (2004) Jasmonate-insensitive1 encodes a MYC transcription factor essential to discriminate between different jasmonate-regulated defense responses in *Arabidopsis*. Plant Cell 16:1938–1950

Ma H (2001) Plant G proteins: the different faces of GPA1. Curr Biol 11:R869–R871

Ma H, Yanofsky MF, Meyerowitz EM (1990) Molecular cloning and characterization of GPA1, a G protein a subunit gene from *Arabidopsis thaliana*. Proc Natl Acad Sci USA 87:3821–3825

Mahady GB, Liu C, Beecher CW (1998) Involvement of protein kinase and G proteins in the signal transduction of benzophenanthridine alkaloid biosynthesis. Phytochemistry 48:93–102

Maleck K, Dietrich RA (1999) Defense on multiple fronts: how do plants cope with diverse enemies? Trends In Plant Science 4:215–219

Mason MG, Botella J (2000) Completing the heterotrimer: Isolation and characterization of an *Arabidopsis thaliana* G protein g-subunit cDNA. Proc Natl Acad Sci USA 97:14784–14788

Mason MG, Botella J (2001) Isolation of a novel G-protein g-subunit from *Arabidopsis thaliana* and its interaction with Gb. Biochim Biophys Acta 1520:147–153

Mayrose M, Bonshtien A, Sessa G (2004) LeMPK3 is a mitogen-activated protein kinase with dual specificity induced during tomato defense and wounding responses. J Biol Chem 279:14819–14827

McDowell JM, Dangl JL (2000) Signal transduction in the plant immune response. Trends Biochem Sci 25:79–82

Menke FL, Van Pelt JA, Pieterse CM, Klessig DF (2004) Silencing of the mitogen-activated protein kinase MPK6 compromises disease resistance in *Arabidopsis*. Plant Cell 16:897–907

Miles GP, Samuel MA, Jones AM, Ellis BE (2004) Mastoparan rapidly activates plant MAP kinase signaling independent of heterotrimeric G proteins. Plant Physiol 134:1332–1336

Molendijk AJ, Ruperti B, Singh MK, Dovzhenko A, Ditengou FA, Milia M, Westphal L, Rosahl S, Soellick TR, Uhrig J, Weingarten L, Huber M, Palme K (2008) A cysteine-rich receptor-like kinase NCRK and a pathogen-induced protein kinase RBK1 are Rop GTPase interactors. Plant J 53:909–923

Morel J, Fromentin J, Blein JP, Simon-Plas F, Elmayan T (2004) Rac regulation of NtrbohD, the oxidase responsible for the oxidative burst in elicited tobacco cell. Plant J 37:282–293

Munnik T, Arisz SA, De Vrije T, Musgrave A (1995) G protein activation stimulates phospholipase D signaling in plants. Plant Cell 7:2197–2210

Navarro L, Dunoyer P, Jay F, Arnold B, Dharmasiri N, Estelle M, Voinnet O, Jones JD (2006) A plant miRNA contributes to antibacterial resistance by repressing auxin signaling. Science 312:436–439

Nibau C, Wu HM, Cheung AY (2006) RAC/ROP GTPases: 'hubs' for signal integration and diversification in plants. Trends Plant Sci 11:309–315

Ono E, Wong HL, Kawasaki T, Hasegawa M, Kodama O, Shimamoto K (2001) Essential role of the small GTPase Rac in disease resistance of rice. Proc Natl Acad Sci USA 98:759–764

Opalski KS, Schultheiss H, Kogel KH, Huckelhoven R (2005) The receptor-like MLO protein and the RAC/ROP family G protein RACB modulate actin reorganization in barley attacked by the biotrophic powdery mildew fungus *Blumeria graminis* f.sp. hordei. Plant J 41:291–303

Ortega X, Polanco R, Castaneda P, Perez LM (2002) Signal transduction in lemon seedlings in the hypersensitive response against *Alternaria alternata*: participation of calmodulin, G-protein and protein kinases. Biol Res 35:373–383

Overmyer K, Brosche M, Kangasjarvi J (2003) Reactive oxygen species and hormonal control of cell death. Trends Plant Sci 8:335–342

Patel S, Caplan J, Dinesh-Kumar SP (2006) Autophagy in the control of programmed cell death. Curr Opin Plant Biol 9:391–396

Pathuri IP, Zellerhoff N, Schaffrath U, Hensel G, Kumlehn J, Kogel KH, Eichmann R, Huckelhoven R (2008) Constitutively activated barley ROPs modulate epidermal cell size, defense reactions and interactions with fungal leaf pathogens. Plant Cell Rep 27:1877–1887

Pei ZM, Murata Y, Benning G, Thomine S, Klusener B, Allen GJ, Grill E, Schroeder JI (2000) Calcium channels activated by hydrogen peroxide mediate abscisic acid signalling in guard cells. Nature 406:731–734

Perekhod EA, Chalenko GI, Il'inskaya LI, Vasyukova NI, Gerasimova NG, Babakov AV, Usov AI, Mel'nikova TM, Ozeretskovskaya OL (1998) Modulation of potato resistance by means of xyloglucan fragments. Appl Biochem Microbiol 34:91–96

Philip B, Levin DE (2001) Wsc1 and Mid2 are cell surface sensors for cell wall integrity signaling that act through Rom2, a guanine nucleotide exchange factor for Rho1. Mol Cell Biol 21: 271–280

Pieterse CM, Van Loon LC (2004) NPR1: the spider in the web of induced resistance signaling pathways. Curr Opin Plant Biol 7:456–464

Rajasekhar VK, Lamb C, Dixon RA (1999) Early events in the signal pathway for the oxidative burst in soybean cells exposed to avirulent *Pseudomonas syringae* pv glycinea. Plant Physiol 120:1137–1146

Ren D, Liu Y, Yang KY, Han L, Mao G, Glazebrook J, Zhang S (2008) A fungal-responsive MAPK cascade regulates phytoalexin biosynthesis in Arabidopsis. Proceedings of the National Academy of Sciences of the United States of America 105:5638–5643

Rentel MC, Lecourieux D, Ouaked F, Usher SL, Petersen L, Okamoto H, Knight H, Peck SC, Grierson CS, Hirt H, Knight MR (2004) OXI1 kinase is necessary for oxidative burst-mediated signalling in *Arabidopsis*. Nature 427:858–861

Robert-Seilaniantz A, Navarro L, Bari R, Jones JD (2007) Pathological hormone imbalances. Current Opinion in Plant Biology 10:372–379

Roos W, Dordschbal B, Steighardt J, Hieke M, Weiss D, Saalbach G (1999) A redox-dependent, G-protein-coupled phospholipase A of the plasma membrane is involved in the elicitation of alkaloid biosynthesis in *Eschscholtzia californica*. Biochim Biophys Acta 1448:390–402

Ropers HH, Hamel BCJ (2005) X-linked mental retardation. Nat Rev Genet 6:46–57

Sagi M, Fluhr R (2001) Superoxide production by plant homologues of the gp91(phox) NADPH oxidase. Modulation of activity by calcium and by tobacco mosaic virus infection. Plant Physiol 126:1281–1290

Sánchez-Rodríguez C, Estévez JM, Llorente F, Hernández-Blanco C, Jordá L, Pagán I, Berrocal-Lobo M, Marco Y, Somerville S, Molina A (2009) The ERECTA receptor-like kinase regulates cell wall-mediated resistance to pathogens in *Arabidopsis thaliana*. Mol Plant-Microbe Interact 22(8):953–963

Schenk PM, Kazan K, Wilson I, Anderson JP, Richmond T, Somerville SC, Manners JM (2000) Coordinated plant defense responses in Arabidopsis revealed by microarray analysis. Proceedings of the National Academy of Sciences of the United States of America 97:11655–11660

Schenk PM, Kazan K, Manners JM, Anderson JP, Simpson RS, Wilson IW, Somerville SC, Maclean DJ (2003) Systemic gene expression in Arabidopsis during an incompatible interaction with Alternaria brassicicola. Plant Physiology 132:999–1010

Schiene K, Puhler A, Niehaus K (2000) Transgenic tobacco plants that express an antisense construct derived from a *Medicago sativa* cDNA encoding a Rac-related small GTP-binding protein fail to develop necrotic lesions upon elicitor infiltration. Mol Gen Genet 263:761–770

Schopfer P, Liszkay A, Bechtold M, Frahry G, Wagner A (2002) Evidence that hydroxyl radicals mediate auxin-induced extension growth. Planta 214:821–828

Schroder M, Kaufman RJ (2005) The mammalian unfolded protein response. Ann Rev Biochem 74:739–789

Schultheiss H, Preuss J, Pircher T, Eichmann R, Huckelhoven R (2008) Barley RIC171 interacts with RACB in planta and supports entry of the powdery mildew fungus. Cell Microbiol 10:1815–1826

Schutz I, Gus-Mayer S, Schmelzer E (2006) Profilin and Rop GTPases are localized at infection sites of plant cells. Protoplasma 227:229–235

Schwechheimer C, Calderon Villalobos LI (2004) Cullin-containing E3 ubiquitin ligases in plant development. Curr Opin Plant Biol 7:677–686

Seifert R, Rosenthal W, Schultz G (1986) Guanine nucleotides stimulate NADPH oxidase in membranes of human neutrophils. FEBS Letts 205:161–165

Sessa G, D'ascenzo M, Loh YT, Martin GB (1998) Biochemical properties of two protein kinases involved in disease resistance signaling in tomato. J Biol Chem 273:15860–15865

Sharp GW, Hynie S (1971) Stimulation of intestinal adenyl cyclase by cholera toxin. Nature 229:266–269

Shiu SH, Karlowski WM, Pan R, Tzeng YH, Mayer KF, Li WH (2004) Comparative analysis of the receptor-like kinase family in *Arabidopsis* and rice. Plant Cell 16:1220–1234

Shoji T, Ogawa T, Hashimoto T (2008) Jasmonate-induced nicotine formation in tobacco is mediated by tobacco COI1 and JAZ genes. Plant Cell Physiol 49:1003–12

Spiegel AM (1996) Defects in G protein-coupled signal transduction in human disease. Ann Rev Physiol 58:143–170

Suh PG, Park JI, Manzoli L, Cocco L, Peak JC, Katan M, Fukami K, Kataoka T, Yun S, Ryu SH (2008) Multiple roles of phosphoinositide-specific phospholipase C isozymes. BMB Rep 41:415–434

Suharsono U, Fujisawa Y, Kawasaki T, Iwasaki Y, Satoh H, Shimamoto K (2002) The heterotrimeric G protein alpha subunit acts upstream of the small GTPase Rac in disease resistance of rice. Proc Natl Acad Sci USA 99:13307–13312

Tao LZ, Cheung AY, Nibau C, Wu HM (2005) RAC GTPases in tobacco and *Arabidopsis* mediate auxin-induced formation of proteolytically active nuclear protein bodies that contain AUX/IAA proteins. Plant Cell 17:2369–2383

Takahashi H, Kanayama Y, Zheng MS, Kusano T, Hase S, Ikegami M, Shah J (2004) Antagonistic interactions between the SA and JA signaling pathways in Arabidopsis modulate expression of defense genes and gene-for-gene resistance to cucumber mosaic virus. Plant Cell Physiol 45:803–809

Thao NP, Chen L, Nakashima A, Hara S, Umemura K, Takahashi A, Shirasu K, Kawasaki T, Shimamoto K (2007) RAR1 and HSP90 form a complex with Rac/Rop GTPase and function in innate-immune responses in rice. Plant Cell 19:4035–4045

Thatcher LF, Manners JM, Kazan K (2009) Fusarium oxysporum hijacks COI1-mediated jasmonate signaling to promote disease development in Arabidopsis. Plant J 58:927–939

Thomma BPHJ, Eggermont K, A., PIAM, Mauch-Mani B, Vogelsang R, A., CBP, Broekaert WF (1998) Separate jasmonate-dependent and salicylate-dependent defense response pathways in Arabidopsis are essential for resistance to distinct microbial pathogens. Proceedings of the National Academy of Sciences of the United States of America 95:15107–15111

Thomma BPHJ, Eggermont K, Tierens KFMJ, Broekaert WF (1999) Requirement of functional ethylene-insensitive 2 gene for efficient resistance of Arabidopsis to infection by Botrytis cinerea. Plant Physiology 121:1093–1101

Thomma BPHJ, Penninckx IAMA, Cammue BPA, Broekaert WF (2001a) The complexity of disease signaling in Arabidopsis. Current Opinion in Immunology 13:63–68

Thomma BPHJ, Tierens KFM, Penninckx IAMA, Mauch-Mani B, Broekaert WF, Cammue BPA (2001b) Different micro-organisms differentially induce Arabidopsis disease response pathways. Plant Physiology and Biochemistry 39:673–680

Torres MA, Dangl JL (2005) Functions of the respiratory burst oxidase in biotic interactions, abiotic stress and development. Curr Opin Plant Biol 8:397–403

Trotochaud AE, Hao T, Wu G, Yang Z, Clark SE (1999) The CLAVATA1 receptor-like kinase requires CLAVATA3 for its assembly into a signaling complex that includes KAPP and a Rho-related protein. Plant Cell 11:393–406

Trusov Y, Rookes JE, Chakravorty D, Armour D, Schenk PM, Botella JR (2006) Heterotrimeric G-proteins facilitate *Arabidopsis* resistance to necrotrophic pathogens and are involved in jasmonate signaling. Plant Physiol 140:210–220

Trusov Y, Rookes JE, Tilbrook K, Chakravorty D, Mason MG, Anderson D, Chen JG, Jones AM, Botella JR (2007) Heterotrimeric G protein g subunits provide functional selectivity in G bg dimer signaling in *Arabidopsis*. Plant Cell 19:1235–1250

Trusov Y, Sewelam N, Rookes JE, Kunkel M, Nowak E, Schenk PM, Botella JR (2009) Heterotrimeric G proteins-mediated resistance to necrotrophic pathogens includes mechanisms independent of salicylic acid-, jasmonic acid/ethylene- and abscisic acid-mediated defense signaling. Plant J 58:69–81

Trusov Y, Zhang W, Assmann SM, Botella JR (2008) Gg1+Gg2 not equal Gb: heterotrimeric G protein Gg-deficient mutants do not recapitulate all phenotypes of Gb-deficient mutants. Plant Physiol 147:636–649

Ullah H, Chen J-G, Temple B, Boyes DC, Alonso JM, Davis KR, Ecker JR, Jones AM (2003) The b-Subunit of the *Arabidopsis* G protein negatively regulates auxin-induced cell division and affects multiple developmental processes. Plant Cell 15:393–409

Van Wees SCM, Chang HS, Zhu T, Glazebrook J (2003) Characterization of the early response of Arabidopsis to *Alternaria brassicicola* infection using expression profiling. Plant Physiology 132:606–617

Vera-Estrella R, Barkla BJ, Higgins VJ, Blumwald E (1994a) Plant defense response to fungal pathogens (activation of host-plasma membrane H^+-ATPase by elicitor-iInduced enzyme dephosphorylation). Plant Physiol 104:209–215

Vera-Estrella R, Higgins VJ, Blumwald E (1994b) Plant defense response to fungal pathogens 2. G-protein-mediated changes in host plasma-membrane redox reactions. Plant Physiol 106: 97–102

Wang D, Weaver ND, Kesarwani M, Dong X (2005) Induction of protein secretory pathway is required for systemic acquired resistance. Science 308:1036–1040

Wang S, Narendra S, Fedoroff N (2007) Heterotrimeric G protein signaling in the *Arabidopsis* unfolded protein response. Proc Natl Acad Sci USA 104:3817–3822

Wei Q, Zhou W, Hu G, Wei J, Yang H, Huang J (2008) Heterotrimeric G-protein is involved in phytochrome A-mediated cell death of *Arabidopsis* hypocotyls. Cell Res 18:949–960

Weiss C, Garnaat C, Mukai K, Hu Y, Ma H (1994) Isolation of cDNAs encoding guanine nucleotide-binding protein b-subunit homologues from maize (ZGB1) and *Arabidopsis* (AGB1). Proc Natl Acad Sci USA 91:9554–9558

Wengier D, Valsecchi I, Cabanas ML, Tang WH, Mccormick S, Muschietti J (2003) The receptor kinases LePRK1 and LePRK2 associate in pollen and when expressed in yeast, but dissociate in the presence of style extract. Proc Natl Acad Sci USA 100:6860–6865

Winge P, Brembu T, Bones AM (1997) Cloning and characterization of rac-like cDNAs from *Arabidopsis thaliana*. Plant Mol Biol 35:483–495

Wong HL, Pinontoan R, Hayashi K, Tabata R, Yaeno T, Hasegawa K, Kojima C, Yoshioka H, Iba K, Kawasaki T, Shimamoto K (2007) Regulation of rice NADPH oxidase by binding of Rac GTPase to its N-terminal extension. Plant Cell 19:4022–4034

Wong HL, Sakamoto T, Kawasaki T, Umemura K, Shimamoto K (2004) Down-regulation of metallothionein, a reactive oxygen scavenger, by the small GTPase OsRac1 in rice. Plant Physiol 135:1447–1456

Xin Z, Zhao Y, Zheng ZL (2005) Transcriptome analysis reveals specific modulation of abscisic acid signaling by ROP10 small GTPase in *Arabidopsis*. Plant Physiol 139:1350–1365

Yang KY, Liu Y, Zhang S (2001) Activation of a mitogen-activated protein kinase pathway is involved in disease resistance in tobacco. Proc Natl Acad Sci USA 98:741–746

Yang Z, Fu Y (2007) ROP/RAC GTPase signaling. Curr OpinPlant Biol 10:490–494

Yang Z, Watson JC (1993) Molecular cloning and characterization of rho, a ras-related small GTP-binding protein from the garden pea. Proc Natl Acad Sci USA 90:8732–8736

Zeng Q, Wang X, Running MP (2007) Dual lipid modification of *Arabidopsis* Gg-subunits is required for efficient plasma membrane targeting. Plant Physiol 143:1119–1131

Zhang S, Liu Y (2001) Activation of salicylic acid-induced protein kinase, a mitogen-activated protein kinase, induces multiple defense responses in tobacco. Plant Cell 13:1877–1889

Zhang X, Zhang L, Dong F, Gao J, Galbraith DW, Song CP (2001) Hydrogen peroxide is involved in abscisic acid-induced stomatal closure in *Vicia faba*. Plant Physiol 126:1438–1448

Zhao J, Sakai K (2003) Multiple signalling pathways mediate fungal elicitor-induced b-thujaplicin biosynthesis in *Cupressus lusitanica* cell cultures. J Exp Bot 54:647–656

Zheng Z, Qamar SA, Chen Z, Mengiste T (2006) *Arabidopsis* WRKY33 transcription factor is required for resistance to necrotrophic fungal pathogens. Plant J 48:592–605

Zheng ZL, Nafisi M, Tam A, Li H, Crowell DN, Chary SN, Schroeder JI, Shen J, Yang Z (2002) Plasma membrane-associated ROP10 small GTPase is a specific negative regulator of abscisic acid responses in *Arabidopsis*. Plant Cell 14:2787–2797

Zheng ZL, Yang Z (2000) The Rop GTPase: an emerging signaling switch in plants. Plant Mol Biol 44:1–9

Zhou J, Loh YT, Bressan RA, Martin GB (1995) The tomato gene Pti1 encodes a serine/threonine kinase that is phosphorylated by Pto and is involved in the hypersensitive response. Cell 83:925–935

Bioinformatics of Seven-Transmembrane Receptors in Plant Genomes

Etsuko N. Moriyama and Stephen O. Opiyo

Abstract Seven-transmembrane receptors (7TMRs; also known as G protein-coupled receptors, GPCRs) constitute the largest receptor superfamily in metazoa. In striking contrast, very few numbers of 7TMRs are reported in plants. Comparative analysis revealed that many of 7TMR proteins found in plants are in fact unique to the plant kingdom. More interestingly, some 7TMR proteins appear to have acquired GPCR functions independently during their evolution. Furthermore, the origin(s) of 7TMR proteins goes back to the level of eukaryote–prokaryote divergence. In order to understand such deep divergence, powerful and sensitive bioinformatics tools are necessary. In this chapter, we first overview the plant 7TMR proteins and how they are distinct from metazoan counterparts. We review various computational methods that are used for classifying 7TMR proteins and their strengths and weaknesses when they are applied for this divergent protein family. We describe our recent efforts to provide a computational tool that facilitates identifying 7TMR candidates from diverse genomes.

1 Seven-Transmembrane Receptors: Overview

Seven-transmembrane (7TM) region containing receptors constitute the largest receptor superfamily in metazoa (Fredriksson et al. 2005; Horn et al. 2003). These proteins are activated by a diverse array of ligands and are involved in various signaling processes such as cell proliferation, neurotransmission, metabolism, smell, taste, and vision. They are the central players in eukaryotic signal transduction. These receptors have been referred to as G protein-coupled receptors (GPCRs) because most transduce extracellular signals into cellular physiological

E.N. Moriyama (✉) and S.O. Opiyo
School of Biological Sciences and Center for Plant Science Innovation, University of Nebraska-Lincoln, 403 Manter Hall, Lincoln, NE 68588-0118, USA
e-mail: emoriyama2@unl.edu

S. Yalovsky et al. (eds.), *Integrated G Proteins Signaling in Plants*,
Signaling and Communication in Plants,
DOI 10.1007/978-3-642-03524-1_13,

responses through the activation of heterotrimeric guanine nucleotide-binding proteins (G proteins) (Gilman 1987; Lefkowitz 2007; Pierce et al. 2002). However, an increasing number of alternative "G protein-independent" signaling mechanisms have been associated with groups of these 7TM receptors (Chen et al. 2004; Kimmel and Parent 2003; Lefkowitz and Shenoy 2005; Luttrell 2008; Smart et al. 2008). Thus, it is no longer appropriate to call this entire protein group as GPCRs. Throughout this chapter, therefore, we refer to these proteins as 7TM receptors (7TMRs). When we refer to "G protein-dependent" 7TMRs specifically, we call the proteins "GPCRs."

1.1 Architecture of the Seven-Transmembrane Receptors

All 7TMR proteins are known to share a common architecture having seven transmembrane regions (Gilman 1987). Although some exceptional cases have been found in several proteins as described later, this architecture is held by the majority of the proteins belonging to this superfamily. Figure 1 shows residue-based diagrams of two 7TMR proteins. Seven membrane-spanning α-helices (I–VII) are linked by three alternating intercellular and extracellular loops (e1–e3 and i1–i3). The extracellular and transmembrane regions are involved in ligand-binding, while the intracellular regions are important in G-protein binding and signal transduction. The majority of 7TMRs have two Cys residues that form a disulfide bridge between e1 and e2, which is considered to be important in proper protein folding. In the common 7TMR architecture, N- and C-terminal regions are located extracellularly and intracellularly, respectively. This orientation has a significant meaning in the function of these receptors. Many 7TMRs are known to have ligand-binding domains in the N-terminals (e.g., hormone-binding domains in luteinizing hormone receptors). In addition to being involved in G-protein coupling, the C-terminal region of 7TMRs are also important in desensitization of the receptor (Luttrell 2008).

1.2 Classification of the Seven-Transmembrane Receptor Superfamily

The largest number of 7TMR-encoding genes are found in vertebrate, especially in mammalian, genomes. For example, the human genome encodes approximately 800 7TMRs, and twice as many are found in the mouse and rat genomes (Gloriam et al. 2007). Figure 2 shows the frequency distribution of transmembrane-protein-encoding genes among various vertebrate and related animal genomes. These animals clearly have a greater representation of 7TM proteins among those with multiple TM regions. Among invertebrates, close to 1,500 *Caenorhabditis elegans*

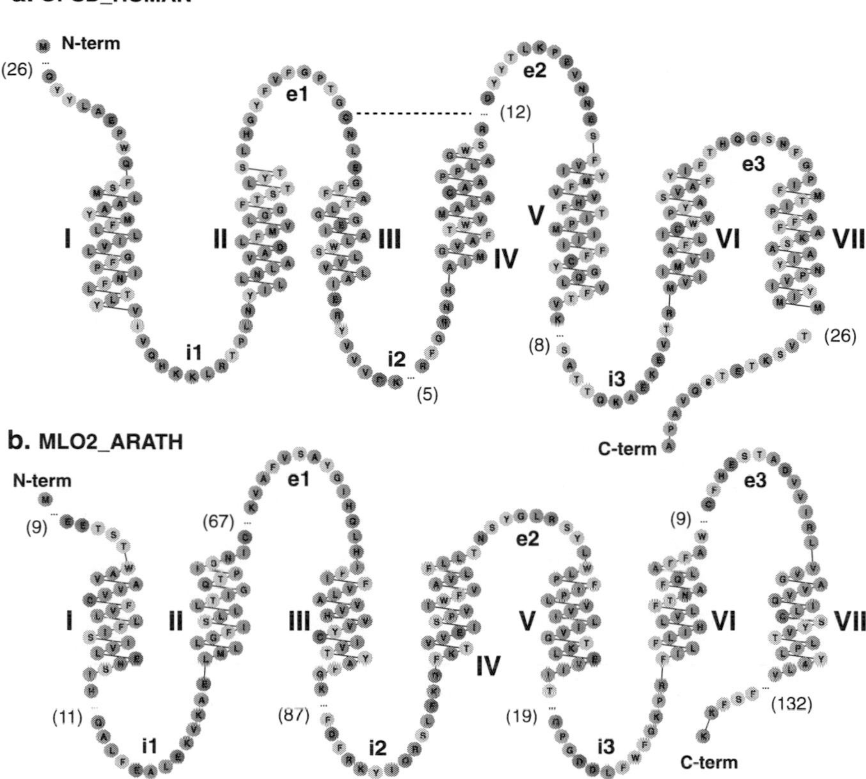

Fig. 1 Residue-based diagram showing the seven-transmembrane structure of 7TMR proteins. Human rhodopsin (**a**; UniProt P08100) and *Arabidopsis thaliana* MLO2 (**b**; UniProt Q9SXB6) proteins are shown. Some residues are omitted from the diagram where an ellipsis symbol (...) is shown. The number next to the ellipsis shows the number of residues omitted. A dashed line connecting two Cys residues in OPSD_HUMAN illustrates a disulfide bridge (the second Cys residue is not shown in the figure). Seven TM regions (I–VII) are connected by three extracellular (e1–e3) and intercellular (i1–i3) loops. "N-term" and "C-term" show the N- and C-terminals of the proteins, respectively. RbDe Web service was used to draw these diagrams (Skrabanek et al. 2003; http://icb.med.cornell.edu/crt/RbDe/Rbde.xml)

genes, or 5% of the genome, are predicted to encode 7TMRs (Fredriksson and Schiöth 2005; Thomas and Robertson 2008). On the other hand, only >200 7TMR genes have been recognized in the *Drosophila melanogaster* genome. Such large numbers of 7TMR-encoding genes in metazoa are mainly correlated with the high degree of expansions of chemoreceptor genes in these genomes. Olfactory receptor genes, for example, constitute 60% of rat 7TMR-encoding genes, and 90% of *C. elegans* 7TMR genes encode chemoreceptors. In *D. melanogaster*, only ~130 genes encode odorant and gustatory receptors.

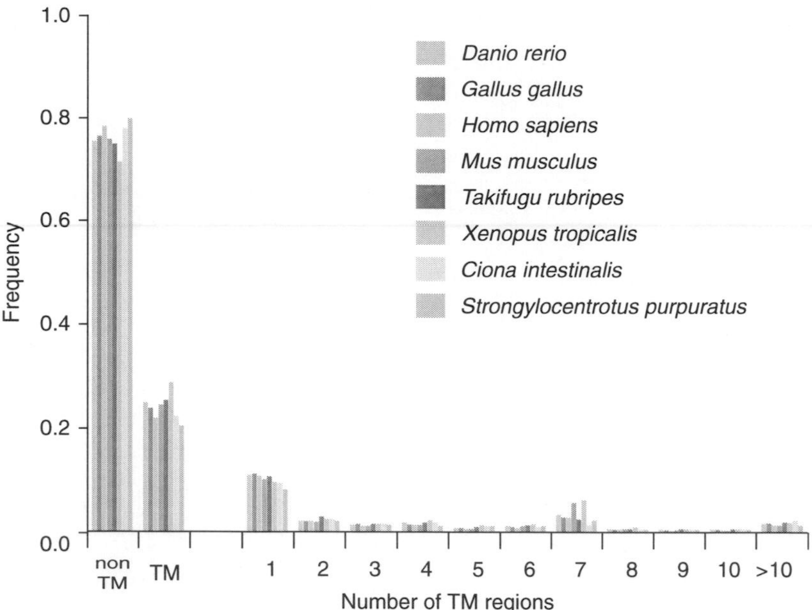

Fig. 2 Frequency distribution of transmembrane-protein encoding genes among various animal genomes. Eight deuterostome species including six vertebrates, one urochordate, and one echinoderm are used for the analysis. TM-region prediction was done by Phobius (Käll et al. 2004, 2007). Proteins predicted to have no (0) TM region are shown above "nonTM." Proteins predicted to have one or more (>0) TM regions are shown above "TM." For each TM number category, the order of the bars from left to right is from the top (*Danio rerio*) to the bottom (*Strongylocentrotus purpuratus*) of the species legend. Histograms from other organisms and by another TM-prediction method are available on the *7TMRmine* website (http://bioinfolab.unl.edu/emlab/7tmr/index.php)

Table 1 shows the classification system used in the Information System for G Protein-Coupled Receptors (GPCRDB) (Horn et al. 2003).[1] The superfamily is divided into six major classes, and six other more divergent groups are listed as "putative families." There are also many proteins whose specific ligands or classification have not been established. These proteins are listed as "Putative/unclassified" or *orphan*.

Fredriksson and his group (Bjarnadóttir et al. 2006; Fredriksson et al. 2003; Schiöth and Fredriksson 2005) used slightly different classification system, the

[1]The original GPCRDB was updated until June 2006. Although the new GPCRDB resumed the development and updating later, as of this writing, the new GPCRDB still has only a part of the original content. Therefore, all statistics from GPCRDB in this chapter were obtained from the original GPCRDB (available from http://www.gpcr.org/7tm_old/). The new GPCRDB is available from http://www.gpcr.org/7tm/.

Table 1 Classification of the 7TMR protein superfamily

Classes in GPCRDB	GRAFS system	Examples	Plant 7TMR
A: Rhodopsin-like	R (Rhodopsin)	Rhodopsin, adrenoceptor, olfactory receptor (vertebrates), thyrotropin receptor, platelet-activating factor receptor, melatonin receptor	
B: Secretin-like	S (Secretin)	Calcitonin receptor, glucagon receptor, parathyroid hormone receptor, secretin receptor, diuretic hormone receptor, Methuselah-like proteins	GCR1
	A (Adhesion)	EMR1, latrophilin, cadherin EGF LAG	
C: Metabotropic glutamate/pheromone	G (Glutamate)	Metabotropic glutamate receptor, calcium-sensing receptor, pheromone receptor, GABA type B receptor, taste receptor type 1 (vertebrates)	
D: Fungal pheromone		Fungal A-pheromone receptor, STE2, STE3, pheromone-B alpha-1 receptor	
E: cAMP receptors		*Dictyostelium* cAMP receptor	
Frizzled/Smoothened family	F (Frizzled/Taste2)	Frizzled, Smoothened	
Putative families			
Ocular albinism proteins		Ocular albinism type 1 (GPR143)	
Insect odorant receptors		Gustatory receptor, odorant receptor	
Plant Mlo receptors		MLO	MLO1–15
Nematode chemoreceptors		Serpentine receptor, str	
Vomeronasal receptors (V1R & V3R)		Vomeronasal receptor V1R	
Taste receptors T2R	F (Frizzled/Taste2)	Taste receptor T2R	

GRAFS system, based on phylogenetic analysis of mammalian GPCRs. In this system, there are five main families: Glutamate (G), Rhodopsin (R), Adhesion (A), Frizzled/Taste2 (F), and Secretin (S). In both systems, Class A: Rhodopsin-like (R family of GRAFS) is by far the largest, containing almost 7,000 entries in GPCRDB. GPCRDB also lists one non-GPCR family, Class Z: Archaeal/bacterial/fungal opsins, which includes archaeal bacteriorhodopsins, proteorhodopsins, and fungal opsin-like proteins. These proteins are photoreceptor/proton-pumps, and share the same topology of 7TMs with GPCRs. However, they do not activate G-proteins and are not considered to be GPCRs. Their ancestral relationships to eukaryotic 7TMRs are not known.

1.3 Seven-Transmembrane Receptors in Plants

Compared with a large number of 7TMRs found in animal genomes, very few 7TMRs have been reported in plants and fungi. Only 24 *Arabidopsis* 7TMRs have been described so far. Fifteen of them constitute the "mildew resistance O" (MLO) family. Although their 7TM topology was biochemically confirmed (Devoto et al. 1999), direct interaction with G-protein α subunit (Gα) has not been shown to date (Chen et al. 2006; AM Jones, personal communication). GCR1, a putative *Arabidopsis* GPCR, has been found to directly interact with the plant Gα subunit GPA1 (Josefsson and Rask 1997; Pandey and Assmann 2004). However, GCR1 was also shown to act independently of the heterotrimeric G-protein complex (Chen et al. 2004). The *Arabidopsis* regulator of G-signaling protein (AtRGS1) is unusual in that it has an RGS-box in the C-terminal region and also predicted 7TM regions in the N-terminal half (Chen et al. 2003). RGS proteins function as GTPase activating proteins (GAP) to desensitize signaling by deactivating the Gα subunits of the heterotrimeric complex. It has been suggested that AtRGS1 is a novel D-glucose receptor having an agonist-regulated GAP function (Chen et al. 2003, 2004; Johnston et al. 2007a). Hsieh and Goodman (2005) reported five expressed proteins predicted to have 7TM regions (heptahelical transmembrane proteins 1–5 or HHP1–5), but these, like the others, do not have candidate ligands. HHP proteins are predicted to have an intracellular N-terminus and an extracellular C-terminus, which is the opposite topology of known GPCRs. Two novel GPCR-type G proteins (GTG1 and GTG2) were characterized recently (Pandey et al. 2009). These proteins interact with GPA1, but also have intrinsic GTP-binding and GTPase activity, which is classic Gα subunit activity. They also appear to function as membrane-localized abscisic acid (ABA) receptors. These proteins are also unique among GPCRs in that they have nine TM regions predicted. Recently, Liu et al. (2007) proposed that another candidate protein GCR2 functions as an ABA-signaling GPCR. However, several later studies demonstrated that GCR2 does not act as an ABA receptor nor in G-protein signaling (Guo et al. 2008; Illingworth et al. 2008; Johnston et al. 2007b).

1.4 Conservation of 7TMR Proteins Among and Between Plants and Animals

The majority of the reported *Arabidopsis* 7TMR proteins are known to share no substantial sequence similarity to known metazoan 7TMRs. It indicates that plant 7TMRs have dramatically diverged from the major groups of metazoan 7TMRs since the plant and metazoan lineages separated more than 1 billion years ago (Douzery et al. 2004; Hedges et al. 2004). Genomic information from a wide variety of organisms has been becoming available at an astonishing rate in recent years. Taking advantage of this advancement, in Table 2, we summarized sequences similar to plant 7TMR proteins across kingdoms at the time of this writing. The majority of the 7TMR proteins found in *Arabidopsis* were also identified in green algae (*Chlamydomonas reinhardtii* and *Ostreococcus* species) as well as in a moss (*Physcomitrella patens subsp. Patens*), indicating that these 7TMR proteins have been conserved throughout plant evolution. A single exception was RGS1. RGS1 was found exclusively in dicot plants.

1.4.1 GCR1

Previous studies reported that *Arabidopsis* GCR1 protein has significant similarities with Class B (Secretin)-like proteins and a *Dictyostelium discoideum* (a slime mold) cyclic AMP receptor (Chen et al. 2004; Josefsson 1999; Josefsson and Rask 1997). Table 2 confirms a wide distribution of GCR1-type 7TMRs across kingdoms. Highly significant similarities were found from ciliates (*Tetrahymena thermophila* and *Paramecium tetraurelia*; 50% similarity[2]), *Monosiga brevicollis* (Choanoflagellida; 34% similarity), *Branchiostoma floridae* (a lancelet, Cephalochordata; 45% similarity), and *Ciona intestinalis* (Tunicata; 47% similarity), in addition to *D. discoideum* (44% similarity). It is noteworthy that except for a few vertebrate (e.g., *Danio rerio*; 47% similarity) proteins, all highly significant hits found were from primitive Chordates or the organisms whose divergence was as deep as or even older than plant/metazoan divergence (Yoon et al. 2008). Note also that there are also hits from fungal species. GCR1 may represent one of the most ancient members of 7TMR proteins.

1.4.2 AtRGS1

As mentioned earlier, AtRGS1 has two domains: 7TM and RGS. Using each region individually as well as the entire sequence as the query, we found a very few similar sequences even from plants. At the time of this writing, we found RGS1 homologue

[2] All % similarities are based on blastp search results. These values are obtained using only the highly similar regions aligned with blastp. The % similarity would be lower if the entire protein regions are aligned.

Table 2 Plant 7TMR proteins and their homologue candidates in various organisms[a]

Plant 7TMR	TAIR[b] locus ID	Length (amino acids)	Plant genomes[c]	Non-plant genomes[d]
GCR1	At1G48270	326	Dicots: B Lo Po Ps R V Monocots: Or Z Moss: Ph Green algae: Ot Ol (dicot/monocot/moss hits: $E < 10^{-104}$) (green algal hits: $E < 10^{-3}$)	$E < 10^{-20}$: *Tetrahymena thermophila*, *Paramecium tetraurelia* $E < 10^{-13}$: *Branchiostoma floridae*, *Ciona intestinalis*, *Monosiga brevicollis*, *Dictyostelium discoideum* $10^{-13} < E < 0.01$: *Trichoplax adhaerens*, *Nematostella vectensis*, *Danio rerio*/other vertebrates, *Trichoderma atroviride*/other fungi
AtRGS1	At3G26090	459	Dicots: Po R V (all plant hits: $E < 10^{-72}$)	$E < 0.01$: *Trichomonas vaginalis*, *C. intestinalis*, *Brugia malayi*, *D. rerio* $0.01 < E^e$: vertebrates, invertebrates $4.0 < E^e$: fungi
		7TM[f]: 300	Dicots: Po R V	$E < 0.01$: no hit $E = 0.22^e$: *T. thermophila* (only 94 amino-acid region) (no vertebrate hit with $E < 10$)
		RGS[f]: 159	Dicots: Po R V	$E < 10^{-4}$: *C. intestinalis* $E < 0.01$: *Brugia malayi*, *Trichomonas vaginalis*, *D. rerio* $0.01 < E^e$: vertebrates, invertebrates $4.0 < E^e$: fungi
GTG1 GTG2	At1g64990 At4g27630	468 467	Dicots: Po R V Monocots: Or Z Green algae: Ch Ot Ol (all plant hits: $E < 10^{-161}$)	$E < 10^{-30}$: vertebrates, invertebrates, *T. adhaerens*, *M. brevicollis*, *D. discoideum*, *Toxoplasma gondii* $E < 10^{-20}$: *T. thermophila*, *P. tetraurelia* $E < 0.02$: fungi
MLO1 MLO2 MLO3 MLO4 MLO5 MLO6 MLO7 MLO8 MLO9 MLO10 MLO11	At4G02600 At1G11310 At3G45290 At1G11000 At2G33670 At1G61560 At2G17430 At2G17480 At1G42560 At5G65970 At5G53760	526 573 508 573 501 583 542 593 460 569 573	Dicots: B Ca Li Lo M Ol Pc Po Ps R S V Monocots: H Or T Z Moss: Ph Green algae: Ot Ol (all plant hits: $E < 10^{-5}$)	$E < 0.01$: no hit (no vertebrate hit with $E < 10$)

(*continued*)

Table 2 (continued)

Plant 7TMR	TAIR[b] locus ID	Length (amino acids)	Plant genomes[c]	Non-plant genomes[d]
MLO12	At2G39200	576		
MLO13	At4G24250	478		
MLO14	At1G26700	554		
MLO15	At2G44110	497		
HHP1	At5G20270	332	Dicots: Li M Pc Po R V	$E < 10^{-30}$: vertebrates, invertebrates
HHP2	At4G30850	358	Monocots: Or Z	$E < 10^{-15}$: *P. tetraurelia*, *T. thermophila*, *T. brucei*, *T. adhaerens*, *M. brevicollis*, *D. discoideum*
HHP3	At2G24150	344	Moss: Ph	$E < 10^{-8}$: fungi
HHP4	At4G37680	385	Green algae: Ch Ot Ol (all plant hits: $E < 10^{-36}$)	$E < 0.1$: *Theileria parva*
HHP5	At4G38320	374		$10^{-5} < E < 0.01$[g]: *Desulfotomaculum reducens*, *Bacillus coahuilensis* $0.01 < E$[e,g]: *Deinococcus geothermalis*, *Bacillus cereus*/other *Bacillus*, other bacteria

[a]Protein similarity search was done with blastp against the non-redundant (nr) protein database available at National Center for Biotechnology Information (NCBI; http://www.ncbi.nlm.nih.gov/). The word size of 2 and BLOSUM45 scoring matrix were chosen for more sensitive search. The search was done in March 2009

[b]The *Arabidopsis* Information Resource (TAIR; http://www.arabidopsis.org) (Swarbreck et al. 2008)

[c]Plant genomes where significant hits were found. Species names are coded as follows:
[Monocots] H *Hordeum* species (barley), Or *Oryza sativa* (rice), T *Triticum aestivum* (common wheat), Z *Zea mays* (corn)
[Dicots] B *Brassica napus* (rape) or *Brassica rapa* (field mustard), Ca *Capsicum annuum* (cayenne pepper), Li *Linum usitatissimum* (common flax), Lo *Lotus japonicus*, M *Medicago truncatula* (barrel medic), Ol *Olea europaea* (olive), Pc *Picea sitchensis* (Sitka spruce) or *Picea abies* (Norway spruce), Ps *Pisum sativum* (garden pea), Po *Populus trichocarpa* (black cottenwood), R *Ricinus communis* (castorbean), S *Solanum lycopersicum* (garden tomato), V *Vitis vinifera* (wine grape)
[Moss] Ph *Physcomitrella patens*
[Green algae] Ch *Chlamydomonas reinhardtii*, Ol *Ostreococcus lucimarinus*, Ot *Ostreococcus tauri*

[d]Non-plant genomes where significant hits were found. Only representative species names are listed. E expect value

[e]These hits with E-values higher than 0.01 are not considered to be significant

[f]For AtRGS1, in addition to the entire protein sequence, sequence regions from 7TM and RGS domains were used separately for blastp search

[g]Bacterial hits

candidates only from four plants: *A. thaliana*, wine grape (*Vitis vinifera*), castorbean (*Ricinus communis*), and black cottonwood (*Populus trichocarpa*). All these four dicot RGS1s have the 7TM + RGS1 structure. No hit was identified from monocots, moss, nor green algae. The search using 7TM or RGS region individually showed no similar sequence from any other plants. In animals, on the other hand, proteins carrying only the RGS domain have been found from a wide range of species including an anaerobic protist *Trichomonas vaginalis* (Metamonada, parabasalia; 49% similarity), although the majority of hits from vertebrates and invertebrates had E-values[3] higher than the threshold of 0.01. The ancestral RGS1 protein may have acquired a 7TM region of unknown origin in the dicot lineage, while in all other plant lineages, these proteins may have been lost.

1.4.3 GTG1/GTG2

As Pandey et al. (2009) reported, these GPCR-type G proteins are highly conserved during the eukaryotic evolution. These proteins are highly conserved throughout plants, including green algae. Highly similar proteins are found in almost all major eukaryotic groups, including metazoa/fungi, *M. brevicollis* (Choanoflagellida), *D. discoideum*, ciliates (e.g., *T. thermophila*), apicomplexans (e.g., *Toxoplasma gondii*), and kinetoplastids (e.g., *Trypanosoma cruzi*). Note that the human homologue candidate protein, G protein-coupled receptor 89 (GPR89), had no proven function. Pandey et al. (2009) reported that the human GPR89 does not have a conserved ATP-/GTP-binding region motif, while the motif is highly conserved in GTG1 and GTG2. GTPase activity was also confirmed with GTG1 and GTG2, but not with the human GPR89. Maeda et al. (2008) recently clarified that the human GPR89 gene encodes a Golgi-resident anion channel, and renamed it as Golgi pH regulator (GPHR). Therefore, although the GTG1/GTG2 group appears to be the most conserved among the 7TMR superfamily, its GPCR function does not seem to be conserved during their evolution.

1.4.4 MLOs

In *Arabidopsis*, the MLO multigene family is composed of 15 genes. As Table 2 shows, MLO proteins are conserved throughout the plant lineage, including a moss and green algae. The number of the MLO proteins, however, varies among the plant genomes due to species-specific duplication and loss events (Devoto et al. 2003; Liu and Zhu 2008). Liu and Zhu (2008) identified possible positive selection at several points of gene duplications, which is consistent with the functional diversification observed among different MLOs (Chen et al. 2006). MLOs appear to be

[3]An E-value or expect value for similarity search is the number of hits you can expect to see by chance when searching the database. The E-value is related to the P-value (the probability), but not the same. An E-value of 1 means that you can expect to see one hit with the same score or higher simply by chance by searching the same size of a database. Similar to the P-value, lower the E-values, especially closer they are to 0, the more significant the matches are. You cannot obtain such hits purely by chance.

unique to plants. No apparent homologue candidate has been identified from the outside of the plant lineage.

1.4.5 HHPs

HHP proteins are conserved throughout the plant lineage, including a moss and green algae (Table 2). As seen in the MLO family, the HHP multigene family also shows species-specific duplication and loss (Hsieh and Goodman 2005). HHP proteins share strong similarities with projestin and adipoQ (adiponectin) receptors (PAQR) in animals. These highly conserved proteins are found in almost all major eukaryotic groups, including metazoa/fungi (as high as 56% similarity), *D. discoideum*, ciliates (e.g., *T. thermophila*), apicomplexans (e.g., *Theileria parva*), and kinetoplastids (e.g., *Trypanosoma cruzi*) (Table 2). In addition to having 7TM regions, the PAQR proteins share sequence similarity with bacterial hemolysin 3 (HLY3) proteins (Tang et al. 2005; Thomas et al. 2007). This is shown in the significant similarities found with many bacterial hemolysin 3 family channel proteins (up to ~50% similarity) (in Table 2, bacterial hits are marked with (f)). The PAQR family is composed of three main subgroups: the adiponectin-receptor (AdipoR) related, the membrane progestin-receptor (mPR) related, and the hemolysin 3 (HLY3)-receptor related (Fernandes et al. 2005). mPR proteins function by activating G proteins and display many other functional characteristics of GPCRs (Thomas et al. 2007; Zhu et al. 2003a,b). Plant HHP proteins belong to the AdipoR group. The orientation of the N- and C-terminals of the AdipoR proteins has been experimentally confirmed to be opposite to that of the regular 7TMR proteins: intracellular N-terminus and extracellular C-terminus (Yamauchi et al. 2003), which seems to be the case also with plant HHP proteins (Hsieh and Goodman 2005). Although more studies are required, GPCR activity appears to be absent in AdipoR receptors. Bacterial hemolysin 3 proteins are also predicted to have 7TM regions with an intracellular N-terminus. Based on their significant sequence similarity, the eukaryotic PAQR family and the prokaryotic HLY3 family appear to have a common bacterial ancestor. Considering that the sequence similarity between the PAQR proteins and other 7TMR proteins is very low, the bacterial origin of the PAQR family appears to differ from that of the other 7TMR superfamily. It implies that GPCR functions may have arisen more than once during eukaryotic evolution (Thomas et al. 2007).

1.5 Are There More 7TMRs in Plants?

The G-protein complex components include heterotrimeric G-proteins, 7TMRs, and the effectors. For about 800 of 7TMRs, the human genome encodes 23 Gα, 5 Gβ, at least 12 Gγ, and dozens of effectors. In striking contrast, both fungi and plants have much simpler G-protein coupled signaling systems. For example, both of *Arabidopsis* and rice genomes contain one canonical Gα, one Gβ, and two Gγ genes (Chen 2008; Jones and Assmann 2004; Temple and Jones 2007).[4] Similarly, a small number of G-proteins are found in fungi; there are two Gα, one Gβ, and one Gγ in *Saccharomyces cerevisiae* (Nakafuku et al. 1987, 1988; Whiteway et al. 1989), while *Neurospora crassa* and other fungi have more (Baasiri et al. 1997; Galagan et al. 2003; Turner and Borkovich 1993). Therefore, on the one hand, it may be reasonable to assume that plants and fungi have fewer GPCRs than humans. On the other hand, close to 200 *Arabidopsis* proteins were predicted to have 7TM regions (see the next section and also Schwacke et al. 2003). Furthermore, at least 61 novel GPCR-like 7TMRs have been recently predicted from the plant pathogenic fungus *Magnaporthe grisea* genome (Kulkarni et al. 2005). Therefore, more divergent groups of 7TMR proteins likely remain undiscovered in non metazoan taxa.

2 Bioinformatics Methods for Protein Classification

In this section, we overview several bioinformatics methods used to classify protein families. We describe the advantage and disadvantage of using such methods for mining 7TMR proteins from diverse genomes.

2.1 Pairwise-Alignment-Based Method: BLAST

Computational protein classification relies on finding similarities between the query (new protein) sequence and those in databases with known (preferably experimentally confirmed) functions. The simplest and most popularly used is the Basic Local Alignment Search Tool (BLAST) (Altschul et al. 1990, http://blast.ncbi.nlm.nih.gov/, 1997). It searches sequence databases for local similarity to the query. The most often used BLAST program is blastp, which uses protein sequences as the query and searches against protein databases. When blastp is used against newly completed genome data, the search needs to be done against the translated CDS (coding sequence) data set. This is problematic since genome annotation processes rely on gene prediction methods. There are many ways the predicted gene structures are incorrect. Some genes, particularly short genes, could have been completely missed by the prediction methods. Some genes may have only a part of the exons or introns identified correctly. Other genes may have chimera structures concatenated with exons belonging to other nearby genes. "Alternatively spliced" genes are most likely predicted to produce a single long transcript including all exons at once. Therefore, the predicted gene sets need to be constantly updated until all genes are correctly identified. For example, the number of CDS included in each release of The Arabidopsis database (TAIR) has increased from 30,690 in 2006 (TAIR6) to

[4]See other chapters of this book for more details on plant G proteins.

31,921 in 2007 (TAIR7) and to 32,825 in 2008 (TAIR8). In order to find all possible 7TMR gene candidates, tblastn, which searches using protein queries against translated nucleotide databases, needs to be used directly against the genomic data. It requires finding and assembling all exons manually to obtain the entire coding regions (see Schiöth et al. 2007 for some examples for mining 7TMRs from genomic data).

One disadvantage of using pairwise-alignment strategy is that the search needs to be done using the queries one by one as we did for Table 2. Another disadvantage is that for each search, similarity can be found only against one single sequence. Information from other similar sequences cannot be incorporated.

2.2 *Conserved Motif Matching Methods*

2.2.1 PROSITE Patterns

Regular expression patterns used in PROSITE (Hulo et al. 2008) takes advantage of multiple alignments of conserved regions obtained from related sequences. Since functionally important regions (e.g., catalytic domains, binding-sites) are under stronger selective constraints, multiple alignments of proteins with known functions are expected to contain conserved regions related to those functions. When distantly related sequences are compared, only functionally crucial sites or short regions might be conserved. Some amino acids, even within such critical sites, may be substituted with other biochemically similar amino acids as long as the protein function is maintained. PROSITE regular expression patterns attempt to represent the information of amino acid conservation in functional regions. Figure 3 shows the PROSITE pattern entry for the "G-protein-coupled receptors family 1 signature" (PS00237) obtained from Class A Rhodopsin-like GPCR sequences.

In general, patterns are derived from relatively short regions. If the query is only a short fragment, as in the case of expressed sequence tag (EST) sequences, and if it does not contain the region where the pattern was derived, this method fails to identify the query correctly. The pattern syntax allows multiple amino-acid possibilities for a site, but all are with equal frequencies. The sequence logo in Fig. 3 clearly shows that in many amino acid sites, some amino acids are often predominant even if other amino acids can be still used. The pattern syntax allows only a limited flexibility in the amino acid substitution pattern.

2.2.2 PRINTS

PRINTS also uses very short conserved motifs called protein fingerprints (Attwood et al. 2003). It overcomes the problems found with PROSITE patterns in two ways. Multiple conserved regions can be identified from a multiple alignment, and used as composite fingerprints, consequently covering larger regions than using a single

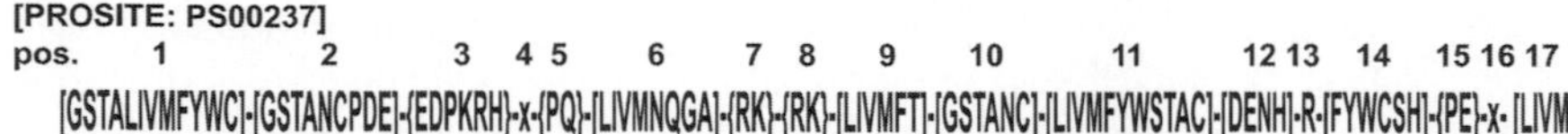

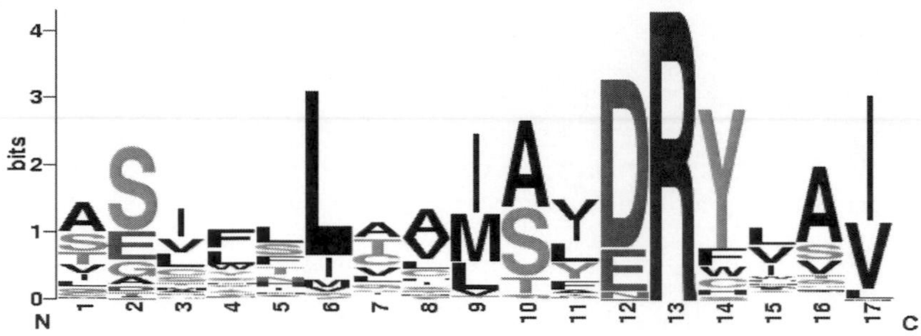

Fig. 3 PROSITE regular expression pattern entry for "G-protein-coupled receptors family 1 signature" (PS00237). This pattern corresponds to 17 amino acid positions covering from the second half of the TM III to the beginning of the loop i2 in the Class A Rhodopsin-like GPCRs. In the OPSD_HUMAN entry shown in Fig. 1, this pattern corresponds to the following amino acid sequence: IALWSLVVLAIERYVVV. The PROSITE pattern syntax is available at http://www.expasy.ch/prosite/prosuser.html. The sequence logo was obtained from the multiple alignment of the Swiss-Prot true positive hits based on this pattern. It represents the amino acid conservation at each of the 17 amino acid positions. The height of each letter in a logo position is proportional to the observed frequency of the corresponding amino acid in the alignment column

motif. Motif searching is done using profile information from the multiple alignment. A motif profile includes amino acid substitution patterns from each alignment column. It provides higher search sensitivity compared with rigid PROSITE patterns. Figure 4a shows the result of the fingerprint search using the *Arabidopsis* GCR1 sequence. The top hit was found against the fingerprint entry for Secretin-like GPCR superfamily signature (GPCRSECRETIN), although the E-value of the top hit (0.0021) is slightly higher than the default threshold (1×10^{-4}). It shows that GPCRSECRETIN is a composite motif consisting of seven elements, and four of the seven motifs are conserved in the GCR1 sequence.

2.3 *Profile-Based Search Methods*

2.3.1 Position-Specific Iterated Blast (PSI-Blast)

As explained in the previous section, profiles incorporate difference in amino acid substitutions along the multiple alignment. Profiles are expressed in Position-Specific Scoring Matrices (PSSMs), which are a series of substitution matrices each specific to an alignment position. PSSMs are used for the profile entries of PROSITE and also in PSI-blast (Altschul et al. 1997). These profiles are often built

a

Ten top scoring fingerprints for your query								
Fingerprint	No. of Motifs	SumId	AveId	PfScore	Pvalue	Evalue	GRAPHScan	
GPCRSECRETIN	4 of 7	82	20	984	2.5e-09	0.0021	i.i.i.i	Graphic

Ten top scoring fingerprints for your query. Detailed by motif										
FingerPrint Name	Motif Number	IdScore	PfScore	Pval	Sequence	Length	low	Pos	high
GPCRSECRETIN	1 of 7	21.15	173	1.24e-02	AINTGASSLSFVGSAFIVLCYCLFK	25	0	21	0
	3 of 7	16.67	231	5.38e-02	AQGYTTHFFCVASFLWTTTIAFTL	24	0	82	0
	5 of 7	22.82	350	1.31e-04	LTFYAPLWGAILYNGFTYFQVIRMLR	26	0	165	0
	7 of 7	21.34	230	2.81e-02	SVLDVGTAALMGLFNSIAYGFN	22	0	251	0

b

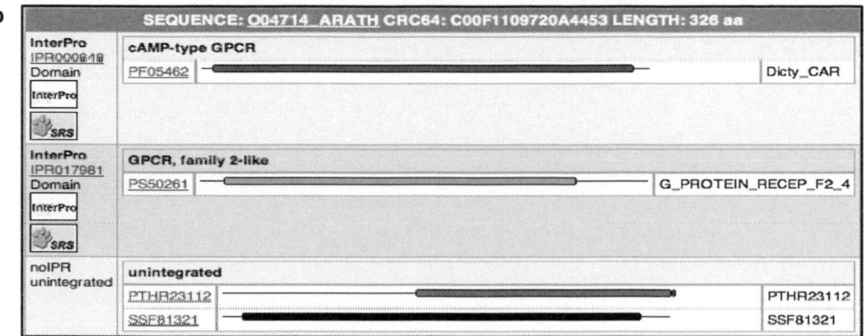

Fig. 4 Search results for *A. thaliana* GCR1. (**a**) The search result by FingerPRINTScan (http://www.bioinf.manchester.ac.uk/fingerPRINTScan/). The protein sequence (Uniprot O04714_ARATHT) was used as the query. BLOSUM45 scoring matrix was selected to find distantly related sequences. (**b**) The search result by InterProScan (http://www.ebi.ac.uk/interpro/) using the same GCR1 sequence. Matches are shown with a Pfam family (PF05462), a PROSITE profile (PS50261), a Panther family (PTHR23112), and a Superfamily family (SSF81321)

from the entire region of the alignments, much longer than the regions covered by PROSITE/patterns and PRINTS/motifs.

PSI-blast starts as a simple protein–protein blast (blastp). From the blastp result, significantly similar sequences (the default E-value threshold = 0.005) are aligned and used to build a PSSM. In the second iteration, this PSSM instead of the single query sequence are used for searching similar sequences. This process is iterated until no more new sequences are found. PSI-blast is more sensitive than blastp and useful for finding very distantly related proteins. However, there are two pit falls the users need to be aware of. One problem is called as *profile corruption*. This happens when unrelated sequences are accidentally included in the PSSM building process. Such PSSMs become useless since they match against unrelated sequences. Further iterations with "corrupted" PSSMs make the situation even worse. Another problem is how to interpret E-values given in PSI-blast results. After the first iteration, E-values are calculated between the PSSM and the hit sequence, which usually improves with iterations. These E-values, however, do not represent the similarity level between the original query and the hit sequences.

2.3.2 Pofile-Hidden Markov Models (Profile HMMs)

Profile HMMs are similar to PSSMs, but they are full probabilistic representations of sequence profiles (Eddy 1998). It provides probabilistic models for multiple alignments. This is currently the state-of-the-art of protein classification methods. For example, the Pfam database is a large collection of protein domain families (Sammut et al. 2008). Each family in Pfam is represented by multiple alignments and profile HMMs. Other examples of protein family and profile HMM databases include: SMART (Letunic et al. 2009), Panther (Mi et al. 2005), and Superfamily (Wilson et al. 2009).

In order to facilitate the use of many classification methods, InterPro integrates various methods/databases and serves as a one-stop search interface (Mulder and Apweiler 2008). Currently, 14 methods are incorporated, including PROSITE, PRINTS, Pfam, SMART, Panther, Superfamily, as well as transmemebrane and signal-peptide prediction methods. Figure 4b shows the result of the InterPro search using the *Arabidopsis* GCR1 sequence. It shows the significant matches with the "*Dictyostelium* cyclic AMP receptor" family (PF05462) and the "G-protein-coupled receptors family 2" (PS50261). These are consistent with the result obtained by PRINTS (Fig. 4a). Note that although InterPro search includes PRINTS search as one of the applications, PRINTS search in InterPro is done with the default parameters (E-value threshold of 10^{-4} and BLOSUM62). Because of this, the InterPro search result does not show any hit by PRINTS. Therefore, although InterPro conveniently assembles many protein classification methods, especially when no hit is found, the users are advised to use these methods directly in their own sites.

PRED-GPCR is a profile-HMM-based method specifically developed for GPCR classification (Papasaikas et al. 2003, 2004). A collection of 265 profile HMMs is derived from 67 GPCR families. Similar to PRINTS using multiple protein fingerprints, in PRED-GPCR multiple short profile HMMs (each built from a short, highly conserved region) represent each family. The six main GPCR classes (A–E and Frizzled/Smoothened) are included in their profile-HMM library. Since plant 7TMR proteins are not well represented in their library, it limits the use of PRED-GPCR for plant-specific 7TMR search. GCR1 or none of the currently known *Arabidopsis* 7TMR proteins is recognized by PRED-GPCR.

2.3.3 Weakness in Alignment-Based Methods

One inherent problem in all of the methods mentioned earlier, however, is that they rely on multiple alignments for generating *models* (e.g., patterns, profiles, profile HMMs). Generating reliable multiple alignments becomes difficult or practically impossible when extremely diverged sequences are involved. Diagnostic patterns and profiles cannot be identified easily from highly diverged sequences, either. Another problem shared by the existing methods is that their models need to be built (trained) from already known protein sequences (*positive samples*). No information from *negative samples* (unrelated protein sequences) is incorporated to the models. Since subsequently found proteins are classified based on these models, possible initial sampling bias is kept and possibly reinforced. Such rigidness of the models would affect the ability of these methods to identify divergent 7TMRs from diverse genomes or to discover new families/groups of 7TMRs yet to be identified.

2.4 Alignment-Free Methods

Extremely diverged 7TMR sequences challenge methods' ability to identify remote similarity in protein classification. Limitation in alignment-based methods described previously prompted computational biologists to develop *alignment-free* methods for protein identification. Instead of using aligned sequences, in alignment-free methods, various sequence descriptors are extracted from each sequence. Most often used descriptors include amino acid composition, dipeptide composition, and various physico-chemical indices. Many methods have been developed particularly for identifying and classifying 7TMR proteins. Approaches used include: discriminant function analyses (Kim et al. 2000; Moriyama and Kim 2005), support vector machines (SVM; Bhasin and Raghava 2004, 2005; Karchin et al. 2002; Liao and Noble 2003; Strope and Moriyama 2007), and partial least squares regression (PLS; Gunnarsson et al. 2003; Lapinsh et al. 2002; Opiyo and Moriyama 2007).

For example, after unsuccessful application of various alignment-based methods, an alignment-free method successfully identified odorant and gustatory receptors from the *D. melanogaster* genome for the first time (Clyne et al. 1999, 2000; Kim et al. 2000). SVM classifiers based on dipeptide composition (GPCRpred and GPCRsclass) showed high accuracies in GPCR family/subfamily prediction (Bhasin and Raghava 2004, 2005). Our recent comparative analyses showed that alignment-free classifiers are in fact more sensitive to remote similarities than profile HMMs (Moriyama and Kim 2005; Opiyo and Moriyama 2007; Strope and Moriyama 2007). These classifiers can also identify weak similarities from short subsequences. Furthermore, we observed that these alignment-free classifiers have an advantage over profile HMMs when a sufficient training set is unavailable (Opiyo and Moriyama 2007).

GPCRpred (Bhasin and Raghava 2004), GPCRsIdentifier (Gao and Wang 2006), and GPCRTree (Davies et al. 2007, 2008) all perform family/subfamily-level GPCR classifications. While GPCRpred and GPCRTree are trained with the five main classes (A–E), GPCRsIdentifier includes also the sixth class (Frizzled/Smoothened) in their training. GPCRpred and GPCRsIdentifier use amino acid and dipeptide compositions as the descriptors. SVMs (in GPCRpred) and the nearest neighbor algorithm (in GPCRsIdentifier) are used for GPCR classification. In GPCRTree, five principal components derived from amino acid properties and transformed with auto cross covariance (originally developed by Lapinsh et al. 2002) are used as descriptors. Eight classification algorithms, including Naīve

Bayes, SVM, and decision list, are combined as a top-down classification approach. At odds with our current understanding of known plant 7TMRs described earlier, both GPCRpred and GPCRTree predicted *Arabidopsis* GCR1 as Class A GPCRs. Furthermore, GPCRpred and GPCRTree predicted MLO and HHP proteins to be either Class A or B or C GPCRs depending on the protein. Since no plant 7TMR was included in their training sets, use of these classifiers for plant 7TMR research needs to be done with caution.

2.5 Transmembrane Prediction Methods

The topology of canonical 7TMR (or GPCR) proteins is to have seven TM regions and an extracellularly located N-terminus. However, no single TM prediction method predicts exactly seven TM regions from all known 7TMRs. Among known GPCR sequences in the GPCRDB, less than 90% are predicted to have exactly seven TM regions (Moriyama et al. 2006). Choosing the TM number ranging from five to nine, for example, covered the majority (99% or more) of the known GPCRs. Note also that as mentioned in the earlier section, GTG1/GTG2 appear to have nine TM regions. Furthermore, HHP proteins have their N-termini located intracellularly (Hsieh and Goodman 2005). Therefore, it is advisable to use a range in the number of predicted TM regions for identification purpose.

Many secreted proteins contain short N-terminal signal peptides, which often have strongly hydrophobic segments. Many TM prediction methods make errors by misidentifying these signal peptides as one of the TM regions. Phobius (Käll et al. 2004, 2007) addressed this problem by combining signal peptide prediction with TM prediction. Figure 5 and Table 3 summarize the comparison of the number of TM regions predicted by Phobius and HMMTOP2.1 (Tusnády and Simon 1998, 2001). Although HMMTOP2.1 is considered as one of the best TM prediction methods (Chen et al. 2002; Cuthbertson et al. 2005), it does not discriminate signal peptide regions. A large difference in the prediction by Phobius and HMMTOP is found in the numbers of 1 and 2–4 TM proteins. These two groups of TM proteins are predicted twice more often by HMMTOP than by Phobius, which results in the

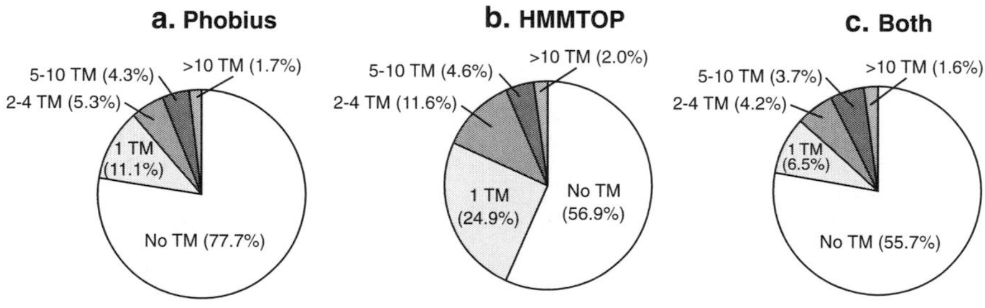

Fig. 5 Predicted transmembrane proteins in the *A. thaliana* genome. The predictions are done by Phobious (**a**), HMMTOP (**b**), or both methods (**c**)

Table 3 Transmembrane proteins predicted from plant genomes

Genomes (proteins)[a]	TM proteins[b]	7TM[c]	5–10TM[c]
Dicots			
Arabidopsis thaliana (32,690)	22%:43%:21% (7,294:14,091:6,916)	245:252:134	1,393:1,499:1,212
Arabidopsis lyrata (32,669)	21%:41%:20% (6,930:13,338:6,560)	209:212:111	1,176:1,273:1,002
Glycine max (75,045)	24%:41%:22% (17,978:30,784:16,866)	482:498:253	2,771:3,044:2,403
Populus trichocarpa (45,555)	20%:36%:19% (9,105:16,435:8,593)	327:297:153	1,704:1,792:1,434
Vitis vinifera (29,526)	24%:42%:23% (7,192:12,300:6,703)	218:216:111	1,220:1,289:1,027
Monocots			
Oryza sativa (26,742)	21%:47%:20% (5,720:12,590:5,372)	194:183:92	1,067:1,150:911
Sorghum bicolor (35,888)	18%:41%:17% (6,624:14,890:6,222)	228:190:106	1,207:1,287:1,012
Zea mays (75,257)	20%:42%:19% (14,959:31,920:14,013)	470:465:227	2,776:3,012:2,383
Fern/moss			
Selaginella moellendorffii (34,697)	18%:39%:17% (6,214:13,424:5,798)	234:230:107	1,385:1,496:1,172
Physcomitrella patens (35,938)	16%:34%:14% (5,609:12,075:5,180)	153:160:76	909:1,022:779
Green algae			
Chlorella vulgaris C-169 (9,994)	18%:37%:17% (1,793:3,679:1,703)	76:77:42	452:482:380
Chlorella sp. NC64A (9,791)	19%:36%:17% (1,838:3,556:1,673)	95:83:41	506:527:424
Chlamydomonas reinhardtii (16,688)	17%:34%:15% (2,784:5,675:2,483)	91:83:27	577:612:467
Ostreococcus lucimarinus (7,651)	17%:33%:16% (1,325:2,524:1,238)	63:46:18	356:373:296
Ostreococcus tauri (7,725)	17%:32%:16% (1,289:2,467:1,216)	56:53:15	340:365:294
Volvox carteri (15,544)	19%:39%:17% (2,907:6,126:2,643)	117:115:55	581:644:485
Diatoms			
Phaeodactylum tricornutum (10,402)	24%:43%:23% (2,496:4,440:2,361)	99:101:45	588:641:487
Thalassiosira pseudonana (11,776)	21%:38%:20% (2,505:4,487:2,338)	102:92:42	605:682:519

[a]Data source for the plant genomes:
The Arabidopsis Information Resource (*A. thaliana*) (TAIR8; Swarbreck et al. 2008), National Center for Biotechnology Information (*O. sativa*), Genoscope (*V. vinifera*), The Maize Genome Sequencing Project (*Z. mays*), and the U.S. Department of Energy Joint Genome Institute for all other species
The number of proteins used for each genome is shown in parentheses. Protein sequences shorter than 35 amino acids and those included more than 30% of "X" (unknown residue) were excluded
[b]The percent proteins predicted to have one or more TM regions in each genome by Phobius, HMMTOP, and both the methods. The numbers of predicted TM proteins are shown in parentheses
[c]The number of proteins predicted to have 7 or 5–10TM regions in each genome by Phobius, HMMTOP, and both the methods

reduced number of non-TM proteins in HMMTOP prediction. Proteins with higher numbers of TMs also show consistent but much smaller differences between Phobius and HMMTOP. Further examinations showed that, for example, among 7,175 *A. thaliana* proteins predicted as non-TM (0) by Phobius and TM (>0) by HMMTOP, 2,847 proteins (39.7%) were predicted to have signal peptides by Phobius. Among 18,221 *A. thaliana* proteins predicted to be non-TM by both methods, on the other hand, only 1,177 (6.5%) were predicted to have signal peptides by Phobius. These observations clearly show that Phobius takes advantage of signal peptide prediction to avoid misidentifying signal peptide regions as TM regions. Proteins predicted to have no (0) TM by both the methods constitute 55.7% of the *A. thaliana* genome; they are most likely truly non-TM proteins. The maximum proportion of non-TM proteins could be as many as 78.7% (predicted as 0 TM by Phobius).

As shown in Table 3, plant genomes appear to contain roughly 200–300 7TM proteins except for the *Glycine max* (soybean) genome, whose larger genome contains close to 500 7TM proteins. Smaller green algal and diatom genomes have 100 or fewer 7TM proteins. The most conservative estimates (predicted by both the methods) of 7TM proteins are 250 or fewer for the majority of the plants (~250 in *G. max*) and fewer than 100 in green algae and diatoms.

2.6 GPCRHMM

This unique GPCR-specific classifier was developed by Wistrand et al. (2006). It combines the HMM-based transmembrane prediction with GPCR-unique feature extraction. Their compartmentalized HMM incorporates distinct loop length patterns and differences in amino acid composition between cytosolic loops, extracellular loops, and membrane regions based on a diverse set of GPCR sequences. Their training set included eleven of 13 PFAM GPCR protein families. The two divergent families excluded from their training set as the outliers are: *Drosophila* odorant receptor family 7tm_6 (PF02949) and the plant family Mlo (PF03094). Due to this rather limited training dataset, similar to other GPCR-specific classification methods (GPCRpred, GPCRsIdentifier, and GPCRTree), except for GCR1, GPCRHMM does not identify any of the currently known plant 7TMRs as positives.

3 Mining 7TMR Proteins from Plant Genomes

As we described in the previous section, many protein classifiers are available. Different classifiers have different strength and weakness. For example, profile-HMM classifiers are accurate in identifying well-established protein family, and few false-positives are given. While alignment-free classifiers are more sensitive to remote similarities, one disadvantage of these classifiers is their relatively high false-positive rate. Therefore, to achieve a thorough mining of 7TMR proteins both types of methods need to be combined.

In Moriyama et al. (2006), we showed a power of hierarchically combining multiple classifiers, including traditional alignment-based and newer alignment-free methods as well as transmembrane prediction methods. Based on the older version of the *A. thaliana* genome, we identified 394 proteins as 7TMR candidates and selected 54 proteins as those prioritized for further investigation. More recently, Gookin et al. (2008) used a similar strategy by combining several methods hierarchically and identified a small number of GPCR candidates from three plant genomes including *A. thaliana*. They confirmed experimentally that seven of the *Arabidopsis* proteins predicted to be GPCR candidates interacted with Gα. It should be noted that Gookin et al.'s focus was to identify only regular GPCR's. Therefore, they used a strict filtering criteria based on the "positive GPCRHMM prediction" as well as "predicted to have 7TM regions." Their GPCR candidates were composed of 16 *Arabidopsis* proteins, including GCR1, RGS1, and one of HHPs (HHP2).

We have recently retrained the same set of classifiers using larger training datasets and performed the classification analysis using the most recent *A. thaliana* genome data (TAIR8; http://www.arabidopsis.org). All of the 24 known *Arabidopsis* 7TMRs are found to have 7–9 TM regions either by Phobius or HMMTOP. GPCRHMM predicted only GCR1 as positive. A profile HMM classifier (SAM2) recognized only GCR1 and MLOs. All the six alignment-free classifiers recognized all but one (GTG2) of the 24 7TMRs as positive (the six classifiers include LDA, QDA, KNN20, SVM_AA, SVM_di, and PLS_ACC; see the *7TMRmine* website, http://bioinfolab.unl.edu/emlab/7tmr/index.php, for more details). GTG2 protein was missed only by PLS_ACC. Based on these results, we combined eight classifiers and performed the 7TMR classification analysis against 18 plant genomes.

Figure 6 summarizes the classification results in Venn diagrams. The eight classifiers are grouped into three meta-classifiers: "5CLASS" is the intersection of the five alignment-free classifiers, "HMM" combines GPCRHMM and profile HMM classifier, and "7–9TM" includes proteins predicted to have 7–9TM regions by either Phobius or HMMTOP. Proteins positively identified by at least two of the three meta-classifiers would most likely include all the 7MTR candidates. For the *A. thaliana* genome, this combination includes 475 (21 + 43 + 410 + 1) 7TMR candidates. Using more limited criteria, we can produce a shorter prioritized list. For example, using a stricter TM-number criteria (7–8TMs by both Phobius and HMMTOP), the number of *A. thaliana* 7TMR candidates identified by the same meta-classifier combination becomes194 (21 + 24 + 148 + 1). Although this list is shorter and easier to handle, it may not include irregular proteins that have more than 7–8TM regions as GTG1 and GTG2. This shorter list should thus provide the first prioritized 7TMR-candidate list. Candidate lists can be expanded using more relaxed criteria for subsequent studies.

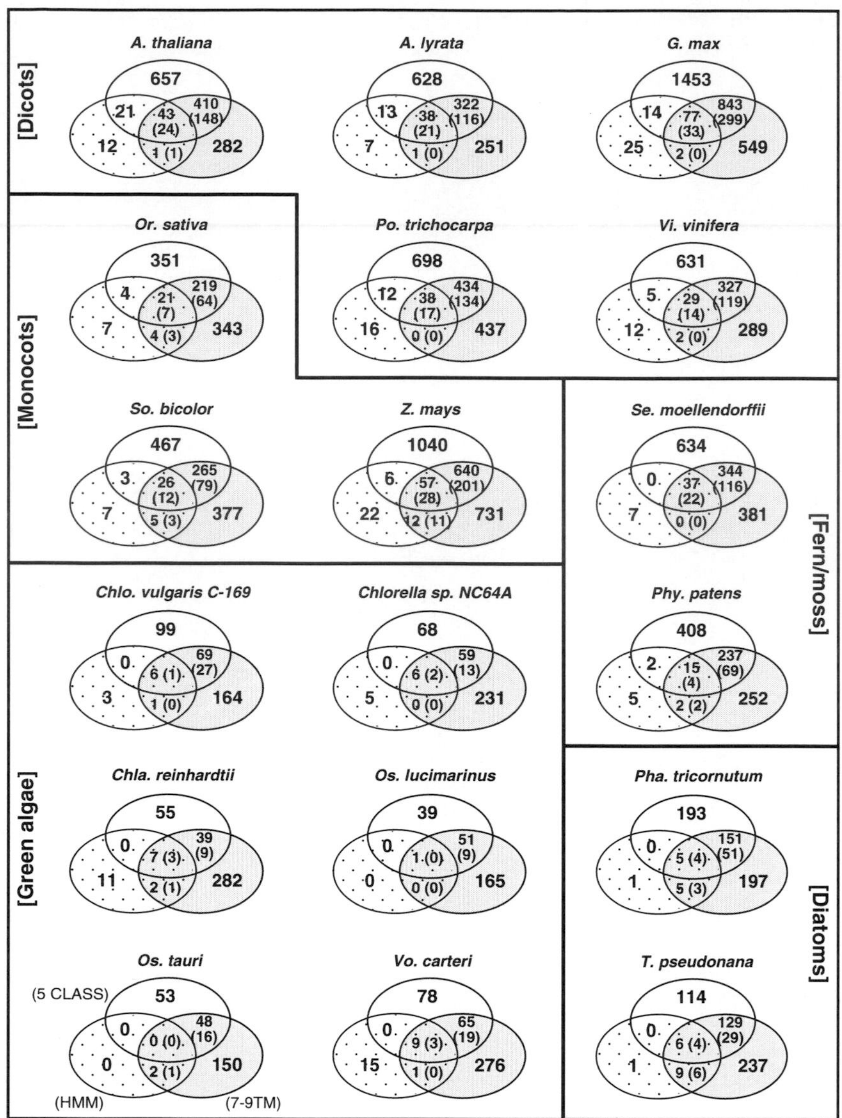

Fig. 6 7TMR protein mining from plant genomes. Each Venn diagram shows the results of 7TMR mining performed by three meta-classifiers (5CLASS, HMM, and 7–9TM). Open circles show the positives identified by the 5CLASS meta-classifier, which is the intersection of five alignment-free classifiers (LDA, QDA, KNN20, SVM_AA, and SVM_di). Circles filled with dots show the positives identified by the HMM meta-classifier, which is the union of GPCRHMM and a profile HMM classifier. Gray circles show the proteins predicted to have 7–9TMs either by Phobius or HMMTOP (the meta-classifier 7–9TMs). Classification results with a more limited TM criterion (7–8TMs by both Phobius and HMMTOP) are also shown in parentheses. See *7TMRmine* website (http://bioinfolab.unl.edu/emlab/7tmr/index.php) for more details of classifiers, more statistics, and lists of predicted plant 7TMR proteins

4 Conclusion

Plants have a simpler G-protein signaling repertoire compared with metazoan systems. Comparative analysis revealed that many of 7TMR proteins found in plants are in fact unique to the plant kingdom. GCR1 and HHPs are the only plant 7TMRs their homologues clearly exist in metazoa. Even more interestingly, some of the 7TMR proteins (e.g., RGA1 and GTGs) appear to have acquired GPCR functions independently after their ancestral proteins parted their evolution from the metazoan lineage. In order to understand how each of the plant 7TMR families has evolved, further investigation is needed.

The origin(s) of 7TMR proteins go(es) back to the level of eukaryote–prokaryote divergence. In order to understand such a deep divergence, more powerful and sensitive bioinformatics tools are necessary. We recently developed a Web server, *7TMRmine* (http://bioinfolab.unl.edu/emlab/7tmr; Lu et al. 2009). Fourteen classifiers (four alignment-based and ten alignment-free) and two transmembrane-prediction methods are incorporated. Using *7TMRmine*, users can dynamically build a multilevel filtering process to generate reduced prioritized candidate lists. This Web server would facilitate exploring all possible 7TMR candidates from diverse plant genomes.

Acknowledgments The authors thank Dr. Alan Jones (University of North Carolina) for discussion and unpublished information. This work was in part funded by the grant number R01LM009219 from the National Library of Medicine to E.N.M.

References

Altschul SF, Gish W, Miller W, Myers EW, Lipman DJ (1990) Basic local alignment search tool. J Mol Biol 215:403–410

Altschul SF, Madden TL, Schaffer AA, Zhang J, Zhang Z, Miller W, Lipman DJ (1997) Gapped BLAST and PSI-BLAST: a new generation of protein database search programs. Nucleic Acids Res 25:3389–3402

Attwood TK, Bradley P, Flower DR, Gaulton A, Maudling N, Mitchell AL, Moulton G, Nordle A, Paine K, Taylor P, Uddin A, Zygouri C (2003) PRINTS and its automatic supplement, prePRINTS. Nucleic Acids Res 31:400–402

Baasiri RA, Lu X, Rowley PS, Turner GE, Borkovich KA (1997) Overlapping functions for two G protein alpha subunits in *Neurospora crassa*. Genetics 147:137–145

Bhasin M, Raghava GP (2004) GPCRpred: an SVM-based method for prediction of families and subfamilies of G-protein coupled receptors. Nucleic Acids Res 32:W383–W389

Bhasin M, Raghava GP (2005) GPCRsclass: a web tool for the classification of amine type of G-protein-coupled receptors. Nucleic Acids Res 33:W143–W147

Bjarnadóttir TK, Gloriam DE, Hellstrand SH, Kristiansson H, Fredriksson R, Schiöth HB (2006) Comprehensive repertoire and phylogenetic analysis of the G protein-coupled receptors in human and mouse. Genomics 88:263–273

Chen CP, Kernytsky A, Rost B (2002) Transmembrane helix predictions revisited. Protein Sci 11:2774–2791

Chen J-G, Willard FS, Huang J, Liang J, Chasse SA, Jones AM, Siderovski DP (2003) A seven-transmembrane RGS protein that modulates plant cell proliferation. Science 301:1728–1731

Chen JG (2008) Heterotrimeric G-proteins in plant development. Front Biosci 13:3321–3333

Chen JG, Pandey S, Huang J, Alonso JM, Ecker JR, Assmann SM, Jones AM (2004) GCR1 can act independently of heterotrimeric G-protein in response to brassinosteroids and gibberellins in *Arabidopsis* seed germination. Plant Physiol 135:907–915

Chen Z, Hartmann HA, Wu MJ, Friedman EJ, Chen JG, Pulley M, Schulze-Lefert P, Panstruga R, Jones AM (2006) Expression analysis of the AtMLO gene family encoding plant-specific seven-transmembrane domain proteins. Plant Mol Biol 60:583–597

Clyne PJ, Warr CG, Carlson JR (2000) Candidate taste receptors in *Drosophila*. Science 287:1830–1833

Clyne PJ, Warr CG, Freeman MR, Lessing D, Kim JH, Carlson JR (1999) A novel family of divergent seven-transmembrane proteins: candidate odorant receptors in *Drosophila*. Neuron 22:327–338

Cuthbertson JM, Doyle DA, Sansom MS (2005) Transmembrane helix prediction: a comparative evaluation and analysis. Protein Eng Des Sel 18:295–308

Davies MN, Secker A, Freitas AA, Mendao M, Timmis J, Flower DR (2007) On the hierarchical classification of G protein-coupled receptors. Bioinformatics 23:3113–3118

Davies MN, Secker A, Halling-Brown M, Moss DS, Freitas AA, Timmis J, Clark E, Flower DR (2008) GPCRTree: online hierarchical classification of GPCR function. BMC Res Notes 1:67

Devoto A, Hartmann HA, Piffanelli P, Elliott C, Simmons C, Taramino G, Goh CS, Cohen FE, Emerson BC, Schulze-Lefert P, Panstruga R (2003) Molecular phylogeny and evolution of the plant-specific seven-transmembrane MLO family. J Mol Evol 56:77–88

Devoto A, Piffanelli P, Nilsson I, Wallin E, Panstruga R, von Heijne G, Schulze-Lefert P (1999) Topology, subcellular localization, and sequence diversity of the Mlo family in plants. J Biol Chem 274:34993–35004

Douzery EJ, Snell EA, Bapteste E, Delsuc F, Philippe H (2004) The timing of eukaryotic evolution: does a relaxed molecular clock reconcile proteins and fossils? Proc Natl Acad Sci USA 101:15386–15391

Eddy SR (1998) Profile hidden Markov models. Bioinformatics 14:755–763

Fernandes MS, Pierron V, Michalovich D, Astle S, Thornton S, Peltoketo H, Lam EW, Gellersen B, Huhtaniemi I, Allen J, Brosens JJ (2005) Regulated expression of putative membrane progestin receptor homologues in human endometrium and gestational tissues. J Endocrinol 187:89–101

Fredriksson R, Lagerström MC, Lundin LG, Schiöth HB (2003) The G-protein-coupled receptors in the human genome form five main families. Phylogenetic analysis, paralogon groups, and fingerprints. Mol Pharmacol 63:1256–1272

Fredriksson R, Lagerström MC, Schiöth HB (2005) Expansion of the superfamily of G-protein-coupled receptors in chordates. Ann NY Acad Sci 1040:89–94

Fredriksson R, Schiöth HB (2005) The repertoire of G-protein-coupled receptors in fully sequenced genomes. Mol Pharmacol 67:1414–1425

Galagan JE, Calvo SE, Borkovich KA, Selker EU, Read ND, Jaffe D, FitzHugh W, Ma LJ, Smirnov S, Purcell S, Rehman B, Elkins T, Engels R, Wang SG, Nielsen CB, Butler J, Endrizzi M, Qui DY, Ianakiev P, Pedersen DB, Nelson MA, Werner-Washburne M, Selitrennikoff CP, Kinsey JA, Braun EL, Zelter A, Schulte U, Kothe GO, Jedd G, Mewes W, Staben C, Marcotte E, Greenberg D, Roy A, Foley K, Naylor J, Stabge-Thomann N, Barrett R, Gnerre S, Kamal M, Kamvysselis M, Mauceli E, Bielke C, Rudd S, Frishman D, Krystofova S, Rasmussen C, Metzenberg RL, Perkins DD, Kroken S, Cogoni C, Macino G, Catcheside D, Li WX, Pratt RJ, Osmani SA, DeSouza CPC, Glass L, Orbach MJ, Berglund JA, Voelker R, Yarden O, Plamann M, Seller S, Dunlap J, Radford A, Aramayo R, Natvig DO, Alex LA, Mannhaupt G, Ebbole DJ, Freitag M, Paulsen I, Sachs MS, Lander ES, Nusbaum C, Birren B (2003) The genome sequence of the filamentous fungus *Neurospora crassa*. Nature 422: 859–868

Gao QB, Wang ZZ (2006) Classification of G-protein coupled receptors at four levels. Protein Eng Des Sel 19:511–516

Gilman AG (1987) G proteins: transducers of receptor-generated signals. Annu Rev Biochem 56:615–649

Gloriam DE, Fredriksson R, Schiöth HB (2007) The G protein-coupled receptor subset of the rat genome. BMC Genomics 8:338

Gookin TE, Kim J, Assmann SM (2008) Whole proteome identification of plant candidate G-protein coupled receptors in *Arabidopsis*, rice, and poplar: computational prediction and in-vivo protein coupling. Genome Biol 9:R120

Gunnarsson I, Andersson P, Wikberg J, Lundstedt T (2003) Multivariate analysis of G protein-coupled receptors. J Chemometrics 17:82–92

Guo J, Zeng Q, Emami M, Ellis BE, Chen JG (2008) The GCR2 gene family is not required for ABA control of seed germination and early seedling development in *Arabidopsis*. PLoS ONE 3:e2982

Hedges SB, Blair JE, Venturi ML, Shoe JL (2004) A molecular timescale of eukaryote evolution and the rise of complex multicellular life. BMC Evol Biol 4:2

Horn F, Bettler E, Oliveira L, Campagne F, Cohen FE, Vriend G (2003) GPCRDB information system for G protein-coupled receptors. Nucleic Acids Res 31:294–297

Hsieh M-H, Goodman HM (2005) A novel gene family in *Arabidopsis* encoding putative hepta-helical transmembrane proteins homologous to human adiponectin receptors and progestin receptors. J Exp Bot 56:3137–3147

Hulo N, Bairoch A, Bulliard V, Cerutti L, Cuche BA, de Castro E, Lachaize C, Langendijk-Genevaux PS, Sigrist CJ (2008) The 20 years of PROSITE. Nucleic Acids Res 36:D245–D249

Illingworth CJ, Parkes KE, Snell CR, Mullineaux PM, Reynolds CA (2008) Criteria for confirming sequence periodicity identified by Fourier transform analysis: application to GCR2, a candidate plant GPCR? Biophys Chem 133:28–35

Johnston CA, Taylor JP, Gao Y, Kimple AJ, Grigston JC, Chen JG, Siderovski DP, Jones AM, Willard FS (2007a) GTPase acceleration as the rate-limiting step in Arabidopsis G protein-coupled sugar signaling. Proc Natl Acad Sci USA 104:17317–17322

Johnston CA, Temple BR, Chen JG, Gao Y, Moriyama EN, Jones AM, Siderovski DP, Willard FS (2007b) Comment on "A G protein coupled receptor is a plasma membrane receptor for the plant hormone abscisic acid." Science 318: 914, author reply 914

Jones AM, Assmann SM (2004) Plants: the latest model system for G-protein research. EMBO Rep 5:572–578

Josefsson LG (1999) Evidence for kinship between diverse G-protein coupled receptors. Gene 239:333–340

Josefsson LG, Rask L (1997) Cloning of a putative G-protein-coupled receptor from *Arabidopsis thaliana*. Eur J Biochem 249:415–420

Käll L, Krogh A, Sonnhammer EL (2004) A combined transmembrane topology and signal peptide prediction method. J Mol Biol 338:1027–1036

Käll L, Krogh A, Sonnhammer EL (2007) Advantages of combined transmembrane topology and signal peptide prediction–the Phobius web server. Nucleic Acids Res 35:W429–W432

Karchin R, Karplus K, Haussler D (2002) Classifying G-protein coupled receptors with support vector machines. Bioinformatics 18:147–159

Kim J, Moriyama EN, Warr CG, Clyne PJ, Carlson JR (2000) Identification of novel multi-transmembrane proteins from genomic databases using quasi-periodic structural properties. Bioinformatics 16:767–775

Kimmel AR, Parent CA (2003) The signal to move: *D. discoideum* go orienteering. Science 300:1525–1527

Kulkarni R, Thon M, Pan H, Dean R (2005) Novel G-protein-coupled receptor-like proteins in the plant pathogenic fungus *Magnaporthe grisea*. Genome Biol 6:R24

Lapinsh M, Gutcaits A, Prusis P, Post C, Lundstedt T, Wikberg JES (2002) Classification of G-protein coupled receptors by alignment-independent extraction of principal chemical properties of primary amino acid sequences. Protein Sci 11:795–805

Lefkowitz RJ (2007) Seven transmembrane receptors: something old, something new. Acta Physiol (Oxf) 190:9–19

Lefkowitz RJ, Shenoy SK (2005) Transduction of receptor signals by beta-arrestins. Science 308:512–517

Letunic I, Doerks T, Bork P (2009) SMART 6: recent updates and new developments. Nucleic Acids Res 37:D229–D232

Liao L, Noble WS (2003) Combining pairwise sequence similarity and support vector machines for detecting remote protein evolutionary and structural relationships. J Comput Biol 10:857–868

Liu Q, Zhu H (2008) Molecular evolution of the MLO gene family in Oryza sativa and their functional divergence. Gene 409:1–10

Liu X, Yue Y, Li B, Nie Y, Li W, Wu WH, Ma L (2007) A G protein-coupled receptor is a plasma membrane receptor for the plant hormone abscisic acid. Science 315:1712–1716

Lu G, Wang Z, Jones AM, Moriyama EN (2009) 7TMRmine: A Web server for hierarchical mining of 7TMR proteins. BMC Genomics 10:275

Luttrell LM (2008) Reviews in molecular biology and biotechnology: transmembrane signaling by G protein-coupled receptors. Mol Biotechnol 39:239–264

Maeda Y, Ide T, Koike M, Uchiyama Y, Kinoshita T (2008) GPHR is a novel anion channel critical for acidification and functions of the Golgi apparatus. Nat Cell Biol 10:1135–1145

Mi H, Lazareva-Ulitsky B, Loo R, Kejariwal A, Vandergriff J, Rabkin S, Guo N, Muruganujan A, Doremieux O, Campbell MJ, Kitano H, Thomas PD (2005) The PANTHER database of protein families, subfamilies, functions and pathways. Nucleic Acids Res 33:D284–D288

Moriyama EN, Kim J (2005) Protein family classification with discriminant function analysis. In: Gustafson JP, Shoemaker R, Snape JW (eds) Genome exploitation: data mining the genome. Springer, New York, pp 121–132

Moriyama EN, Strope PK, Opiyo SO, Chen Z, Jones AM (2006) Mining the *Arabidopsis thaliana* genome for highly-divergent seven transmembrane receptors. Genome Biology 7:R96

Mulder NJ, Apweiler R (2008) The InterPro database and tools for protein domain analysis. Curr Protoc Bioinformatics Chapter 2, Unit 2 7

Nakafuku M, Itoh H, Nakamura S, Kaziro Y (1987) Occurrence in *Saccharomyces cerevisiae* of a gene homologous to the cDNA coding for the alpha subunit of mammalian G proteins. Proc Natl Acad Sci USA 84:2140–2144

Nakafuku M, Obara T, Kaibuchi K, Miyajima I, Miyajima A, Itoh H, Nakamura S, Arai K, Matsumoto K, Kaziro Y (1988) Isolation of a second yeast Saccharomyces cerevisiae gene (GPA2) coding for guanine nucleotide-binding regulatory protein: studies on its structure and possible functions. Proc Natl Acad Sci USA 85:1374–1378

Opiyo SO, Moriyama EN (2007) Protein family classification with partial least squares. J Proteome Res 6:846–853

Pandey S, Assmann SM (2004) The *Arabidopsis* putative G protein-coupled receptor GCR1 interacts with the G protein alpha subunit GPA1 and regulates abscisic acid signaling. Plant Cell 16:1616–1632

Pandey S, Nelson DC, Assmann SM (2009) Two novel GPCR-type G proteins are abscisic acid receptors in *Arabidopsis*. Cell 136:136–148

Papasaikas PK, Bagos PG, Litou ZI, Hamodrakas SJ (2003) A novel method for GPCR recognition and family classification from sequence alone using signatures derived from profile hidden Markov models. SAR QSAR Environ Res 14:413–420

Papasaikas PK, Bagos PG, Litou ZI, Promponas VJ, Hamodrakas SJ (2004) PRED-GPCR: GPCR recognition and family classification server. Nucleic Acids Res 32:W380–W382

Pierce KL, Premont RT, Lefkowitz RJ (2002) Seven-transmembrane receptors. Nat Rev Mol Cell Biol 3:639–650

Sammut SJ, Finn RD, Bateman A (2008) Pfam 10 years on: 10, 000 families and still growing. Brief Bioinform 9:210–219

Schiöth HB, Fredriksson R (2005) The GRAFS classification system of G-protein coupled receptors in comparative perspective. Gen Comp Endocrinol 142:94–101

Schiöth HB, Nordström KJ, Fredriksson R (2007) Mining the gene repertoire and ESTs for G protein-coupled receptors with evolutionary perspective. Acta Physiol (Oxf) 190:21–31

Schwacke R, Schneider A, van der Graaff E, Fischer K, Catoni E, Desimone M, Frommer WB, Flugge UI, Kunze R (2003) ARAMEMNON, a novel database for *Arabidopsis* integral membrane proteins. Plant Physiol 131:16–26

Skrabanek L, Campagne F, Weinstein H (2003) Building protein diagrams on the web with the residue-based diagram editor RbDe. Nucleic Acids Res 31:3856–3858

Smart R, Kiely A, Beale M, Vargas E, Carraher C, Kralicek AV, Christie DL, Chen C, Newcomb RD, Warr CG (2008) Drosophila odorant receptors are novel seven transmembrane domain proteins that can signal independently of heterotrimeric G proteins. Insect Biochem Mol Biol 38:770–780

Strope PK, Moriyama EN (2007) Simple alignment-free methods for protein classification: a case study from G-protein-coupled receptors. Genomics 89:602–612

Swarbreck D, Wilks C, Lamesch P, Berardini TZ, Garcia-Hernandez M, Foerster H, Li D, Meyer T, Muller R, Ploetz L, Radenbaugh A, Singh S, Swing V, Tissier C, Zhang P, Huala E (2008) The Arabidopsis Information Resource (TAIR): gene structure and function annotation. Nucleic Acids Res 36:D1009–D1014

Tang YT, Hu T, Arterburn M, Boyle B, Bright JM, Emtage PC, Funk WD (2005) PAQR proteins: a novel membrane receptor family defined by an ancient 7-transmembrane pass motif. J Mol Evol 61:372–380

Temple BR, Jones AM (2007) The plant heterotrimeric G-protein complex. Annu Rev Plant Biol 58:249–266

Thomas JH, Robertson HM (2008) The *Caenorhabditis* chemoreceptor gene families. BMC Biol 6:42

Thomas P, Pang Y, Dong J, Groenen P, Kelder J, de Vlieg J, Zhu Y, Tubbs C (2007) Steroid and G protein binding characteristics of the seatrout and human progestin membrane receptor alpha subtypes and their evolutionary origins. Endocrinology 148:705–718

Turner GE, Borkovich KA (1993) Identification of a G protein alpha subunit from *Neurospora crassa* that is a member of the Gi family. J Biol Chem 268:14805–14811

Tusnády GE, Simon I (1998) Principles governing amino acid composition of integral membrane proteins: application to topology prediction. J Mol Biol 283:489–506

Tusnády GE, Simon I (2001) The HMMTOP transmembrane topology prediction server. Bioinformatics 17:849–850

Whiteway M, Hougan L, Dignard D, Thomas DY, Bell L, Saari GC, Grant FJ, O'Hara P, MacKay VL (1989) The STE4 and STE18 genes of yeast encode potential beta and gamma subunits of the mating factor receptor-coupled G protein. Cell 56:467–477

Wilson D, Pethica R, Zhou Y, Talbot C, Vogel C, Madera M, Chothia C, Gough J (2009) SUPERFAMILY–sophisticated comparative genomics, data mining, visualization and phylogeny. Nucleic Acids Res 37:D380–D386

Wistrand M, Kall L, Sonnhammer EL (2006) A general model of G protein-coupled receptor sequences and its application to detect remote homologs. Protein Sci 15:509–521

Yamauchi T, Kamon J, Ito Y, Tsuchida A, Yokomizo T, Kita S, Sugiyama T, Miyagishi M, Hara K, Tsunoda M, Murakami K, Ohteki T, Uchida S, Takekawa S, Waki H, Tsuno NH, Shibata Y, Terauchi Y, Froguel P, Tobe K, Koyasu S, Taira K, Kitamura T, Shimizu T, Nagai R, Kadowaki T (2003) Cloning of adiponectin receptors that mediate antidiabetic metabolic effects. Nature 423:762–769

Yoon HS, Grant J, Tekle YI, Wu M, Chaon BC, Cole JC, Logsdon JM Jr, Patterson DJ, Bhattacharya D, Katz LA (2008) Broadly sampled multigene trees of eukaryotes. BMC Evol Biol 8:14

Zhu Y, Bond J, Thomas P (2003a) Identification, classification, and partial characterization of genes in humans and other vertebrates homologous to a fish membrane progestin receptor. Proc Natl Acad Sci USA 100:2237–2242

Zhu Y, Rice CD, Pang Y, Pace M, Thomas P (2003b) Cloning, expression, and characterization of a membrane progestin receptor and evidence it is an intermediary in meiotic maturation of fish oocytes. Proc Natl Acad Sci USA 100:2231–2236

Unconventional GTP-Binding Proteins in Plants

Lei Ding, Timothy E. Gookin, and Sarah M. Assmann

Abstract G proteins, including monomeric G proteins and heterotrimeric G proteins composed of Gα, Gβ, and Gγ subunits, are molecular switches in cellular signaling. In addition to these classic G proteins, plants have several types of unconventional G proteins, such as extra-large G proteins (XLGs), developmentally regulated G proteins (DRGs), and GPCR-type G proteins (GTGs). XLGs are nuclear-localized proteins with Gα-like *C*-termini and large, unique, *N*-terminal extensions. XLGs are involved in regulation of primary root growth, root waving and skewing, and plant responses to sugars, osmotic stress, pathogens, and hormones. DRGs have all the conserved GTPase domain motifs found in conventional G proteins but do not have any other sequence similarities with conventional G proteins. The functions of DRGs in plants remain unknown. GTGs have nine predicted transmembrane domains and exhibit GTP-binding and GTPase activities, as well as ABA-binding. As one class of ABA receptors, GTGs mediate most classic ABA responses. These unconventional G proteins diversify signaling pathways mediated by G proteins in plants.

1 Introduction

GTP-binding proteins, also known as G proteins, are a large group of proteins that bind and, typically, hydrolyze GTP (Bourne et al. 1990, 1991; Bourne 1995). Based on their subunit number and molecular mass, G proteins are classified into two

L. Ding (✉)
Department of Biology, Indiana University, Bloomington, IN 47405, USA
e-mail: ding3@indiana.edu

T.E. Gookin and S.M. Assmann (✉)
Department of Biology, 208 Mueller Laboratory, Pennsylvania State University, University Park, PA 16802, USA
e-mail: tegookin@psu.edu; sma3@psu.edu

S. Yalovsky et al. (eds.), *Integrated G Proteins Signaling in Plants*,
Signaling and Communication in Plants,
DOI 10.1007/978-3-642-03524-1_14,

major groups: heterotrimeric G proteins, which are composed of a GTP-binding and GTP-hydrolyzing α subunit along with β and γ subunits, and monomeric G proteins (often called small G proteins), which are comprised of a single subunit and have relatively lower molecular mass than the heterotrimeric G proteins. Despite these differences, heterotrimeric and monomeric G proteins share conserved structure of the GTPase domain, and molecular signaling via both types of G proteins is conserved in higher eukaryotes (Bourne et al. 1991; Sprang 1997).

Like metazoans and yeast, plants maintain both heterotrimeric and small G proteins. Molecular, genetic, biochemical, and physiological evidence demonstrates the presence of heterotrimeric G proteins in plants and their involvement in plant cell signaling (Ma 1994; Fujisawa et al. 2001; Assmann 2002; Jones 2002; Jones and Assmann 2004; Perfus-Barbeoch et al. 2004; Temple and Jones 2007; Chen 2008; Ding et al. 2008a). However, plants seem to have a small complement of prototypical heterotrimeric G protein subunits: one or two Gαs, one Gβ, and on the order of two Gγs (Jones and Assmann 2004; Temple and Jones 2007). In contrast, humans have 23 Gαs, 5 Gβs, and 12 Gγs. Similar to metazoan organisms, plants also contain most categories of small G proteins, such as Rho, Rab, Arf, and Ran (Takai et al. 2001; Yang 2002; Vernoud et al. 2003; Nibau et al. 2006; Meier 2007; Nielsen et al. 2008), although plants appear not to have Ras GTPases (Yang 2002; Vernoud et al. 2003).

The classic heterotrimeric G proteins and the typical small G proteins described above comprise the majority of the G proteins. However, there exist a few unconventional types of G proteins in animals and fungi, which can be assigned to neither of the two conventional types of G proteins. For instance, mammals have extra-large Gαs (XLαs) proteins, which are large alternative splicing variants of the Gsα gene *GNAS1* (Kehlenbach et al. 1994; Weinstein et al. 2004). Although expression of the *XLαs* occurs mainly in neural and endocrine tissues and thus is much more restricted than the *Gsα* (Pasolli et al. 2000; Pasolli and Huttner 2001; Plagge et al. 2004), like Gsα, XLαs interacts with the $\beta\gamma$ dimer and couple receptor-mediated adenylyl cyclase activation (Klemke et al. 2000; Bastepe et al. 2002; Linglart et al. 2006). Gh, originally identified as a GTP-binding protein (Im and Graham 1990; Im et al. 1990), turned out also to be a transglutaminase (TGase II) (Nakaoka et al. 1994). Thus, Gh/TGase II is a dual function protein with both transglutaminase activity and signaling activity as a G protein (Im et al. 1997; Mhaouty-Kodja 2004). Gh/TGase II has a unique GTP-binding motif, which is different from those common in heterotrimeric G proteins and small G proteins (Iismaa et al. 1997, 2000). Animals and fungi have another group of unconventional GTP-binding proteins called the developmentally regulated GTP-binding proteins (DRGs). The DRGs share with the heterotrimeric and small G proteins a conserved GTPase domain but harbor no similarity with the other classes of G proteins outside the GTPase domain (Sazuka et al. 1992; Li and Trueb 2000).The function of DRGs is not well understood, although they appear to play a regulatory role in genesis of blood cells (Mahajan et al. 1996; Zhao and Aplan 1998).

Similar to animals and fungi, in addition to canonical heterotrimeric G proteins and small G proteins, plants also have some unconventional GTP-binding proteins, including extra-large GTP-binding proteins (XLGs), DRGs, and G-protein-coupled receptor (GPCR)-type GTP-binding proteins (GTGs). These unconventional G proteins have mainly been studied, to date, in the model species *Arabidopsis*, and their study is still in the early stages. In this chapter, we review recent progress on the structure and function of these unconventional plant G proteins.

2 Extra-Large GTP-Binding Proteins

Prior to the completion of *Arabidopsis* genome sequencing, the Assmann laboratory identified an *Arabidopsis* expressed sequence tag (EST) clone with homology to GPA1 and used this EST clone as a probe to screen an *Arabidopsis* cDNA library and identify a corresponding full-length cDNA (Lee and Assmann 1999). The protein translated from the cDNA bore similarity to *Arabidopsis* GPA1, but was much longer, and was named extra-large GTP-binding protein 1 (XLG1). XLG1 (At2g23460) has a Gα-like domain at its *C*-terminus (Lee and Assmann 1999). Two additional family members, *XLG2* (At4g34390) and *XLG3* (At1g31930), were identified from the sequenced *Arabidopsis* genome by homology-based searches using the XLG1 protein sequence (Assmann 2002; Ding et al. 2008b). Like XLG1, each of XLG2 and XLG3 also contains a Gα-like domain at its *C*-terminus (Fig. 1). The identity and similarity between the Gα-like domains of XLGs and *Arabidopsis* GPA1 are 32% identity and 54% similarity for XLG1, 27% and 48% for XLG2, and 32% and 52% for XLG3, respectively.

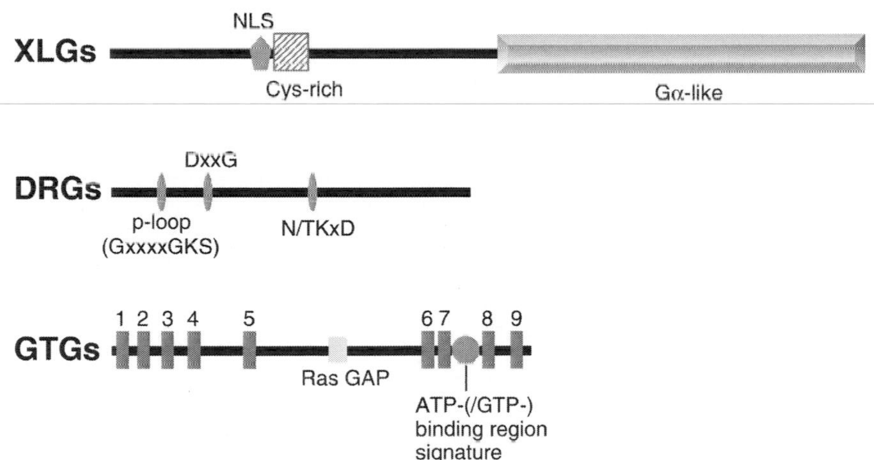

Fig. 1 Structural features of unconventional G proteins. Conserved domains of XLGs (Lee and Assmann 1999; Ding et al. 2008b), DRGs (Devitt et al. 1999; Etheridge et al. 1999), and GTGs (Pandey et al. 2009) are illustrated. NLS, predicted nuclear localization signal. 1–9: nine transmembrane domains of GTGs. Ras GAP: the consensus sequence of Ras GTPase-activating proteins

2.1 Structural features of Arabidopsis XLGs

Generally speaking, the Gα-like domains of XLGs maintain all three motifs essential for GTP binding and GTP hydrolysis, although some typically conserved residues in these motifs are not conserved in the XLGs (Lee and Assmann 1999; Temple and Jones 2007; Ding et al. 2008a; Ding et al. 2008b). The p-loop motif, featured by GxxxxGKS/T for Gα proteins, is required for NTP binding (Saraste et al. 1990; Bourne et al. 1991; Sprang 1997; Temple and Jones 2007). The p-loop of XLG1 and XLG3 is characterized by GxxxxGTS, whereas XLG2 has GxxxxGAT in its p-loop. The DxxGQ motif is involved in GTP hydrolysis (Bourne et al. 1991; Sprang 1997; Temple and Jones 2007). However, the region of the XLGs that corresponds to the DxxGQ motif shows significant variation: the conserved D is replaced by R in all three XLGs and none of the XLGs has conserved amino acid residues G and Q. The third motif is the NKxD motif, which is essential for guanine recognition (Bourne et al. 1991; Sprang 1997; Temple and Jones 2007). XLG1 and XLG3 maintain all three conserved residues in this motif; however, XLG2 has one variation in the motif, with the N replaced by a T while keeping the conserved K and D. Despite these changes in the conserved residues, recombinant XLG1 has been demonstrated to specifically bind GTP (Table 1) (Lee and Assmann 1999). Another structural feature of conventional Gα proteins is that upon GTP binding or hydrolysis, Gα proteins undergo conformational changes at three regions named switch I, II, and III respectively (Sprang 1997). Compared with conventional Gαs, the switch regions of XLGs have significant alterations and sequence divergence (Temple and Jones 2007; Ding et al. 2008a).

Besides the *C*-terminal Gα-like domain, each of the *Arabidopsis* XLGs has an additional *N*-terminal region of more than 400 amino acids, which is not present in *Arabidopsis* GPA1, in any other conventional Gα protein, or in the mammalian XLαs. In these *N*-terminal regions, there exist two conserved motifs: one is a predicted nuclear localization signal (NLS) and the other is a cysteine-rich region (Fig. 1) (Lee and Assmann 1999; Ding et al. 2008b). The region extending from the cysteine-rich region to the beginning of the Gα region bears even higher sequence conservation among the three XLGs than their Gα-like domains, probably indicating an important functional role of the *N*-terminal region of XLGs. The cysteine-rich regions of XLGs show a $CX_2CX_{10-11}CX_2CX_4CX_2CX_{13}CX_2C$ spacing pattern, which resembles those of zinc fingers proteins (Leon and Roth 2000). A unique structural feature of XLG1 is that it harbors a TonB box consensus sequence (^{90}D–S–I–T–V–S–P–T^{97}) at its *N*-terminus (Lee and Assmann 1999), a sequence feature of transporter proteins in the bacterial outer membrane which are targeted by the TonB protein (Postle 1993; Postle and Larsen 2007). Since the *Arabidopsis* genome does not appear to encode a homolog of the bacterial TonB protein, the function of this TonB box consensus sequence in XLG1 remains unknown.

Table 1 Biochemical characteristics of the *Arabidopsis* unconventional G proteins[a]

Unconventional G proteins	GTP-binding activity	GTPase activity	Cation requirement	Interaction with heterotrimeric G protein subunits
XLG1	Yes (GTP-γ-^{35}S labeling)	–	–	–
XLG2	–	–	–	Interacts with AGB1 in planta (coimmuno precipitation)
XLG3	–	–	–	–
GTG1	Yes (BODIPY-GTPγS, BODIPY-GTP, [^{35}S] GTPγS-binding)	Yes (BODIPY-GTP, ENZchek phosphate assay, [^{32}P] GTP hydrolysis)	Mg^{2+} (BODIPY-GTP)	Interacts with GPA1 (coimmuno precipitation, split-ubiquitin, biochemical interaction)
GTG2	Yes (BODIPY-GTPγS, BODIPY-GTP, [^{35}S] GTPγS-binding)	Yes (BODIPY-GTP, ENZchek phosphate assay, [^{32}P] GTP hydrolysis)	Mg^{2+} (BODIPY-GTP)	Interacts with GPA1 (coimmuno precipitation, split-ubiquitin, biochemical interaction)
DRG1	–	–	–	–
DRG2	–	–	–	–
DRG3	–	–	–	–

[a]Methods used to test respective biochemical characteristics are indicated in parentheses

2.2 XLGs are Plant-Specific Proteins

When XLG-specific *N*-termini are used to search the Genbank (http://www.ncbi.nlm.nih.gov/BLAST) and SUPERFAMILY (http://supfam.mrc-lmb.cam.ac.uk/SUPERFAMILY) databases, to date, XLG homologs are only found in plant species, and not in non-plant species, including those species whose genomes have been completely sequenced, such as human, mouse, rat, *Caenorhabditis elegans*, *Drosophila melanogaster*, *Saccharomyces cerevisiae*, and *E. coli* (Ding et al. 2008b). EST homologs of *XLGs* appear to be widely present in most, if not all, plant species, including dicots, monocots, gymnosperms, and nonvascular plants such as mosses (Ding et al. 2008b). XLGs, thus, may be plant-specific proteins even though they contain a Gα-like domain that is conserved in both plants and metozoans.

It is worth mentioning that searches of the sequenced rice (*Oryza sativa*) genome (http://rice.plantbiology.msu.edu/) reveal the presence of three *XLG* genes in rice:

OsXLG1 (Os06g02130), *OsXLG2* (Os11g10050), and *OsXLG3* (Os12g40190), which encode proteins of 896, 856, and 867 amino acids, respectively. In an early version of the rice genome annotation, there existed a fourth *XLG* gene, *OsXLG4* (Os10g02810), which was predicted to encode a protein of 625 amino acids, but this annotation was later deemed obsolete due to the likelihood that this sequence is a pseudogene. Each of the OsXLGs has a *C*-terminal Gα-like region and a conserved *N*-terminal region homologous to those of the *Arabidopsis* XLGs. OsXLG1, OsXLG2, and OsXLG3 are 61%, 61%, and 62% similar to *Arabidopsis* XLG1 at the amino acid level, respectively. Like *Arabidopsis* XLGs, the *N*-terminus of each OsXLG contains a cysteine-rich region and a putative NLS at positions corresponding to those in the *Arabidopsis* XLGs. The cysteine-rich regions of OsXLGs have a $CX_2CX_{10/14}CX_2CX_4CX_2CX_{13}CX_2C$ spacing pattern, which is almost identical to those of *Arabidopsis* XLGs. None of the OsXLGs has the TonB-box consensus sequence.

2.3 XLGs are Ubiquitously Expressed Nuclear Proteins

Like the conventional heterotrimeric G-protein subunit genes *GPA1* and *AGB1* (Weiss et al. 1993; Huang et al. 1994; Weiss et al. 1994), *XLGs* are expressed in all organs that have been examined, including roots, hypocotyls, cotyledons, rosette leaves, cauline leaves, stems, flowers, and green siliques of *Arabidopsis* (Lee and Assmann 1999; Ding et al. 2008b). At the tissue level, all three *XLGs* are mainly expressed in vascular tissues, shoot and root meristem, and lateral root meristems (Ding et al. 2008b). Several differences were noted in the expression patterns of the three *XLGs* as determined from reporter gene analyses: (1) *XLG2* and *XLG3* express in petals, stigma, and pollen, whereas *XLG1* does not express in these organs; (2) *XLG2* and *XLG3* are generally expressed in the entire root tip, whereas *XLG1* appeared to be limited to quiescent center and columnella cells. Consistent with these expression patterns, a 1.9-fold enrichment of *XLG1* expression was detected in the quiescent center by microarray (Nawy et al. 2005).

Arabidopsis GPA1 localizes to the plasma membrane and endoplasmic reticulum (ER) (Weiss et al. 1993; Chen et al. 2004; Wang et al. 2008), while AGB1 localizes at the plasma membrane and ER and in the nucleus (Anderson and Botella 2007; Wang et al. 2007). In agreement with the presence of a predicted NLS in the *N*-terminus of each XLG, several pieces of experimental evidence support nuclear localization of XLGs. First, the fluorescence of fusion proteins between either full-length or *N*-terminal truncations of XLGs and GFP reporter proteins was detected in the nucleoplasm regardless of whether the GFP reporter was fused to the *N*-termini or *C*-termini of the XLGs (Ding et al. 2008b). Second, biochemical fractionation and immunoblotting experiments demonstrated enrichment of XLG2 in the nuclear extract as compared with total protein extracts (Zhu et al. 2009). And third, *Arabidopsis XLGs*, when cloned into the Cub (C-terminal ubiquitin) vector in the split-ubiquitin yeast two-hybrid system (Obrdlik et al. 2004), can bring the artificial transcription factor PLV (protein A-LexA-VPA6) to the yeast nucleus even in the absence of an interaction partner (Ding et al. 2008b). This nuclear localization is presumably not due to simple diffusion but due to nuclear localization of XLGs as the XLG proteins are large-molecular weight proteins (~100 kDa).

2.4 Physiological Processes Regulated by XLGs

The biological functions of XLGs have been explored in the model species *Arabidopsis*. By drawing on the availability of T-DNA insertional mutants for the *XLG* genes, XLGs have been implicated in an array of biological processes, including negative regulation of primary root growth, promotion of root waving and skewing, modulation of plant responses to sugars, osmotic stress, and plant hormones, and involvement in plant defense responses (Table 2) (Ding et al. 2008b; Pandey et al. 2008; Zhu et al. 2009). Some of these physiological processes are collectively regulated by all three XLGs, whereas others obligately involve only one XLG.

2.4.1 Negative Regulation of Primary Root Growth

Corresponding to the expression of *XLGs* in the primary root meristems, the *xlg1-1 xlg2-1 xlg3-1* triple mutant displayed increased primary root length relative to wild-type plants when grown in darkness, while the *xlg1-1*, *xlg2-1*, and *xlg3-1* single mutants appeared wild type in this regard (Ding et al. 2008b). This phenotype of the *xlg1-1 xlg2-1 xlg3-1* triple mutant was completely or partially complemented by any of the three *XLG* cDNAs, which substantiated that the primary root phenotype in the *xlg* triple mutant is due to loss of function in each of the three *XLG* genes. Therefore, the three *XLG* genes function together in negatively modulating primary root growth of *Arabidopsis* in darkness. Since there is no dramatic difference in sizes of mature root cells in the *xlg1-1 xlg2-1 xlg3-1* triple mutant as compared to wild type plants, and the proportional sizes of the elongation and meristematic zones also appear to be wild type (Ding et al. 2008b), the increased root length of this *xlg* triple mutant presumably results from an increased cell division rate and/or a lengthened time that cells remain in a mitotic state (Beemster and Baskin 1998, 2000).

In early observations, no primary root growth phenotype was observed in the *xlg1-1 xlg2-1 xlg3-1* triple mutant under illumination (Ding et al. 2008b). In these experiments, the seedlings were grown on agar plates supplied with Murashige and Skoog (MS) salts (Murashige and Skoog 1962) and wrapped with parafilm. However, when agar plates were supplied with Linsmaier and Skoog (LS) salts (Linsmaier and Skoog 1965) and wrapped with surgical tape, the *xlg1-1 xlg2-1 xlg3-1* triple mutant exhibited lengthened primary roots under light as well (Pandey et al. 2008). Overexpression of any XLG in the *xlg1-1 xlg2-1 xlg3-1* triple mutant partially but not completely rescued this long primary root phenotype. Thus, all

Table 2 Major responses mediated by the unconventional G proteins in *Arabidopsis*[a]

Unconventional G proteins	Physiological responses		Responsible proteins	Regulatory roles	References
XLGs	Redundantly regulated by more than one XLG	Primary root growth	XLG1, XLG2, XLG3	Negatively regulate primary root growth both in darkness and under light	Ding et al. (2008b) Pandey et al. (2008)
		Responses to sugars	XLG1, XLG2, XLG3	Negatively modulate promotion of primary root elongation by sucrose as a nutrient	Ding et al. (2008b)
		Response to osmotic stress	XLG1, XLG2, XLG3	Negatively modulate inhibition of primary root elongation and seed germination by osmotic stress	Ding et al. (2008b)
		ABA responsiveness	XLG1, XLG2, XLG3	Negatively modulate ABA sensitivity of seed germination, but positively regulate ABA inhibition of primary root growth in both darkness and light	Ding et al. (2008b)
		Response of primary roots to ethylene and/or ACC[b]	XLG1, XLG2, XLG3	Negatively modulate the sensitivity of primary roots to ACC and/or ethylene in the dark	Ding et al. (2008b)
		Response to auxin transport	XLG1, XLG2, XLG3	Positively modulate sensitivity of primary root growth to changed auxin transport by NPA	Pandey et al. (2008)
	Regulated primarily by a single XLG	Responses to *P. syringae*	XLG2	Required for full resistance to *P. syringae*	Zhu et al. (2009)
		Root waving	XLG3	Promote root waving	Pandey et al. (2008)

		Root skewing	XLG3	Promote root skewing under 3 μM propyzamide or 100 nM oryzalin	Pandey et al. (2008)
		Response of hypocotyl to ethylene and/or ACC[b]	XLG3	Negatively modulate the sensitivity of hypocotyls to ACC and/or ethylene.	Pandey et al. (2008)
GTGs	Redundantly regulated by two GTGs	Seed germination	GTG1, GTG2	Participate in ABA inhibition of seed germination	Pandey et al. (2009)
		Postgermination growth	GTG1, GTG2	Participate in prevention of cotyledon greening by ABA	Pandey et al. (2009)
		Primary root elongation	GTG1, GTG2	Participate in retardation of primary root elongation by ABA	Pandey et al. (2009)
		Induction of ABA responsive genes	GTG1, GTG2	Participate in induction of ABA responsive genes, such as *RAB18*, *RD29B*, *DREB2A*, *DREB2B*, and *ERD10*	Pandey et al. (2009)
		Stomatal movement	GTG1, GTG2	Participate in promotion of stomatal closure by ABA	Pandey et al. (2009)
DRGs	Not yet characterized		–	–	–

[a]Information is derived from extant phenotypic analyses of loss-of-function mutants

[b]ACC (1- aminocyclopropane-1-carboxylic acid) is the immediate precursor of ethylene

three XLGs cooperatively function to regulate primary root elongation. The varied growth behaviors of the *xlg1-1 xlg2-1 xlg3-1* triple mutant under the two seemingly similar light growth conditions turned out likely to be due to different atmospheric conditions within the sealed plates rather than to slight variations in salt components and concentrations (Pandey et al. 2008). Parafilm is much less porous than surgical tape and thus causes higher accumulation of ethylene in plates wrapped with parafilm (Buer et al. 2003). Considering that the *xlg1-1 xlg2-1 xlg3-1* triple mutant is hypersensitive to ethylene (Ding et al. 2008b), which inhibits primary root growth, the primary root growth of the *xlg1-1 xlg2-1 xlg3-1* triple mutant was likely more inhibited by ethylene than wild type in the parafilm-wrapped plates; therefore, under this condition, the primary roots of the *xlg1-1 xlg2-1 xlg3-1* triple mutant had similar length as those of wild type. In contrast, when wrapped with porous surgical tape, a high level of ethylene does not accumulate in the agar plates, and thus the *xlg1-1 xlg2-1 xlg3-1* triple mutant experienced less inhibition and showed longer primary root length than the wild type.

2.4.2 Promotion of Root Waving and Root Skewing

When grown on the surface of hard agar plates tilted at an angle, *Arabidopsis* roots show wavy growth, which is called root waving (Okada and Shimura 1990). In contrast, under the same conditions, the *xlg1-1 xlg2-1 xlg3-1* triple mutant and *xlg3* single mutants exhibited much more relaxed root waving with fewer waves of longer wavelength and diminished amplitudes (Pandey et al. 2008). The facts that *xlg3* single mutants showed attenuated root waving and that the reduced root waving in the *xlg1-1 xlg2-1 xlg3-1* triple mutant can be rescued by *XLG3* complementation, but not by introduction of *XLG1* or *XLG2* to the triple mutant background, indicate that XLG3 is the key player in the regulation of root waving by XLGs.

When *Arabidopsis* is grown on firm agar surfaces oriented vertically, the primary roots, depending on the ecotype, tend to show slanted rather than vertical growth, and this is called root skewing. Columbia (Col) does not exhibit root skewing under normal conditions, but does so when treated with low concentrations of chemicals that influence dynamic turnover of microtubules, such as propyzamide (3 μM) or oryzalin (100 nM) (Rutherford and Masson 1996; Nakamura et al. 2004). Under these treatment conditions, the *xlg3* single mutants and the *xlg1-1 xlg2-1 xlg3-1* triple mutant, which are in the Col background, showed significantly decreased root slanting compared with wild type. The reduced slanting of the *xlg1-1 xlg2-1 xlg3-1* triple mutant could be rescued only by complementation with *XLG3* and not by *XLG1* or *XLG2*, implying that XLG3 participates in the regulation of root skewing under these conditions (Pandey et al. 2008).

It is generally thought that root waving involves interaction of gravitropic responses of roots sensed at the tip and touch responses of roots against the hard agar surface (Okada and Shimura 1990; Thompson and Holbrook 2004), while root skewing is due to an interaction of root touch response, microtubule orientation, and direction of cell expansion (Furutani et al. 2000; Hashimoto 2002; Sedbrook et al. 2004). Although *XLG3* plays a role in both root waving and skewing, it appears that *xlg3* mutants do not show defects in either gravitropic responses or touch responses when these responses are tested separately (Pandey et al. 2008). Hence, it seems to be less likely that XLG3 might regulate root waving or root skewing via modulating gravitropic responses and/or touch responses, and it has been hypothesized that the role of XLG3 may be in the modulation of hormonal responses that affect waving and skewing (Pandey et al. 2008).

2.4.3 Modulation of Responses to Sugars and Osmotic Stress

Sucrose plays dual roles in the regulation of primary root growth in darkness: a nutritional role for dark-grown plants (dominant at low concentration) and an osmotic effect as an osmolyte (dominant at high concentration). At 0.5–3% sucrose, the *xlg1-1 xlg2-1 xlg3-1* triple mutant showed significantly increased primary root length in darkness relative to wild type, and this difference peaked at 2–3% sucrose (Ding et al. 2008b), indicating possible roles of *XLGs* in negatively modulating sensitivity of primary roots to nutritional sucrose. However, at high concentration (6%) of sucrose, the difference in the primary root length between the *xlg1-1 xlg2-1 xlg3-1* triple mutant and wild type shrank to a nonsignificant level, presumbly due to an osmotic effect created by high concentration of sucrose. Under molarites of mannitol (which does not have a nutritional role) equivalent to 3% sucrose, the *xlg1-1 xlg2-1 xlg3-1* triple mutant actually showed significantly shorter primary roots than did the wild-type plants, suggesting a hypersensitivity of the *xlg1-1 xlg2-1 xlg3-1* triple mutant to osmotic stress. The hypersensitivity of the *xlg* triple mutant to osmotic stress was also evidenced in seed germination assays in which 3% and 6% mannitol each inhibited seed germination of the *xlg1-1 xlg2-1 xlg3-1* triple mutant much more than that of wild type (Ding et al. 2008b). XLGs appear to modulate root growth both positively and negatively, i.e., promotion of primary root elongation by sucrose as a nutrient and inhibition of primary root elongation and seed germination by osmotic stress.

2.4.4 Modulation of Responses to Plant Hormones

Arabidopsis GPA1 and AGB1 modulate responses to plant hormones (Wang et al. 2001a; Ullah et al. 2002, 2003; Chen et al. 2004; Pandey et al. 2006; Fan et al. 2008). Similarly, XLGs also play roles in *Arabidopsis* responses to auxin transport, abscisic acid (ABA), and ethylene although the *xlg1-1 xlg2-1 xlg3-1* triple mutant still maintains responsiveness to these plant hormones (Ding et al. 2008b; Pandey et al. 2008).

ABA: As compared with wild type, the *xlg1-1 xlg2-1 xlg3-1* triple mutant showed ~4-fold increased sensitivity to ABA inhibition of seed germination, but displayed decreased ABA sensitivity in primary root growth in both darkness and light (Ding 2005; Ding et al. 2008b). It, therefore, appears that XLGs negatively modulate ABA sensitivity of seed germination but positively regulate ABA inhibition of primary root growth. It is unclear whether the faster rate of primary root growth in the *xlg1-1 xlg2-1 xlg3-1* triple mutant overcomes the inhibitory effects of ABA and, thus, contributes to the decreased ABA sensitivity of primary root growth in this mutant.

Ethylene: When exposed to ethylene or its immediate precursor 1- aminocyclopropane-1-carboxylic acid (ACC), dark-grown *Arabidopsis* seedlings show the triple response: shortened and thickened hypocotyls, inhibited root elongation, and exaggerated apical hooks (Guzman and Ecker 1990; Kieber et al. 1993). When grown on LS medium with ACC treatment (plates wrapped with surgical tape), the *xlg3* mutants showed a more exaggerated responses in hypocotyl length and width and in root length compared with wild-type plants, an indication of increased sensitivity to ACC and/or ethylene (Pandey et al. 2008). Similarly, an ~5-fold increase in the sensitivity of primary root growth to ACC was observed in the *xlg1-1 xlg2-1 xlg3-1* triple mutant in darkness compared to wild-type plants (Ding et al. 2008b). These data indicate that XLGs negatively modulate the sensitivity of hypocotyls and primary roots to ACC and/or ethylene.

Auxin: When grown on agar plates containing different concentrations of *N*-1-naphthylphthalamic acid (NPA), an inhibitor of auxin transport, the *xlg1-1 xlg2-1 xlg3-1* triple mutant exhibited changed responses. At 10^{-6} and 10^{-5} M NPA, Col seedlings showed 30 and 70% reduction in root length, respectively, compared with the control without NPA, whereas the *xlg1-1 xlg2-1 xlg3-1* triple mutant only showed 20 and 40% reduction compared with this control (Pandey et al. 2008). Thus, XLGs positively modulate sensitivity of primary root growth to alterations in auxin transport by NPA.

2.4.5 Involvement in Defense Responses

XLG2 transcripts were induced following infection with a virulent strain (DC 3,000) and an avirulent strain (*Rpm1*) of *P. syringae* pv. *tomato* as well as with the nonhost strain *P. syringae* pv. *phaseolicola* (Zhu et al. 2009). Consistent with these inductions, *xlg2* single mutants displayed a moderate increase (~5-fold) in susceptibility to *P. syringae*. This increased susceptibility to *P. syringae* in the *xlg2* mutants was abolished when a wild-type copy of *XLG2* genomic DNA was introduced into the mutants. This evidence strongly supports a role of XLG2 in *Arabidopsis* responses to *P. syringae*. Similar to *XLG2*, *XLG3* was also induced by the avirulent strain (*Rpm1*) of *P. syringae* pv. *Tomato*, but *XLG1* seems to be unaffected by infection with this strain (Zhu et al. 2009). *xlg3* and *xlg1* mutants showed wild-type-like resistance to this pathogen. Considering that the *xlg1-1 xlg2-1 xlg3-1* triple mutant showed similar resistance to these *P. syringae* strains as the *xlg2* single mutants, it appears that XLG2 plays a dominant role in responses to *P. syringae* by *XLGs*, while XLG3 and XLG1 play at best minimal roles in *Arabidopsis* resistance to this pathogen.

2.5 *How do XLGs Function: Speculations*

Considering that XLGs have Gα-like domains similar to GPA1, an intriguing hypothesis would be that XLGs might serve as additional G protein α subunits in plants. To evaluate this hypothesis, four key questions need to be answered: one is whether XLG2 and XLG3, like XLG1, have GTP-binding activity; the second is whether the XLGs have GTPase activity; the third is whether XLGs can interact with AGB1; and the fourth is whether XLGs and AGB1 show any functional correlation.

The first two questions await future research. As to the third question, XLG2 and AGB1 are implicated to exist in the same protein complex by coimmunoprecipitation analysis (Table 1) (Zhu et al. 2009), making it tempting to hypothesize that XLG2 and AGB1 might interact in the same way as conventional Gα and Gβ do. However, given the significant structural differences between XLGs and conventional Gαs and negative results from testing AGB1 and XLGs interaction in various yeast two-hybrid assays (Ding et al. 2008b; Zhu et al. 2009), direct physical interactions between XLGs and AGB1, as opposed to indirect interaction in a multiprotein complex, cannot be assumed, and remain to be demonstrated.

Regarding the fourth question, XLGs and AGB1 appear to show functional overlap as indicated by phenotypic characterization of the *xlg1-1 xlg2-1 xlg3-1* triple and *agb1* mutants (Pandey et al. 2006, 2008; Ding et al. 2008b). Both XLGs and AGB1 are involved in negative regulation of primary root growth, promotion of root waving and root skewing, negative modulation of *Arabidopsis* responses to ACC and/or ethylene, negative modulation of ABA inhibition of seed germination, and positive modulation of ABA inhibition of primary root growth. Considering these functional similarities and the overlapping localization of XLGs and AGB1 in the nucleus (Anderson and Botella 2007; Ding et al. 2008b; Wang et al. 2008), these proteins may function together, at least in the nucleus. The possibility also remains that XLGs may relocate to other subcellular compartments harboring AGB1 under some, as yet unknown, conditions.

3 Developmentally Regulated GTP-Binding Proteins

In the early 1990s, a new type of G protein, named DRG for "developmentally regulated G protein," was identified as a transcript that was highly expressed in mouse embryonic brain tissue and downregulated in mature tissue (Kumar et al. 1992). Homologs of mouse DRG1 were identified in systems as diverse as archaea (*Halobacterium*) and insects (*Drosophila*). The GTP-binding activity of mouse DRG1 was confirmed by Sazuka et al. (1992). DRGs are now known to compose one subfamily of the OBG GTP-binding family, which in turn belongs to the OBG-HflX-like superfamily of the TRAFAC class of GTPases (Leipe et al. 2002). As expected, DRGs contain all of the domains characteristic of GTPases (Leipe et al.

2002). In addition to the conserved GTPase domains, DRGs also contain an *N*-terminal glycine-rich motif and a *C*-terminal TGS domain, which is suggested to be an RNA-binding domain although its function has not been experimentally demonstrated. The TGS domain is named as such because it is found in threonyl-tRNA synthetases (ThrRSs), the DRG family of G proteins, and guanosine polyphosphate phosphohydrolases/synthetases (SpoT/RelA). The proposed RNA-binding function of the TGS domain is based on the remarkable structural similarities between the TGS domain of *E. coli* threonyl-tRNA synthetase, the αL motif of the novel *E. coli* RNA-binding heat shock protein Hsp15, and the αL motif of the ribosome protein S4 (Staker et al. 2000). *Xenopus* DRG proteins have been shown to bind RNA (uridylic acid homopolymers) in vitro, but interestingly, this binding appears to be at least partially independent of the TGS domain as a deletion mutant that is lacking the TGS domain still retains the binding activity (Ishikawa et al. 2003).

In non-plant eukaryotic organisms, DRGs are widely conserved as a small family comprised of two proteins, while archeon genomes encode only single DRG which cannot readily be classified as belonging to either the DRG1 or DRG2 clades (Li and Trueb 2000). The archeon DRGs show ~45% similarity to both human DRG1 and DRG2 (Li and Trueb 2000).

3.1 Presence of DRGs in Plants

Although DRGs appear to be widely distributed throughout the plant kingdom, plant DRGs have been studied primarily in *Arabidopsis* and pea, the first two plant species in which DRGs were discovered. *Arabidopsis*, like other eukaryotes, has one protein belonging to each of the two DRG clades (Li and Trueb 2000), but also has a third DRG family member. The *Arabidopsis* DRG1 (At4g39520) and DRG2 (At1g17470) protein sequences are ~56% identical, while the third family member, *Arabidopsis* DRG3 (At1g72660), is 95% identical to DRG2 and thus belongs to the DRG2 clade (Stafstrom 2008). Interestingly, the *Arabidopsis* DRG2 and DRG3 sequences have a 32-amino acid *C*-terminal extension, compared to *Arabidopsis* DRG1, hinting of a functional divergence between the plant DRG1 and DRG2 clades. This extension seems to be a common characteristic of the plant DRG2 clade as pea PsDRG, a DRG homolog with 90% identity to *Arabidopsis* DRG2, also has an extended *C*-terminus (Devitt et al. 1999).

Arabidopsis DRG2 was identified and characterized in nearly simultaneous reports by Devitt et al. (1999) and Etheridge et al. (1999). The encoded protein was originally called AtDRG and AtDRG1 by the two groups, respectively, but has since been renamed as *Arabidopsis* DRG2 (Stafstrom 2008) to follow the naming convention established by the first formal phylogenetic analysis of the DRG family (Li and Trueb 2000).

Whole proteome molecular evolutionary analysis of the sequenced genomes of the nonvascular moss *Physcomitrella*, the monocot rice, the woody dicot poplar, and the herbaceous dicot *Arabidopsis* indicate that the presence of both DRG1 and

DRG2 clade members within a single plant proteome appears to be the rule, with all species having one DRG1 protein and at least one DRG2 protein with the seemingly characteristic *C*-terminal extension (T. Gookin unpublished observation). The *Physcomitrella* proteome appears to have only one DRG2 clade protein while the *Arabidopsis*, rice, and poplar proteomes have two; interestingly, the second rice DRG2 clade protein is produced as a splice variant which does not have the *C*-terminal extension, and is, in fact, even shorter than the rice DRG1 protein (T. Gookin unpublished observation). These observations are intriguing from an evolutionary perspective, and their functional significance awaits investigation.

Plant DRGs contain all of the domains responsible for GTP-binding and GTPase activity, but empirical evidence of these biochemical functions has not been reported. In a specific comparison with human DRG1 and DRG2, the three *Arabidopsis* DRGs and the single known pea DRG show complete conservation of the p-loop motif (GxxxxGKS), the DxxG motif, and the N/TKxD motif, which in all six sequences are represented by GFPSVGKS, DLPG, and NKID, respectively, except for a single substitution of L for K in the N/TKxD motif of PsDRG (Fig. 1).

3.2 Expression Patterns of Plant DRGs

Northern analysis by both Devitt et al. (1999) and Etheridge et al. (1999) showed that *Arabidopsis DRG2* transcripts are present in all of the major *Arabidopsis* plant organs and that this expression is not temporally limited to only young organs, as might have been hypothesized from the expression patterns of the first-identified mouse *DRG* transcript. These expression surveys were extended by Etheridge et al. (1999), who showed by Northern analysis that mRNA homologous to *Arabidopsis DRG2* was also widely present in a variety of plant species, including tomato, macadamia, mango, sugarcane, and the symbiotic unicellular alga *Symbiodinium*.

Detailed *Arabidopsis DRG2* mRNA and protein expression patterns were identified by in situ cDNA hybridization and immunogold localization experiments, respectively. Etheridge et al. showed that *DRG2* mRNA transcripts were most abundant in developing organs and tissues such as apical meristems, immature flower buds, developing pistils and petals, fertilized ovaries, and torpedo-stage embryos, as well as in pericycle, and *Arabidopsis* DRG2 protein was found to be primarily limited to cytosolic vesicles (Etheridge et al. 1999).

Devitt et al. (1999) examined mRNA and protein expression patterns of PsDRG by northern analysis and western blotting, respectively. *PsDRG* mRNA transcript abundance was initially low in axillary buds but increased dramatically after removal of the apical bud. Interestingly, these changes were not mirrored by PsDRG protein levels in the axillary buds as the protein levels remained unchanged. In agreement with the immunolocalization studies by Etheridge et al. (1999), subcellular fractionation experiments showed PsDRG was predominantly located in the microsomal fraction (Devitt et al. 1999), with only a small proportion present in the cytosolic fraction. Subsequent quantitative real-time RT-PCR results confirmed that *Arabidopsis DRG2* was widely expressed in many plant organs, and also showed that while *DRG1* transcripts were expressed at a similar level as *DRG2*, *DRG3* transcripts were expressed at much lower levels, up to 100-fold less in some tissues (Stafstrom 2008).

More detailed examination of *DRG1* and *DRG2* using the reporter gene β-glucuronidase (*GUS*) under the control of the *DRG1* and *DRG2* promoters showed that although *DRG1* and *DRG2* are coexpressed in a number of tissues, there are some distinct differences. Young seedlings showed a general GUS staining of all tissues with both the *DRG1* and *DRG2* promoter fusion constructs, except that *DRG2* was not expressed in hypocotyls or cotyledon petioles. In addition, young seedlings expressing *DRG1* promoter constructs showed strong GUS staining in the leaf vascular tissue while the *DRG2* promoter construct resulted in strong staining in trichomes (Stafstrom 2008). *DRG1* and *DRG2* GUS staining was heavy in pollen, anthers, and stigmas, but only the *DRG2* reporter gene construct showed additional staining in sepals and petals. In roots, both the *DRG1* and *DRG2* reporter constructs gave strong staining in vascular tissue and emerging lateral roots (Stafstrom 2008), but interestingly, GUS expression remained high in the root tip while dropping to nearly undetectable levels in the zone of maturation between the tip of the newly formed root and the parent root. Combined with the earlier observation of pericycle-localized *DRG2* expression by Etheridge et al. (1999), the root tissue localization of *DRG1* and *DRG2* promoter *GUS* staining suggests that both DRGs might play a role in lateral root primordia initiation and growth of the advancing root tip. Reporter gene analyses are not yet available for *DRG3*.

Although Stafstrom (2008) examined *DRG* mRNA and protein expression levels under diverse environmental conditions (including low and high temperature, pH, osmotic, genotoxic, heavy metal and salinity stresses, as well as herbicide treatment), only exposure of plants to 37°C produced any appreciable differences. In response to high temperature, *DRG1* mRNA levels increased tenfold, yet DRG1 protein levels did not rise, whereas *DRG2* mRNA levels remained constant, yet DRG2 protein levels increased. Most dramatically, *DRG3* transcripts increased over 1,000-fold with the high temperature treatment. This may indicate that DRG3 plays a more predominant role in heat responses, but given the disparity between *DRG* mRNA and protein levels in these and previous experiments, it is clear that more analysis is needed.

3.3 Biological Functions and Regulatory Mechanisms of Plant DRGs: Still a Mystery

The DRG family of proteins remains an enigma. Little is known regarding their function, and what is known only raises more questions. Mouse and human DRG proteins have been shown to bind the basic helix–loop–helix proteins TAL-1/SCL,

TAL2, and Lyl1 (Mahajan et al. 1996; Zhao and Aplan 1998), which function in normal hematopoietic development. In the pathogenic yeast, *Candida*, Drg1p was shown to regulate invasive filament growth in vitro, similar to the basic helix–loop–helix protein Efg1p (Chen and Kumamoto 2006). From these clues and the expression data reviewed above, it is certain that the developmentally regulated G-protein family is appropriately named, but their functions in plants will only become clear once their loss-of-function mutants are characterized, their GTP binding and GTPase characteristics are assessed, their protein–protein interaction partners are identified, and their RNA binding targets, if such exist, are isolated.

4 GPCR-Type GTP-Binding Proteins

A third class of unconventional G proteins, GPCR-type G proteins (GTGs), has also been functionally characterized in *Arabidopsis* (Pandey et al. 2009). GTGs are G proteins with multitransmembrane span topology reminiscent of GCPRs, although with nine predicted transmembrane spans instead of the seven characteristic of GPCRs (Fig. 1). The *Arabidopsis* genome encodes two GTG proteins, GTG1 (At1g64990) and GTG2 (At4g27630), which are 90% identical at the amino acid level. Both GTG1 and GTG2 share 45% identity and 68% similarity with an annotated human orphan receptor, G-protein-coupled receptor 89 (GPR89) (Lai et al. 2000; Strausberg et al. 2002; Clark et al. 2003; Matsuda et al. 2003; Oh et al. 2006; Maeda et al. 2008).

4.1 Structural and Biochemical Features of GTGs

A portion of the large third intracellular loop of the GTG1 and GTG2 (the 230–243 amino acids for GTG1) is similar to a consensus sequence of Ras GTPase-activating proteins as annotated by PROSITE (Fig. 1) (Pandey et al. 2009). The two proteins also possess a "protein kinases ATP-binding region signature" sequence (corresponding to the 382–411 amino acids of GTG1) (Fig. 1) (Pandey et al. 2009). This "protein kinases ATP-binding region signature" sequence is different from the p-loop, a motif commonly present in the GTP- and ATP-binding proteins (Saraste et al. 1990; Leipe et al. 2002), but it was hypothesized that it might function as an GTP-binding motif within the GTGs. Using BODIPY-GTPγS or BODIPY-GTP (McEwen et al. 2001; Willard et al. 2005), this hypothesis was experimentally evaluated in real-time assays and the GTG proteins were found to both bind GTP specifically and hydrolyze the bound GTP in a Mg^{2+}-dependent manner (Table 1) (Pandey et al. 2009). These GTP-binding and GTPase activities were verified using the ENZchek phosphate assays and thin layer chromatography analyses of [^{32}P] GTP hydrolysis, and GTP-binding was further experimentally verified by binding assays with [^{35}S]-GTP γS. This evidence demonstrated that the *Arabidopsis* GTGs are de facto G proteins, although these experiments do not specifically address whether the "protein kinases ATP-binding region signature" sequence in GTGs is responsible for the observed GTP-binding and GTPase activities.

4.2 GTGs are Widely Expressed Plasma Membrane Proteins

Homologs of the *Arabidopsis* GTGs are broadly present in both monocot and dicot plants, vertebrates, invertebrates, fungi, and unicellular organisms (Pandey et al. 2009). At the tissue and organ level, *Arabidopsis GTG1* and *GTG2* appear to be universally expressed (Pandey et al. 2009). Expression of the two genes was detected in all organs examined, including cotyledons, both cauline and rosette leaves, stems, roots, flowers, and guard cells via quantitative PCR analyses of their transcripts and staining *GTG* promoter:*GUS* transgenic lines.

Consistent with the presence of nine transmembrane domains in *Arabidopsis* GTG1 and GTG2, the two proteins are enriched in the microsomal fraction compared with the soluble fraction, and GFP-tagged GTG1 and GTG2 transiently expressed in *Arabidopsis* mesophyll protoplasts localized at the cell periphery (Pandey et al. 2009), suggesting that *Arabidopsis* GTG1 and GTG2 are primarily plasma membrane proteins.

4.3 GTGs Interact with Arabidopsis GPA1

Homology of GTGs with human GPR89 led Pandey et al. (2009) to test whether GTGs may interact with heterotrimeric G protein GPA1. As expected, both GTG1 and GTG2 interacted with the *Arabidopsis* GPA1 in yeast split ubiquitin assays. These interactions between GTGs and GPA1 were also confirmed by in vivo coimmunoprecipitation (Table 1) (Pandey et al. 2009). Interestingly, interaction of GPA1 with GTGs boosted GTP-binding activity of GTGs and dramatically reduced their GTPase activity (Pandey et al. 2009).

4.4 GTGs mediate Arabidopsis Responses to ABA

Biological functions of *Arabidopsis* GTGs have been discovered by characterizing the *gtg1* and *gtg2* single mutants and the *gtg1 gtg2* double mutants. While no significant phenotype was found in the *gtg1* or *gtg2* single mutants, the *gtg1 gtg2* double mutant is defective in a variety of classic ABA responses, including inhibition of seed germination, prevention of cytoledon greening, retardation of primary root elongation, induction of ABA responsive genes such as *RAB18*, *RD29B*, *DREB2A*, *DREB2B*, and *ERD10*, and promotion of stomatal closure (Table 2) (Pandey et al. 2009). All these defects were complemented through introduction of either wild-type *GTG1* or wild-type *GTG2* to the *gtg1 gtg2* double mutant, confirming that it was the loss of both *GTG1* and *GTG2* that was responsible for the ABA hyposensitivity.

4.5 GTGs are ABA Receptors

Although three proteins FCA, CHLH, and GCR2 (Razem et al. 2006; Shen et al. 2006; Liu et al. 2007), had been reported as ABA receptors, questions have been raised regarding the identity of FCA, CHLH, and GCR2 as ABA receptors (Gao et al. 2007; Johnston et al. 2007; Guo et al. 2008; Illingworth et al. 2008; Jang et al. 2008; Risk et al. 2008; Muller and Hansson 2009), and, in fact, identification of FCA as an ABA receptor was retracted in 2008 (Razem et al. 2008). Using a binding assay methodology similar to the one used to identify BRI1 as a BR receptor (Wang et al., 2001b; Kinoshita et al. 2005), which is different from the ABA-binding methodology used for FCA, CHLH, and GCR2, Pandey et al. (2009) demonstrated direct, specific, and saturable binding of recombinant GTG proteins to ^{3}H-ABA, a biochemical feature expected for an ABA receptor. These data, in combination with the observations that *Arabidopsis* GTG1 and GTG2 are membrane-localized proteins which couple with GPA1 and that *gtg1 gtg2* mutants show hyposensitivity in classic ABA responses, led Pandey et al. (2009) to propose that GTGs are ABA receptors. Interestingly, the presence of GDP, rather than GTP, faciliated the binding of ABA to GTGs, indicating that GDP-bound GTGs may be the active forms. This is opposite to the classic paradigm of heterotrimeric G-protein signaling, in which it is the GTP-bound form of Gα that actively propagates the signal. Whether the GTG proteins bind apoplastic (external) or symplastic (cytosolic) ABA remains to be determined. In addition, it should be mentioned that the *gtg1 gtg2* double mutant still maintains wild type responses in ABA-inhibition of stomatal opening, suggesting to the authors that additional ABA receptors remain to be identified (Pandey et al. 2009).

5 Conclusions and Perspectives

In addition to conventional G proteins, plants have three types of unconventional G proteins, XLGs, DRGs, and GTGs, which are expressed in almost all plant organs. The presence of signaling pathways mediated by these unconventional G proteins may compensate for the limited number of heterotrimeric G proteins in plants. The functions of XLGs and GTGs have been characterized by analyzing their respective loss-of-function mutants. Despite evidence for tangible biological functions of these unconventional G proteins, their regulatory mechanisms remain elusive. Testing whether XLGs function as Gα subunits, as discussed earlier, and identifying the interacting proteins of XLGs will help to further our understanding of their regulatory mechanisms. DRGs are the group that is least known among the three sets of unconventional G proteins. So far only the expression patterns of *Arabidopsis* and pea *DRGs* have been reported. It remains unknown whether plant DRGs have GTP-binding and GTPase activities, what their biogical functions are, whether plant DRGs have RNA-binding activity as *Xenopus* DRGs do, and, if they do have RNA binding activity, what the RNA targets of plant DRGs are. Although GTGs have been identified as ABA receptors, components downstream of GTGs in the signaling pathway remain mysterious. It is interesting that the human homolog of the GTGs, GPR89, did not show GTP-binding or GTPase activity under identical assay conditions as used for the GTGs (Pandey et al. 2009), leading Pandey et al. to suggest possible functional divergence. Recently, a human Golgi pH regulator (GPHR) was cloned and it turned out to be GPR89 (Maeda et al. 2008). By reconstitution in planar lipid bilayers, the human GPR89 was found to be a voltage-dependent anion channel that modulates acidification of Golgi vesicles. It is of great interest to investigate whether the *Arabidopsis* GTGs may also act as anion channels and whether they may also localize and function in the Golgi, in addition to their localization at the plasma membrane. In addition, since GTGs possess a consensus sequence of Ras GTPase-activating proteins, it is also appealing to investigate whether GTGs may have GTPase activating activity. In conclusion, many aspects of unconventional plant G proteins await intensive investigation in order to expand understanding of their functions and regulatory mechanisms.

Acknowledgments We gratefully acknowledge comments on this chapter from Dr. Sona Pandey. Research on G proteins in the authors' laboratories was supported by NSF grant MCB-0209694 and USDA grant 2006-35100-17254 to S.M.A.

References

Anderson DJ, Botella JR (2007) Expression analysis and subcellular localization of the *Arabidopsis thaliana* G-protein β-subunit AGB1. Plant Cell Rep 26:1469–1480

Assmann SM (2002) Heterotrimeric and unconventional GTP binding proteins in plant cell signaling. Plant Cell (Suppl.) 14:S355–S373

Bastepe M, Gunes Y, Perez-Villamil B, Hunzelman J, Weinstein LS, Juppner H (2002) Receptor-mediated adenylyl cyclase activation through XLαs, the extra-large variant of the stimulatory G protein α-subunit. Mol Endocrinol 16:1912–1919

Beemster GT, Baskin TI (1998) Analysis of cell division and elongation underlying the developmental acceleration of root growth in *Arabidopsis thaliana*. Plant Physiol 116:1515–1526

Beemster GT, Baskin TI (2000) *Stunted plant 1* mediates effects of cytokinin, but not of auxin, on cell division and expansion in the root of *Arabidopsis*. Plant Physiol 124:1718–1727

Bourne HR (1995) GTPases: a family of molecular switches and clocks. Philos Trans R Soc Lond B Biol Sci 349:283–289

Bourne HR, Sanders DA, McCormick F (1990) The GTPase superfamily: a conserved switch for diverse cell functions. Nature 348:125–132

Bourne HR, Sanders DA, McCormick F (1991) The GTPase superfamily: conserved structure and molecular mechanism. Nature 349:117–127

Buer CS, Wasteneys GO, Masle J (2003) Ethylene modulates root-wave responses in *Arabidopsis*. Plant Physiol 132:1085–1096

Chen JG (2008) Heterotrimeric G-proteins in plant development. Front Biosci 13:3321–3333

Chen JG, Pandey S, Huang J, Alonso JM, Ecker JR, Assmann SM, Jones AM (2004) GCR1 can act independently of heterotrimeric G-protein in response to brassinosteroids and gibberellins in *Arabidopsis* seed germination. Plant Physiol 135:907–915

Chen X, Kumamoto CA (2006) A conserved G protein (Drg1p) plays a role in regulation of invasive filamentation in *Candida albicans*. Microbiology 152:3691–3700

Clark HF, Gurney AL, Abaya E, Baker K, Baldwin D, Brush J, Chen J, Chow B, Chui C, Crowley C, Currell B, Deuel B, Dowd P, Eaton D, Foster J, Grimaldi C, Gu Q, Hass PE, Heldens S, Huang A, Kim HS, Klimowski L, Jin Y, Johnson S, Lee J, Lewis L, Liao D, Mark M, Robbie E, Sanchez C, Schoenfeld J, Seshagiri S, Simmons L, Singh J, Smith V, Stinson J, Vagts A, Vandlen R, Watanabe C, Wieand D, Woods K, Xie MH, Yansura D, Yi S, Yu G, Yuan J, Zhang M, Zhang Z, Goddard A, Wood WI, Godowski P, Gray A (2003) The secreted protein discovery initiative (SPDI), a large-scale effort to identify novel human secreted and transmembrane proteins: a bioinformatics assessment. Genome Res 13:2265–2270

Devitt ML, Maas KJ, Stafstrom JP (1999) Characterization of DRGs, developmentally regulated GTP-binding proteins, from pea and *Arabidopsis*. Plant Mol Biol 39:75–82

Ding L (2005) Functional characterization of extra-large G proteins (XLGs) in *Arabidopsis thaliana*. PhD thesis, The Pennsylvania State University

Ding L, Chen J-G, Jones AM, Assmann SM (2008a) Heterotrimeric G-protein-coupled signaling in higher plants. Annual Plant Reviews 33:30–63

Ding L, Pandey S, Assmann SM (2008b) *Arabidopsis* extra-large G proteins (XLGs) regulate root morphogenesis. Plant J 53:248–263

Etheridge N, Trusov Y, Verbelen JP, Botella JR (1999) Characterization of ATDRG1, a member of a new class of GTP-binding proteins in plants. Plant Mol Biol 39:1113–1126

Fan LM, Zhang W, Chen JG, Taylor JP, Jones AM, Assmann SM (2008) Abscisic acid regulation of guard-cell K^+ and anion channels in Gβ- and RGS-deficient *Arabidopsis* lines. Proc Natl Acad Sci USA 105:8476–8481

Fujisawa Y, Kato H, Iwasaki Y (2001) Structure and function of heterotrimeric G proteins in plants. Plant Cell Physiol 42:789–794

Furutani I, Watanabe Y, Prieto R, Masukawa M, Suzuki K, Naoi K, Thitamadee S, Shikanai T, Hashimoto T (2000) The *SPIRAL* genes are required for directional control of cell elongation in *Arabidopsis thaliana*. Development 127:4443–4453

Gao Y, Zeng Q, Guo J, Cheng J, Ellis BE, Chen JG (2007) Genetic characterization reveals no role for the reported ABA receptor, GCR2, in ABA control of seed germination and early seedling development in *Arabidopsis*. Plant J 52:1001–1013

Guo J, Zeng Q, Emami M, Ellis BE, Chen JG (2008) The GCR2 gene family is not required for ABA control of seed germination and early seedling development in *Arabidopsis*. PLoS ONE 3:e2982

Guzman P, Ecker JR (1990) Exploiting the triple response of *Arabidopsis* to identify ethylene-related mutants. Plant Cell 2:513–523

Hashimoto T (2002) Molecular genetic analysis of left-right handedness in plants. Philos Trans R Soc Lond B Biol Sci 357:799–808

Huang H, Weiss CA, Ma H (1994) Regulated expression of the *Arabidopsis* G protein a subunit gene *GPA1*. Int J Plant Sci 155:3–14

Iismaa SE, Chung L, Wu MJ, Teller DC, Yee VC, Graham RM (1997) The core domain of the tissue transglutaminase Gh hydrolyzes GTP and ATP. Biochemistry 36:11655–11664

Iismaa SE, Wu MJ, Nanda N, Church WB, Graham RM (2000) GTP binding and signaling by Gh/transglutaminase II involves distinct residues in a unique GTP-binding pocket. J Biol Chem 275:18259–18265

Illingworth CJ, Parkes KE, Snell CR, Mullineaux PM, Reynolds CA (2008) Criteria for confirming sequence periodicity identified by Fourier transform analysis: application to GCR2, a candidate plant GPCR? Biophys Chem 133:28–35

Im MJ, Graham RM (1990) A novel guanine nucleotide-binding protein coupled to the α_1-adrenergic receptor. I. Identification by photolabeling or membrane and ternary complex preparation. J Biol Chem 265:18944–18951

Im MJ, Riek RP, Graham RM (1990) A novel guanine nucleotide-binding protein coupled to the α_1-adrenergic receptor. II. Purification, characterization, and reconstitution. J Biol Chem 265:18952–18960
Im MJ, Russell MA, Feng JF (1997) Transglutaminase II: a new class of GTP-binding protein with new biological functions. Cell Signal 9:477–482
Ishikawa K, Azuma S, Ikawa S, Morishita Y, Gohda J, Akiyama T, Semba K, Inoue J (2003) Cloning and characterization of *Xenopus laevis drg2*, a member of the developmentally regulated GTP-binding protein subfamily. Gene 322:105–112
Jang YH, Lee JH, Kim JK (2008) Abscisic acid does not disrupt either the *Arabidopsis* FCA-FY interaction or its rice counterpart in vitro. Plant Cell Physiol 49:1898–1901
Johnston CA, Temple BR, Chen JG, Gao Y, Moriyama EN, Jones AM, Siderovski DP, Willard FS (2007) Comment on "A G protein coupled receptor is a plasma membrane receptor for the plant hormone abscisic acid". Science 318:914
Jones AM (2002) G-protein-coupled signaling in *Arabidopsis*. Curr Opin Plant Biol 5:402–407
Jones AM, Assmann SM (2004) Plants: the latest model system for G-protein research. EMBO Rep 5:572–578
Kehlenbach RH, Matthey J, Huttner WB (1994) XLαs is a new type of G protein. Nature 372: 804–809
Kieber JJ, Rothenberg M, Roman G, Feldmann KA, Ecker JR (1993) *CTR1*, a negative regulator of the ethylene response pathway in *Arabidopsis*, encodes a member of the Raf family of protein kinases. Cell 72:427–441
Kinoshita T, Cano-Delgado A, Seto H, Hiranuma S, Fujioka S, Yoshida S, Chory J (2005) Binding of brassinosteroids to the extracellular domain of plant receptor kinase BRI1. Nature 433:167–171
Klemke M, Pasolli HA, Kehlenbach RH, Offermanns S, Schultz G, Huttner WB (2000) Characterization of the extra-large G protein α-subunit XLαs. II. Signal transduction properties. J Biol Chem 275:33633–33640
Kumar S, Tomooka Y, Noda M (1992) Identification of a set of genes with developmentally down-regulated expression in the mouse brain. Biochem Biophys Res Commun 185:1155–1161
Lai CH, Chou CY, Ch'ang LY, Liu CS, Lin W (2000) Identification of novel human genes evolutionarily conserved in *Caenorhabditis elegans* by comparative proteomics. Genome Res 10:703–713
Lee YR, Assmann SM (1999) *Arabidopsis thaliana* 'extra-large GTP-binding protein' (AtXLG1): a new class of G-protein. Plant Mol Biol 40:55–64
Leipe DD, Wolf YI, Koonin EV, Aravind L (2002) Classification and evolution of P-loop GTPases and related ATPases. J Mol Biol 317:41–72
Leon O, Roth M (2000) Zinc fingers: DNA binding and protein-protein interactions. Biol Res 33:21–30
Li B, Trueb B (2000) DRG represents a family of two closely related GTP-binding proteins. Biochim Biophys Acta 1491:196–204
Linglart A, Mahon MJ, Kerachian MA, Berlach DM, Hendy GN, Juppner H, Bastepe M (2006) Coding *GNAS* mutations leading to hormone resistance impair in vitro agonist- and cholera toxin-induced adenosine cyclic 3′, 5′-monophosphate formation mediated by human XLαs. Endocrinology 147:2253–2262
Linsmaier EM, Skoog F (1965) Organic growth factor requirements of tobacco tissue cultures. Physiol Plant 18:100–127
Liu X, Yue Y, Li B, Nie Y, Li W, Wu WH, Ma L (2007) A G protein-coupled receptor is a plasma membrane receptor for the plant hormone abscisic acid. Science 315:1712–1716
Ma H (1994) GTP-binding proteins in plants: new members of an old family. Plant Mol Biol 26:1611–1636
Maeda Y, Ide T, Koike M, Uchiyama Y, Kinoshita T (2008) GPHR is a novel anion channel critical for acidification and functions of the Golgi apparatus. Nat Cell Biol 10:1135–1145
Mahajan MA, Park ST, Sun XH (1996) Association of a novel GTP binding protein, DRG, with TAL oncogenic proteins. Oncogene 12:2343–2350

Matsuda A, Suzuki Y, Honda G, Muramatsu S, Matsuzaki O, Nagano Y, Doi T, Shimotohno K, Harada T, Nishida E, Hayashi H, Sugano S (2003) Large-scale identification and characterization of human genes that activate NF-κB and MAPK signaling pathways. Oncogene 22: 3307–3318

McEwen DP, Gee KR, Kang HC, Neubig RR (2001) Fluorescent BODIPY-GTP analogs: real-time measurement of nucleotide binding to G proteins. Anal Biochem 291:109–117

Meier I (2007) Composition of the plant nuclear envelope: theme and variations. J Exp Bot 58:27–34

Mhaouty-Kodja S (2004) Gha/tissue transglutaminase 2: an emerging G protein in signal transduction. Biol Cell 96:363–367

Muller AH, Hansson M (2009) The barley magnesium chelatase 150-kDa subunit is not an abscisic acid receptor. Plant Physiol 150:157–166

Murashige T, Skoog F (1962) A revised medium for rapid growth and bio-assays with tobacco tissue cultures. Physiol Plant 15:473–497

Nakamura M, Naoi K, Shoji T, Hashimoto T (2004) Low concentrations of propyzamide and oryzalin alter microtubule dynamics in *Arabidopsis* epidermal cells. Plant Cell Physiol 45:1330–1334

Nakaoka H, Perez DM, Baek KJ, Das T, Husain A, Misono K, Im MJ, Graham RM (1994) Gh: a GTP-binding protein with transglutaminase activity and receptor signaling function. Science 264:1593–1596

Nawy T, Lee JY, Colinas J, Wang JY, Thongrod SC, Malamy JE, Birnbaum K, Benfey PN (2005) Transcriptional profile of the *Arabidopsis* root quiescent center. Plant Cell 17:1908–1925

Nibau C, Wu HM, Cheung AY (2006) RAC/ROP GTPases: 'hubs' for signal integration and diversification in plants. Trends Plant Sci 11:309–315

Nielsen E, Cheung AY, Ueda T (2008) The regulatory RAB and ARF GTPases for vesicular trafficking. Plant Physiol 147:1516–1526

Obrdlik P, El-Bakkoury M, Hamacher T, Cappellaro C, Vilarino C, Fleischer C, Ellerbrok H, Kamuzinzi R, Ledent V, Blaudez D, Sanders D, Revuelta JL, Boles E, Andre B, Frommer WB (2004) K^+ channel interactions detected by a genetic system optimized for systematic studies of membrane protein interactions. Proc Natl Acad Sci USA 101:12242–12247

Oh DY, Kim K, Kwon HB, Seong JY (2006) Cellular and molecular biology of orphan G protein-coupled receptors. Int Rev Cytol 252:163–218

Okada K, Shimura Y (1990) Reversible root tip rotation in *Arabidopsis* seedlings induced by obstacle-touching stimulus. Science 250:274–276

Pandey S, Chen JG, Jones AM, Assmann SM (2006) G-protein complex mutants are hypersensitive to abscisic acid regulation of germination and postgermination development. Plant Physiol 141:243–256

Pandey S, Monshausen GB, Ding L, Assmann SM (2008) Regulation of root-wave response by extra large and conventional G proteins in *Arabidopsis thaliana*. Plant J 55:311–322

Pandey S, Nelson DC, Assmann SM (2009) Two novel GPCR-type G proteins are abscisic acid receptors in *Arabidopsis*. Cell 136:136–148

Pasolli HA, Huttner WB (2001) Expression of the extra-large G protein α-subunit XLαs in neuroepithelial cells and young neurons during development of the rat nervous system. Neurosci Lett 301:119–122

Pasolli HA, Klemke M, Kehlenbach RH, Wang Y, Huttner WB (2000) Characterization of the extra-large G protein α-subunit XLαs. I. Tissue distribution and subcellular localization. J Biol Chem 275:33622–33632

Perfus-Barbeoch L, Jones AM, Assmann SM (2004) Plant heterotrimeric G protein function: insights from *Arabidopsis* and rice mutants. Curr Opin Plant Biol 7:719–731

Plagge A, Gordon E, Dean W, Boiani R, Cinti S, Peters J, Kelsey G (2004) The imprinted signaling protein XLαs is required for postnatal adaptation to feeding. Nat Genet 36:818–826

Postle K (1993) TonB protein and energy transduction between membranes. J Bioenerg Biomembr 25:591–601

Postle K, Larsen RA (2007) TonB-dependent energy transduction between outer and cytoplasmic membranes. Biometals 20:453–465

Razem FA, El-Kereamy A, Abrams SR, Hill RD (2006) The RNA-binding protein FCA is an abscisic acid receptor. Nature 439:290–294

Razem FA, El-Kereamy A, Abrams SR, Hill RD (2008) Retraction. The RNA-binding protein FCA is an abscisic acid receptor. Nature 456:824

Risk JM, Macknight RC, Day CL (2008) FCA does not bind abscisic acid. Nature 456:E5–E6

Rutherford R, Masson PH (1996) *Arabidopsis thaliana sku* mutant seedlings show exaggerated surface-dependent alteration in root growth vector. Plant Physiol 111:987–998

Saraste M, Sibbald PR, Wittinghofer A (1990) The P-loop–a common motif in ATP- and GTP-binding proteins. Trends Biochem Sci 15:430–434

Sazuka T, Tomooka Y, Ikawa Y, Noda M, Kumar S (1992) DRG: a novel developmentally regulated GTP-binding protein. Biochem Biophys Res Commun 189:363–370

Sedbrook JC, Ehrhardt DW, Fisher SE, Scheible WR, Somerville CR (2004) The *Arabidopsis sku6/spiral1* gene encodes a plus end-localized microtubule-interacting protein involved in directional cell expansion. Plant Cell 16:1506–1520

Shen YY, Wang XF, Wu FQ, Du SY, Cao Z, Shang Y, Wang XL, Peng CC, Yu XC, Zhu SY, Fan RC, Xu YH, Zhang DP (2006) The Mg-chelatase H subunit is an abscisic acid receptor. Nature 443:823–826

Sprang SR (1997) G protein mechanisms: insights from structural analysis. Annu Rev Biochem 66:639–678

Stafstrom JP (2008) Expression patterns of *Arabidopsis DRG* genes: Promoter::GUS fusions, quantitative RT-PCR and patterns of protein accumulation in response to environmental stresses. Int J Plant Sci 169:1046–1056

Staker BL, Korber P, Bardwell JC, Saper MA (2000) Structure of Hsp15 reveals a novel RNA-binding motif. EMBO J 19:749–757

Strausberg RL, Feingold EA, Grouse LH, Derge JG, Klausner RD, Collins FS, Wagner L, Shenmen CM, Schuler GD, Altschul SF, Zeeberg B, Buetow KH, Schaefer CF, Bhat NK, Hopkins RF, Jordan H, Moore T, Max SI, Wang J, Hsieh F, Diatchenko L, Marusina K, Farmer AA, Rubin GM, Hong L, Stapleton M, Soares MB, Bonaldo MF, Casavant TL, Scheetz TE, Brownstein MJ, Usdin TB, Toshiyuki S, Carninci P, Prange C, Raha SS, Loquellano NA, Peters GJ, Abramson RD, Mullahy SJ, Bosak SA, McEwan PJ, McKernan KJ, Malek JA, Gunaratne PH, Richards S, Worley KC, Hale S, Garcia AM, Gay LJ, Hulyk SW, Villalon DK, Muzny DM, Sodergren EJ, Lu X, Gibbs RA, Fahey J, Helton E, Ketteman M, Madan A, Rodrigues S, Sanchez A, Whiting M, Young AC, Shevchenko Y, Bouffard GG, Blakesley RW, Touchman JW, Green ED, Dickson MC, Rodriguez AC, Grimwood J, Schmutz J, Myers RM, Butterfield YS, Krzywinski MI, Skalska U, Smailus DE, Schnerch A, Schein JE, Jones SJ, Marra MA (2002) Generation and initial analysis of more than 15, 000 full-length human and mouse cDNA sequences. Proc Natl Acad Sci USA 99:16899–16903

Takai Y, Sasaki T, Matozaki T (2001) Small GTP-binding proteins. Physiol Rev 81:153–208

Temple BR, Jones AM (2007) The plant heterotrimeric G-protein complex. Annu Rev Plant Biol 58:249–266

Thompson MV, Holbrook NM (2004) Root-gel interactions and the root waving behavior of *Arabidopsis*. Plant Physiol 135:1822–1837

Ullah H, Chen JG, Temple B, Boyes DC, Alonso JM, Davis KR, Ecker JR, Jones AM (2003) The β-subunit of the *Arabidopsis* G protein negatively regulates auxin-induced cell division and affects multiple developmental processes. Plant Cell 15:393–409

Ullah H, Chen JG, Wang S, Jones AM (2002) Role of a heterotrimeric G protein in regulation of *Arabidopsis* seed germination. Plant Physiol 129:897–907

Vernoud V, Horton AC, Yang Z, Nielsen E (2003) Analysis of the small GTPase gene superfamily of *Arabidopsis*. Plant Physiol 131:1191–1208

Wang S, Assmann SM, Fedoroff NV (2008) Characterization of the *Arabidopsis* heterotrimeric G protein. J Biol Chem 283:13913–13922

Wang S, Narendra S, Fedoroff N (2007) Heterotrimeric G protein signaling in the *Arabidopsis* unfolded protein response. Proc Natl Acad Sci USA 104:3817–3822

Wang XQ, Ullah H, Jones AM, Assmann SM (2001a) G protein regulation of ion channels and abscisic acid signaling in *Arabidopsis* guard cells. Science 292:2070–2072

Wang ZY, Seto H, Fujioka S, Yoshida S, Chory J (2001b) BRI1 is a critical component of a plasma-membrane receptor for plant steroids. Nature 410:380–383

Weinstein LS, Liu J, Sakamoto A, Xie T, Chen M (2004) Minireview: *GNAS*: normal and abnormal functions. Endocrinology 145:5459–5464

Weiss CA, Garnaat CW, Mukai K, Hu Y, Ma H (1994) Isolation of cDNAs encoding guanine nucleotide-binding protein β-subunit homologues from maize (ZGB1) and *Arabidopsis* (AGB1). Proc Natl Acad Sci USA 91:9554–9558

Weiss CA, Huang H, Ma H (1993) Immunolocalization of the G protein α subunit encoded by the *GPA1* gene in *Arabidopsis*. Plant Cell 5:1513–1528

Willard FS, Kimple AJ, Johnston CA, Siderovski DP (2005) A direct fluorescence-based assay for RGS domain GTPase accelerating activity. Anal Biochem 340:341–351

Yang Z (2002) Small GTPases: versatile signaling switches in plants. Plant Cell (Suppl.) 14: S375–S388

Zhao XF, Aplan PD (1998) SCL binds the human homologue of DRG in vivo. Biochim Biophys Acta 1448:109–114

Zhu H, Li G-J, Ding L, Cui X, Berg H, Assmann SM, Xia Y (2009) *Arabidopsis* extra large G protein 2 (XLG2) interacts with the Gβ subunit of heterotrimeric G protein and functions in disease resistance. Mol Plant 2:513–525

Evolution of the ROP GTPase Signaling Module

John E. Fowler

Abstract The core ROP GTPase module is composed of ROP along with regulators that directly influence its activity. This module plays an important role in several crucial cellular pathways in plants, such as polarized cell growth and pathogen response. ROP's ability to influence multiple pathways likely derives from the complexity of the module components (e.g., the ROP regulators PRONE GEF, and RopGAP), which have expanded in number and diversified in flowering plants. In this chapter, I use a variety of new sequence resources to investigate the origin of ROP GTPase diversity in angiosperms, thus updating our phylogenetic understanding of ROP evolution. Data from more basal plants help strengthen the notion that diversity in the ROP family first increased in a gymnosperm ancestor and continued to grow in angiosperms. In addition, our understanding of the evolution of the various ROP regulators and downstream effectors will be considered.

1 Introduction

Since the identification of the first ROP GTPase in 1993 (Yang and Watson 1993), a multitude of studies has demonstrated the importance of the ROP family of proteins, along with ROP regulators and effectors, in the development and function of flowering plants. ROP signaling in plants appears to play a role in polarized cell growth, interactions with microbial pathogens, auxin- and ABA-influenced gene expression, and abiotic stress response (reviewed in Yang 2002; Gu et al. 2004; Berken 2006; Nibau et al. 2006, and in this volume). Given the importance of these processes in higher plants, it is not surprising that ROPs and ROP interactors have been detected in all angiosperm genomes that have been scrutinized for their

J.E. Fowler
Department of Botany and Plant Pathology and Center for Genome Research and Biocomputing, 2082 Cordley Hall, Corvallis, OR 97331, USA
e-mail: fowlerj@science.oregonstate.edu

S. Yalovsky et al. (eds.), *Integrated G Proteins Signaling in Plants*,
Signaling and Communication in Plants,
DOI 10.1007/978-3-642-03524-1_15,

presence (e.g., grapes, legumes – Abbal et al. 2007; Yuksel and Memon 2008). Furthermore, the available genome sequences in higher plants indicate that the ROP family has diversified into multiple, conserved subgroups, each with distinct sequence characteristics (Winge et al. 2000; Zheng and Yang 2000; Christensen et al. 2003). The functional complexity documented for ROP signaling (reviewed in Nibau et al. 2006) may be derived, in part, from this sequence diversity and the ensuing structural diversity. Intriguingly, ROP-encoding sequences have been detected in more basal plant lineages, e.g., moss and conifers (Winge et al. 2000; Christensen et al. 2003; Brembu et al. 2006 and see Sect. 15.3), suggesting that ROP function has been integral to plant cells since their origin.

These observations raise questions regarding the evolution of the ROP signaling module. At what point in evolution did ROP-specific characteristics appear? What is the evolutionary relationship among ROP subgroups, and when did the subgroups originate? Are ROP regulators also associated with evolutionary diversity, and if so, what are the evolutionary origins of this diversity? Are specific ROP module genes associated with rapid evolutionary change? Previous analyses strongly support the contention that the ROP family is monophyletic in plants (Winge et al. 2000; Boureux et al. 2007), i.e., it originated as a single gene, which has duplicated and diversified over the course of plant evolution. Thus, the ROP module provides an attractive model for investigating the evolution of a signal transduction mechanism from simple to complex, through addition of new components and gene duplication and diversification. An understanding of ROP module evolution may also inform functional studies of module components, by guiding the application of "phylogeny of function" (Pereira-Leal and Seabra 2001), i.e., the concept that functionally similar proteins group together within clades. Finally, plant science is transitioning from using *Arabidopsis thaliana* as the single pre-eminent model to using it as the primary reference for translational research toward applied goals in other species. Thus, a better understanding of ROP module evolution would allow for more intelligent manipulation of ROP signaling function in agronomically important species.

2 Composition of the ROP Signaling Module

ROP GTPases are a subfamily of the Rho-type small GTPase family found in almost all eukaryotes, which in turn falls within the larger Ras superfamily (Boureux et al. 2007). As a part of this superfamily, ROPs follow the general model for the "switch"-like activity of small GTP-binding proteins, cycling between an active GTP-bound form and an inactive GDP-bound form. In its active form, ROP binds to downstream effector proteins, positively influencing effector activity and/or localization, to achieve a given cellular function. Its intrinsic GTPase activity provides a mechanism to hydrolyze the bound GTP to GDP, thus inactivating ROP and resetting the switch.

Proteins that influence this cycle through direct interaction with ROP appear to play an important role in determining both the relative concentration and the localization of active ROP. The core Rop GTPase signaling module, thus, consists of the ROP GTPase itself, and three classes of direct regulators: Guanine nucleotide Exchange Factors (RopGEFs), GTPase Activating Proteins (RopGAPs), and Guanine nucleotide Dissociation Inhibitors (RhoGDIs). Each class regulates ROP at a different level. RopGEFs catalyze exchange of GDP for GTP, positively regulating ROP activity. In plants, there are at least two structurally distinct families of RopGEFs: those in the CZH family, with representatives in species across the eukaryotic domain (Meller et al. 2005; Basu et al. 2008), and those in the PRONE family, a plant-specific group (Berken et al. 2005; Kaothien et al. 2005; Gu et al. 2006). RopGAPs enhance the slow intrinsic GTPase activity of ROP, negatively regulating ROP activity. All known RopGAPs contain the conserved RhoGAP catalytic domain present across eukaryotes (Jiang and Ramachandran 2006; Grunt et al. 2008). Finally, on the basis of fungal and animal models, RhoGDIs appear to act at two levels, ostensibly as negative regulators: they sequester ROP in the cytosol away from its site of action at the plasma membrane, and they inhibit exchange of GDP for GTP (reviewed in Dovas and Couchman 2005). However, at least one RhoGDI may play a positive role in plants, by recycling inactive ROP through the cytoplasm back to its site of activity at the plasma membrane (Klahre et al. 2006). Like RopGAPs, plant RhoGDIs contain a structural domain conserved across eukaryotes, in this case, a domain thought to be important for ROP binding (Bischoff et al. 2000).

This brief overview highlights two contrasting notions. First, several features of the ROP module are likely similar to those in Rho GTPase signaling modules in other model eukaryotes (e.g., *Saccharomyces cerevisiae*, mammalian tissue culture cells), based on inheritance of an ancestral module (small GTPase, GAP, RhoGDI, CZH-type GEF) from the last common ancestor of these organisms. In contrast, evolutionary change since that divergence has generated novel features associated with the ROP module. For example, the PRONE GEF catalytic structure and its protein family has emerged (Elias 2008), along with novel combinations of the RhoGAP domain with other structural motifs (e.g., formin or pleckstrin-homology domains – Grunt et al. 2008; Hwang et al. 2008). In addition, the GTPase structure itself has been altered, via the gene duplication and divergence seen in higher plant genomes (Winge et al. 2000; Christensen et al. 2003; and see Section 3.2 "Diversification in the ROP GTPase Family"). Duplication provides the evolutionary opportunity for subfunctionalization and/or neofunctionalization of one or both of the gene duplicates (Freeling 2009), potentially subjecting them to distinct selective forces, and therefore, to distinct pathways of evolution. Such novelty and diversification likely provide the raw material that constitutes, at least in part, the plant-specific, ROP-associated signaling mechanisms.

In the following sections, the evolution of each component of the ROP module will be examined, primarily from a sequence perspective. The recent explosion of sequence information from a diversity of plant species (both higher and more basal) has greatly increased our ability to comprehend the overall trends in gene and protein family evolution. Therefore, the module components will be examined using published analyzes, and in the case of ROP GTPases, more recently available sequences, to illustrate how these new resources can improve our understanding.

3 The ROP GTPases

3.1 Origin of the ROP GTPases

ROPs are a monophyletic clade in plants, most closely related to the Rac subfamily present across almost all eukaryotic kingdoms (Boureux et al. 2007). Thus, it has been proposed that Rac is the founding member of the Rho GTPase family, which then gave rise to other GTPase subgroups in various clades of species over the course of evolution (e.g., Rho and Cdc42 in fungi and metazoans) (Boureux et al. 2007). The close relationship between ROP and Rac sequences clearly explains why a number of ROPs were designated as RACs upon isolation (e.g., OsRAC1 from rice Kawasaki et al. 1999). However, as expected for a group more directly derived from the most ancestral form, the ROPs have been ascribed broad functional roles (e.g., cell polarity, pathogen response), relative to the functional specificity seen in the Rac, Rho, and Cdc42 subgroups in fungi and metazoans (Boureux et al. 2007). Thus, using a distinct name (ROP) to differentiate the plant family from those in metazoans appears justified. In this chapter, I use ROP designations to refer to the proteins and genes discussed, except where the RAC designation is the most common in the recent literature. Table 1, modified from Vernoud et al. (2003), cross-references the various names used to designate ROP family members in *A. thaliana*.

Table 1 Names used in the literature for ROP GTPases of *Arabidopsis thaliana*

Name used in this chapter[a]	Alternative names		Systematic gene ID[b]
ROP1	Arac11/AtRAC11[c]		At3g51300
ROP2	Arac4/AtRAC4		At1g20090
ROP3	Arac1/AtRAC1		At2g17800
ROP4	Arac5/AtRAC5		At1g75840
ROP5	Arac6/AtRAC6	AtRac2[d]	At4g35950
ROP6	Arac3/AtRAC3	AtRac1	At4g35020
ROP7	Arac2/AtRAC2		At5g45970
ROP8	Arac9/AtRAC9		At2g44690
ROP9	Arac7/AtRAC7		At4g28950
ROP10	Arac8/AtRAC8		At3g48040
ROP11	Arac10/AtRAC10		At5g62880

[a]Often seen in the literature with "At" as a prefix or suffix (e.g., in Li et al. 1998)
[b]The *Arabidopsis* Genome Initiative nomenclature, as currently administered by TAIR (http://www.arabidopsis.org/portals/nomenclature/guidelines.jsp)
[c]For example, in Winge et al. 2000
[d]For example, in Lemichez et al. 2001

ROP proteins are highly conserved across their entire length (from 66 to 100% amino acid identity, comparing available sequences throughout the land plants – from moss to angiosperms). For example, even AtROP8, the *Arabidopsis* ROP least similar to ROPs of the moss *Physcomitrella patens*, encodes a peptide that is fully 75% identical to those in the moss. The conserved regions include near-invariant residues in the G1–G5 G-box-motifs that make up the nucleotide-binding site of the small GTPases (Bourne et al. 1991; Sørmo et al. 2006). Two additional structural elements are nearly invariant, and present in ROPs, but not in other Rho family members (Berken and Wittinghofer 2008). First is a set of five consecutive amino acid residues, "SYRGA" (Serine–Tyrosine–Argnine–Glycine–Alanine), located just *C*-terminal to the Switch II region of the GTPase. This element may include a recognition site (SYR) for serine-threonine kinases, and appears to date back at least to the origin of land plants, as it is present in the four ROPs in *P. patens*. The element diverges in only one of 150 ROP sequences (see Section 3.2 "Diversification in the ROP GTPase Family") across a broad phylogenetic spectrum in land plants (in the grape VvROP11, which contains SYQGA – Abbal et al. 2007). The second ROP-specific element is in an α-helix-forming, 12 amino acid sequence called the "Rho insert." This insert is specific to Rho-type GTPases, and is consistently two residues shorter in ROPs than in other Rho proteins (Berken and Wittinghofer 2008). Again, this ROP-specific deletion represents an early divergence from the ancestral GTPase, as it is present in all ROPs of *P. patens*. Among analyzed flowering plant sequences, only two ROPs (ROP9 from *A. thaliana* and a ROP from papaya) show an additional deletion of two amino acids. However, sequences in this region show marked variation across the ROP family, and may be involved in providing functionally distinguishing characteristics among different ROP subgroups (Christensen et al. 2003, and see Sect. 14.3.3).

We are precluded from stating that ROPs were present while plants first colonized the land, due to the absence of a fully sequenced genome from the most basal group of land plants (liverworts – Qiu et al. 2006). Nonetheless, it appears that ROPs were in plant genomes in a form similar to their current one early in the evolutionary history of land plants. To assess the earlier stages of ROP evolution, the available genomes of algae have been inspected for Rho-type genes (Brembu et al. 2006; Elias 2008). A Rho molecule closely related to ROPs has been detected in two of the smallest free-living eukaryotes, *Ostreococcus tauri* and *O. lucimarinus*. These phytoplankton species belong to the Prasinophyceae, an early diverging clade in the Chlorophyta (green algae) (Keeling 2007). Neither sequence contains either of the two ROP-specific elements described above. Yet, both genomes do apparently encode highly diverged homologs of the PRONE GEF domain, as does the genome of the red alga *Cyanidioschyzon merolae* (Elias 2008). Thus, current data suggest that at least one ROP-specific component of the ROP module, the PRONE GEF, was present at a fairly early stage in the evolution of the Archaeplastida. (Archaeplastida is the eukaryotic supergroup encompassing organisms associated with the primary endosymbiosis that generated the chloroplast – Adl et al. 2005). Intriguingly, neither *Chlamydomonas reinhardtii* nor *Volvox carteri*, two closely related species of chlorophycean algae, encodes recognizable Rho

GTPase or PRONE GEF proteins, suggesting the loss of this signaling mechanism in at least one clade of the Chlorophyta (Elias 2008). Additional genome sequences, from liverworts and the charophycean algae (the group most closely related to land plants), may address the possibility that the ROP GTPase hallmarks are a later innovation, perhaps associated with colonization of land.

3.2 Diversification in the ROP GTPase Family

All land plant genomes sequenced and analyzed to date encode more than a single ROP, with copy number ranging from 2 (the lycophyte *Selaginella moellendorffii*) to 13 (poplar – *Populus trichocarpa*) (Table 2). Functional characterization of ROPs has been done most extensively for *A. thaliana*'s 11 ROPs, which, with its relatively high copy number, provides an explanation for observed functional redundancy (reviewed in Gu et al. 2004). It has long been recognized that the ROPs in flowering plants can be categorized into a few subgroups, with an early analysis suggesting four groups (I through IV), based on sequence similarity (Zheng and Yang 2000; Yang 2002). Subsequent phylogenetic analyzes with additional

Table 2 Number of ROP-encoding genes in plant genomes

Data from:	Plant group	Species	Number of ROPs	
			Type-I	Type-II
Sequenced genome	Moss	*Physcomitrella patens*	4	0
	Lycophyte	*Selaginella moellendorffii*	2	0
	Dicot – Rosid	Vitis vinifera (grape)	6	2
		Carica papaya (papaya)	4 (+1)[b]	2
		Arabidopsis thaliana	8	3
		Populus trichocarpa (poplar)	10	3
	Monocot – Grass	*Sorghum bicolor*	1	4
		Brachypodium distachyon	3	2
		Zea mays	3	6
		Oryza sativa	3	4
cDNA sequencing (ESTs) and contig assembly[a]	Gymnosperm – Conifer	*Pinus taeda* (loblolly pine)	3	1
		Picea glauca (white spruce)	5	1
	Basal angiosperm	*Amborella trichopoda*	2 (+1) [b]	1 (+1) [b]
	Magnoliid	*Liriodendron tulipifera* (tulip poplar)	4	1 (+1) [b]
	Monocot – Grass	*Panicum virgatum* (switchgrass)	5	4

[a]Because EST sequencing does not cover the entire genome, the numbers are a minimum. Only data for large-scale EST projects are shown

[b]The nucleotide sequence for these genes (denoted by a + 1) is not complete at the 3′ end. Thus, it is not possible to confirm completely that the gene belongs in the assigned category, as this is ultimately dependent on *C*-terminal amino acid sequence. However, characteristics of the available sequence strongly suggest inclusion in the given category

sequences have generated minor revisions to this initial grouping and nomenclature (Winge et al. 2000; Christensen et al. 2003). Due to the paucity of sequences from nonangiosperm plants during these earlier analyzes, the evolutionary origin of the proposed ROP subgroups was not discernible. However, taken together, these analyzes did propose two major points: (1) at least four ROP paralogs were present in the angiosperm progenitor; and (2) a major new type of ROP GTPase, the Type-II ROP, had appeared at least by the time of this progenitor. Type-II ROPs are primarily distinguishable by unique sequence features in the *C*-terminal Hyper-Variable Region or "HVR." These features are derived from an evolutionary change that extended the HVR, due to an additional intron and exon relative to conventional type-I ROPs. Because the HVR plays a central role in subcellular targeting of ROPs, the type-II ROPs localize to the plasma membrane by a mechanism distinct from that of the type-I ROPs, which appears to affect their function (Ivanchenko et al. 2000; Lavy et al. 2002).

The rapid increase in available plant DNA sequences in the past few years, both genomic and from cDNA (i.e., Expressed Sequence Tags – ESTs), provided an incentive to revisit the phylogenetic analysis of the ROP family to determine whether these earlier proposals are better supported with additional data. A new phylogenetic tree for the ROP family (Fig. 1) has been generated from 150 nucleotide sequences using maximum likelihood methods with the RAxML software package (Stamatakis 2006). In this analysis, the addition of many more sequences from nonangiosperm and basal angiosperm lineages allows for higher resolution in defining the steps in plant evolution at which distinct ROP subgroups originated. Of particular importance were the sequencing of the genomes of the moss *P. patens* (Rensing et al. 2008) and the lycophyte *S. moellendorffii* (a basal vascular plant) (Wang et al. 2005), the 454-based ultra high-throughput sequencing of cDNAs from a number of basal lineages (e.g., the basal-most angiosperm *Amborella trichopoda*) by the Ancestral Angiosperm Genome Project (ancangio.uga.edu and Albert et al. 2005), and the large-scale EST sequencing from a number of gymnosperms (including white spruce – *Picea glauca*; and loblolly pine – *Pinus taeda*) (Cairney et al. 2006; Ralph et al. 2008). The PlantGDB comparative genomics portal, which generates transcript assemblies from ESTs across the plant kingdom (Duvick et al. 2008), was also useful as a source of gymnosperm and nonrosid dicot sequences, as no complete genomes are available in these groups; and the phytozome.net project from the Joint Genome Institute was also valuable, as a resource for obtaining sequences from complete genomes. The most important details regarding the computational methods are available in the legend of Fig. 1, with additional information, alignments, and sequence sources available at the website: oregonstate.edu/~fowlerjo/RopEvolution/.

The resulting unrooted phylogram, in which branch length is proportional to the amount of inferred sequence change, shows three major, well-supported clades (Fig. 1). Inclusion of a smaller group of gymnosperm sequences (designated "Gym1a") in one of the three major clades is not well supported, although this analysis places it closest to Clade 3. Both monocot and dicot ROP genes are present

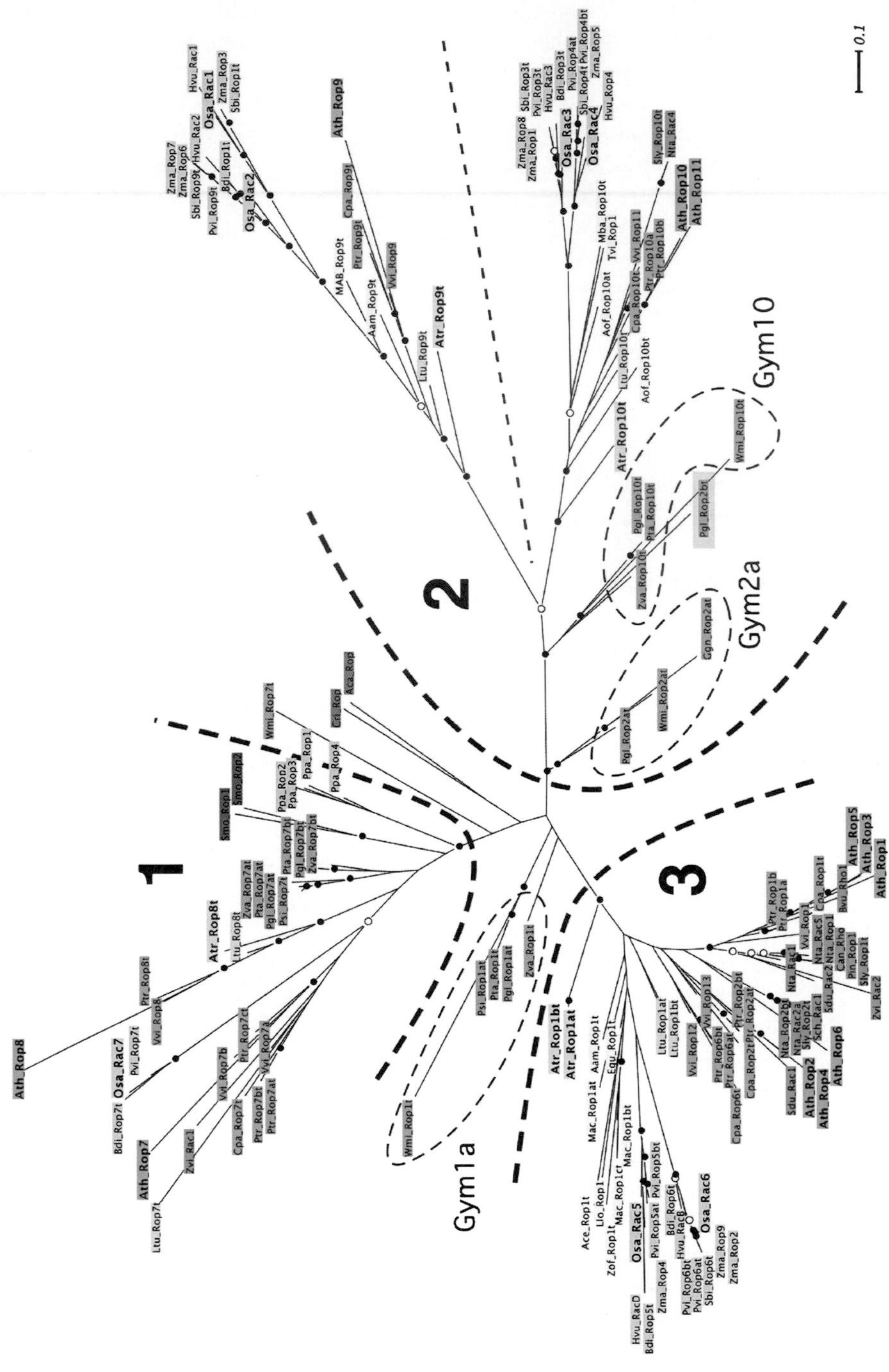
1
2
3
Gym1a
Gym2a
Gym10
0.1

in all three clades. Furthermore, there is some support for bifurcation of Clade 2 into two subgroups (*dotted green line*), each also containing monocot and dicot taxa. Because *Arabidopsis* (designated "Ath" for *A. thaliana*) and rice ("Osa" for *Oryza sativa*) are both reference species, the names for their ROPs are in bold and in a larger font. ROP sequences from *Amborella trichopoda* (Atr) are also in larger boldface, due to the placement of the plant at the base of the angiosperm lineage, and thus its unique informative value. Clades 1 and 3 and the two subgroups of Clade 2 each include at least one *Amborella* sequence. Thus, this analysis strongly supports the proposal that these four lineages were present at the origin of the angiosperms. The grouping is not significantly different from previously published trees (Winge et al. 2000; Christensen et al. 2003), with group II corresponding to Clade 2, group IV corresponding to Clade 3, and groups I and III both contained in Clade 1 (Zheng and Yang 2000; Yang 2002). A representative ROP from *Arabidopsis* can be used to refer to each of these four lineages: ROP7 (Clade 1), ROP9 and ROP10 (subgroups from Clade 2), and ROP1 (Clade 3). Each major clade is discussed in the following sections.

<

Fig. 1 Unrooted phylogram of 150 Rop coding sequences, generated using maximum likelihood analysis with the RAxML software. To aid in interpretation of the phylogram, sequences in particular groups of species are color coded as follows: rosid dicots, dark green; nonrosid dicots, light green; grasses (monocots), orange; nongrass monocots, yellow; basal angiosperms, pale green; gymnosperms, blue; ferns, purple; lycophyte, magenta; and moss, pink. The three major, well-supported clades are denoted with dark red dotted lines, and numbered (1, 2 and 3); other notations are discussed in the text. Species designations: Ath, *Arabidopsis thaliana*; Atr, *Amborella trichopoda*; Bdi, *Brachypodium distachyon*; Bvu, *Beta vulgaris*; Can, *Capsicum annuum*; Cpa, *Carica papaya*; Cri, *Ceratopteris richardii*; Egu, *Elaeis guineensis*; Ggn, *Gnetum gnemon*; Hvu, *Hordeum vulgare*; Llo, *Lilium longiflorum*; Ltu, *Liriodendron tulipifera*; MAB, *Musa* ABB Group; Mac, *Musa acuminata*; Mba, *Musa balbisiana*; Nta, *Nicotiana tabacum*; Osa, *Oryza sativa*; Pgl, *Picea glauca*; Pin, *Petunia integrifolia*; Ppa, *Physcomitrella patens*; Psi, *Picea sitchensis*; Pta, *Pinus taeda*; Ptr, *Populus trichocarpa*; Pvi, *Panicum virgatum*; Sbi, *Sorghum bicolor*; Sch, *Solanum chacoense*; Sdu, *Scoparia dulcis*; Sly, *Solanum lycopersicum*; Smo, *Selaginella moellendorffii*; Tvi, *Tradescantia virginiana*; Vvi, *Vitis vinifera*; Wmi, *Welwitschia mirabilis*; Zma, *Zea mays*; Zof, *Zingiber officinale*; Zva, *Zamia vazquezii*; Zvi, *Zinnia violacea*. If a published name was available for a gene, this name was used (e.g., rice Rac1 = Osa_Rac1). If no published name was available, the gene was named for the most closely related *Arabidopsis* ROP, followed by a lower case letter designation ("a" or "b") to differentiate between likely duplicates; in addition, a "t" was used as a suffix to designate the nomenclature as tentative (e.g., Ptr_Rop7at from poplar). The alignment used for the analysis is 621 nucleotides long, and does not include positions coding for amino acids at the extreme *N*-terminus and in a portion of the HVR region, as these could not be confidently aligned across all taxa. 250 runs with distinct starting trees were used in the default RAxML mode, using a general time reversible model of nucleotide substitution and the gamma model of rate heterogeneity, with estimated base frequencies and rate variation across sites modeled by codon position. 500 replicates of nonparametric bootstrapping were used to measure confidence. Filled circles indicate well-supported nodes (bootstrap values 70% or greater), and open circles represent nodes with some support (bootstrap values 65–69%). Because many details are not easily discernible in this figure, particularly at the "tips" of the tree, a complete tree file in Newick format, along with the original alignment plus sequence source information, is available for download at the website: oregonstate.edu/~fowlerjo/RopEvolution/

3.2.1 Clade 1 ROPs

Clade 1 (the ROP7 group) appears to be most directly derived from the ancestral ROP form. All members have a conventional HVR sequence, and thus are type-I ROPs. Clade 1 includes all four moss and both lycophyte ROPs, and a number of gymnosperm ROPs (all designated in Fig. 1 with Rop7at or Rop7bt, with "t" for tentative nomenclature). Moreover, the two available ROP sequences from ferns are also associated with this clade, although this placement is not well supported. Notably, the genomes of three dicot species (*A. thaliana*, *P. trichocarpa,* and grape – *Vitis vinifera*) and of the basal angiosperm *Liriodendron tulipifera* (a magnoliid) encode at least two phylogenetically distinct ROP types in this clade, grouped with *Arabidopsis* ROP7 and ROP8. This could be explained by the existence of two Clade 1 ROPs at the angiosperm origin (i.e., a total of five paralogs, rather than four). Discovery of a ROP7-type sequence in *Amborella* would provide additional evidence to support this possibility, and would argue for bifurcation of Clade 1 in angiosperms. Intriguingly, however, two sequenced genomes (the grasses *Sorghum bicolor* and *Zea mays*) do not encode a representative of Clade 1. This suggests that, at least in these two species, any essential functions provided by Clade 1 ROPs were supplanted by other genes, most likely by type-I ROPs in Clade 3. ROPs in Clade 1 are the least characterized of the ROPs, although some evidence supports a role for *Arabidopsis* ROP7 in differentiation of xylem vessels (Brembu et al. 2005).

3.2.2 Clade 2 ROPs

Clade 2 (the ROP9 and ROP10 groups) encompasses all of the type-II ROPs. In fact, only three sequences in this clade – all from gymnosperms – encode type-I ROPs, based on their HVR sequences. These include two ROPs from Clade 2's basal-most group denoted "Gym2a" and the highlighted ROP Pgl_Rop2bt. (One Gym2a clade member is not of full length, and so cannot be assessed). Thus, it appears likely that Gym2a ROPs are derived from the type-I ROP that also gave rise to the type-II ROPs. This analysis is the first to document type-II ROPs in gymnosperms, in the group designated "Gym10" (due to the similarity of its members to *Arabidopsis* ROP10). The Gym10 group includes sequences from species across the gymnosperms, including conifers, a gnetophyte, and a cycad, indicating that type-II ROPs probably originated prior to, or coincident with, the gymnosperm origin. All four Gym10 ROPs contain a polybasic region and a variant of the GC–CG box (S/NC–CG) in their HVRs, both elements that are required for plasma membrane localization of *Arabidopsis* type-II ROPs (Lavy and Yalovsky 2006). Thus, it appears that the plasma membrane localization mechanism documented for *Arabidopsis* type-II ROPs operated in the ancestor of both angiosperms and gymnosperms, and could have been present at the origin of this ROP type. The generation of comprehensive sequence resources in the ferns would help address whether the origin of type-II ROPs was coincident with the appearance of seed plants, i.e., the evolutionary transition from ferns to gymnosperms.

The current sequence data indicate that the original type-II ROP gene did not give rise to the ROP9 and ROP10 subgroups in gymnosperms, but rather that those two subgroups are extant only in angiosperms. As noted previously (Winge et al. 2000; Christensen et al. 2003), in grasses, type-II ROPs appear to have preferentially expanded relative to type-I ROPs, in contrast to the preferential expansion of type-I ROPs seen in dicots (Table 2). Remarkably, the *S. bicolor* genome may have only one functional type-I ROP, compared with four of type-II; it shows no evidence of a Clade 1 sequence, and one of its two Clade 3 sequences (omitted from this analysis) appears to be a pseudogene (data not shown). OsRac1, the best-studied member of Clade 2, has roles in the pathogen response in rice, e.g., by influencing the production of reactive oxygen species (reviewed in Nibau et al. 2006).

3.2.3 Clade 3 ROPs

Clade 3 (the ROP1 group) contains the best-characterized ROPs at the functional level, e.g., in pollen tube tip growth, for ROP1 in *Arabidopsis* and Nt-Rac5 in tobacco (reviewed in Kost 2008). In addition, Clade 3 taxa show very little sequence divergence, hence the relatively short branch lengths. This results in a few useful characters for inferring the evolutionary relationships. Therefore, although there are many nodes (i.e., many taxa), few are well supported, particularly near the base of the clade. However, the depicted tree is consistent with an earlier analysis (Christensen et al. 2003) in that monocots and dicots bifurcate at the base of the clade. The many newly added nongrass monocot and nonrosid dicot sequences follow the same pattern. If the tree is correct, a single ROP-encoding gene likely produced Clade 3 over the course of angiosperm evolution. This gene would have undergone a relatively high rate of evolutionary duplication and retention (compared with Clades 1 and 2), particularly in dicots, to generate a large fraction of the modern angiosperm ROP family. For example, Clade 3 contains three genes each from grape and papaya, at least four genes from *Nicotiana tabacum* (tobacco), and six genes each from *Arabidopsis* and poplar. This diversification offers the opportunity for functional specialization within Clade 3. For example, in *Arabidopsis*, *ROP1* is expressed specifically in pollen and appears to have pollen-specific functions (Li et al. 1998), whereas *ROP2* and *ROP4* have redundant vegetative roles in leaf epidermal cells (Fu et al. 2005). However, because monocot and dicot lineages have diversified independently, functional predictions regarding particular *Rop* genes from dicot to monocot (or vice versa) should be made with caution. For example, in maize, none of the three Clade 3 ROPs (*rop2*, *rop4*, and *rop9*) are associated with pollen-specific expression (Christensen et al. 2003), and *rop2* is known to function both in pollen and during the vegetative phase (Arthur et al. 2003 and Fowler, unpublished observations).

To independently assess the evolutionary relationships within Clade 3, possible syntenic relationships among chromosomal regions that encoded Clade 3 ROPs were determined. Synteny is deduced by detection of a set of collinear gene homologs along two different chromosomal segments. It indicates that those chromosomal regions are derived from an ancestral chromosome (i.e., gene order and location have been preserved across evolutionary time) (Lyons and Freeling 2008). Furthermore, synteny can be found either between species, or within a genome, if large-scale genome duplications have occurred in that genome. As evolutionary time passes, gene loss, inversion, or transposition can obscure collinearity, making synteny less obvious or undetectable. Winge et al. (2000) used syntenic relationships to establish that ROP1 through ROP5 in *Arabidopsis* (all in Clade 3) were derived from genome duplication events, suggesting that such an approach could be fruitful for comparisons to other sequenced dicot genomes. The CoGe bioinformatics platform (synteny.cnr.berkeley.edu/CoGe) provides a useful set of tools for assessing synteny among *Arabidopsis*, grape, papaya, and poplar, all dicots in the rosid group (Lyons et al. 2008). Figure 2 shows such an analysis, revealing the relationships among regions of the grape, papaya, and *Arabidopsis* genomes that

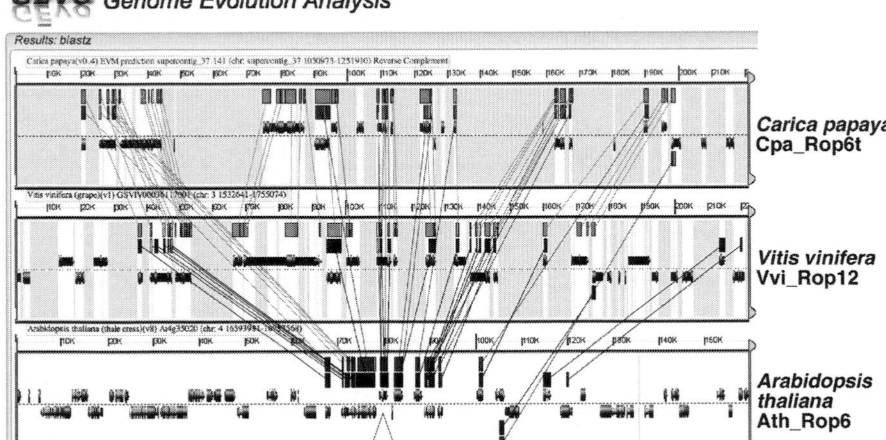

Fig. 2 A diagram generated by the GEvo software tool to illustrate genetic collinearity between chromosomal regions from papaya, grape and *Arabidopsis*. Each of the three horizontal images encodes a ROP protein homolog (designated Cpa_Rop6t, Vvi_Rop12, and Ath_Rop6 in the phylogenetic analysis) at the midpoint of the pictured chromosomal segment (arrowhead). Gene models are shown on either the top (5′ on *left*) or bottom (5′ on *right*) strands (above or below the central line representing the chromosome). Pairwise BLAST hits between genes or gene exons among the chromosomes are represented by blocks of a particular color above each chromosome, with a line drawn between the blocks representing the paired hits. In this case, synteny is present across all three species, due to the extensive shared homology across these segments. Note that the genes are more closely spaced in *Arabidopsis*. This particular image can be regenerated using the url: tinyurl.com/qa8zyt. GEvo is available at: synteny.cnr.berkeley.edu/CoGe/GEvo.pl (Lyons and Freeling 2008)

Table 3 Dicot ROP Clade 3 genes associated with extensive chromosomal collinearity (i.e., synteny)

Species	ROP1 syntenic group	ROP2 syntenic group	ROP6 syntenic group
A. thaliana	Ath_Rop1, Ath_Rop3, Ath_Rop5	Ath_Rop2, Ath_Rop4	Ath_Rop6
Carica papaya	Cpa_Rop1t	Cpa_Rop2t	Cpa_Rop6t
Populus trichocarpa	Ptr_Rop1a, Ptr_Rop1b	Ptr_Rop2at, Ptr_Rop2bt	Ptr_Rop6at, Ptr_Rop6bt
V. vinifera	Vvi_Rop1	Vvi_Rop13	Vvi_Rop12

contain the most closely related homologs of *Arabidopsis ROP6*. Using these tools, I demonstrated clear syntenic relationships across species and within genomes (data not shown, available on the website cited above) to identify three groups of genes that shared extensive collinearity (Table 3). For all genes but one (Ath_Rop6), the synteny-derived grouping is consistent with the phylogenetically predicted clades (Fig. 1); *Arabidopsis* ROP6 maps on an unsupported branch containing no other rosid taxa, perhaps due to unusual nucleotide divergence. Overall, the synteny analysis provides independent support for the phylogram, and suggests that three Clade 3 ROP paralogs were present at the origin of the rosids. Furthermore, although some minimal collinearity could be detected across all three syntenic groups among the rosids, no syntenic relationship could be established between the rosid and the grass ROPs in Clade 3 (data not shown). This is consistent with the scenario supported by the phylogram, in which the monocot/dicot split preceded ROP gene duplication within Clade 3.

3.3 Functional Implications of ROP Diversification

Diversification of sequences in the ROP family allows diversification of protein function, introducing greater complexity and functional possibilities for the ROP signaling system. This has been clearly demonstrated by the neofunctionalized type-II ROPs, which operate under different cellular parameters than type-I ROPs, due to their distinct localization mechanism (reviewed in Gu et al. 2004; Berken 2006; Nibau et al. 2006). As the functions of type-II ROPs in flowering plants become better understood, it will be interesting to see whether those functions are also present in gymnosperms, and whether they offer any advantage relative to the more ancestral plant groups that lack this ROP type.

A second type of functional diversification that may accompany ROP family expansion is its ability to interact with distinct sets of regulators or effectors. This would require that some determinants for interaction (e.g., protein binding affinity) be distributed to unique ROPs or sets of ROPs. If, due to their selective advantage, such determinants became fixed in a progenitor for a particular ROP clade, the sequences corresponding to the determinants would appear highly conserved only in that particular clade. Such "subfamily-specific sequences" have been identified previously for the angiosperm ROP clades discussed here (Christensen et al. 2003). Intriguingly, some of these sequences cluster in the "insert region" (see Section 3.1 "Origin of the ROP GTPases"), which is involved in the binding of at least one ROP/PRONE–GEF pair; thus, it is reasonable to hypothesize that certain ROP subfamily-specific amino acids in this region help determine which PRONE family proteins can bind (reviewed in Berken and Wittinghofer 2008). Manual inspection of the new, larger ROP sequence dataset indicates that it is largely consistent with the consensus sequences for each subfamily defined earlier (data not shown). Testing whether these sequence elements control interaction specificity will be possible using site-directed mutagenesis, as assays for interaction between specific protein pairs (e.g., the in vitro ROP/PRONE guanine nucleotide exchange reaction – Berken et al. 2005; Gu et al. 2006) become available.

4 Evolution of ROP Regulators

4.1 Positive Regulators: The RopGEFs

The most ancient RopGEF known is represented in *Arabidopsis* by the SPIKE1 protein, a member of the CZH family of RhoGEFs (Meller et al. 2005; Basu et al. 2008). In contrast to the expansion of the ROP family (and the PRONE GEF family, see *below*), neither *SPIKE1* (Basu et al. 2008) nor its ortholog in rice has a homolog in its own genome. Remarkably, the rice and *Arabidopsis* proteins are 75% identical across their entire length of ~1,800 amino acids, suggesting a strong selection for maintaining protein structure. Loss-of-function mutations in *SPIKE1* are seedling lethal, with pleiotropic cell morphology defects throughout the plant, as the gene is ubiquitously expressed in the sporophyte (Qiu et al. 2002). In vitro, SPIKE1 can catalyze nucleotide exchange on multiple ROP family members (Basu et al. 2008), and thus may provide a general ROP activation function crucial for growth in all (or most) vegetative plant cells. This crucial role in plant growth may help explain its evolutionary conservation.

The PRONE GEF family is similar to the ROP family in that both have multiple homologs: 6 in *P. patens*, 4 in *S. moellendorffii*, 11 in rice, and 14 in *Arabidopsis* (Berken et al. 2005; Gu et al. 2006; Elias 2008). The protein has a conserved central domain that provides the GEF activity, and variable *N*- and *C*-terminal domains; amino acid identity among the rice and *Arabidopsis* PRONE sequences ranges from 25 to 97% across the entire protein (Berken et al. 2005). Thus, the divergent domains surrounding the central catalytic domain may help provide functional specificity, by conditionally activating an associated ROP (e.g., only when bound to a third protein). Evidence for this type of regulation exists in the pollen tube: autoinhibition of AtRopGEF12 PRONE activity, mediated by its *C*-terminal domain, is apparently relieved upon binding of the *C*-terminal domain to the receptor kinase AtPRK2a (Zhang and McCormick 2007).

As mentioned previously, the origin of the PRONE GEF itself can be traced back at least to the common ancestor of red algae and plants, as shown by an extensive analysis of sequenced eukaryotic genomes (Elias 2008). To address the origin of PRONE family diversity, as well as the relationships among family members, initial gene family phylogenies have been generated, indicating that at least two subgroups of the PRONE family are present in both the *Arabidopsis* and rice genomes (Berken et al. 2005; Zhang and McCormick 2007). The alignment generated by Elias (2008) was used to produce a maximum likelihood tree that included the entire set of PRONE proteins in *P. patens*, *S. moellendorffii*, rice, and *Arabidopsis* (Fig. 3). This phylogram shows five well-supported subgroups containing both dicot and monocot sequences, implying the existence of five PRONE–GEF paralogs near the base of the angiosperms. Intriguingly, this is similar to the number of ROP paralogs at the angiosperm origin (Fig. 1); however, no subgroup shows obvious preferential expansion in dicots vs. monocots. In addition, three of the subgroups (those containing AtRopGEF1 through AtRopGEF7) are more closely related to each other than to the other two subgroups. However, given the lack of

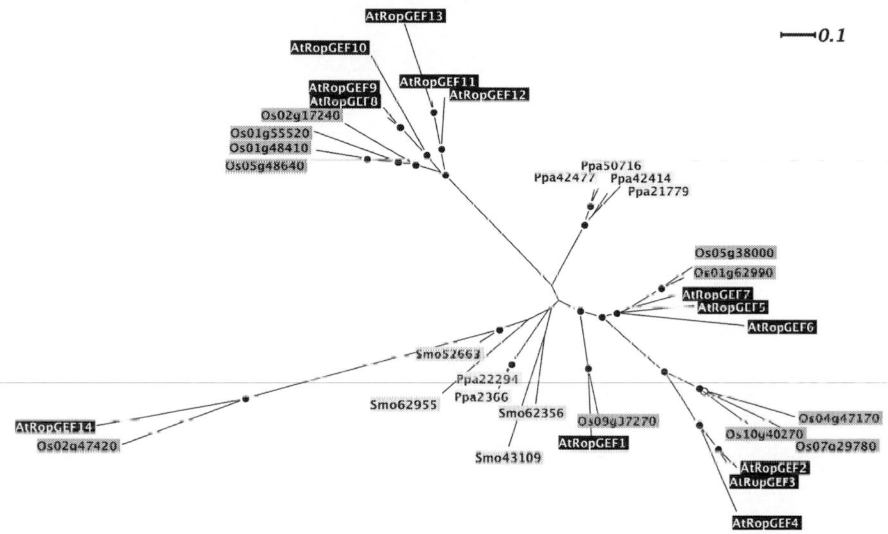

Fig. 3 Unrooted phylogram of 36 PRONE GEF peptide sequences, generated using maximum likelihood analysis with the RAxML software. To aid in interpretation of the phylogram, species are designated and color coded as follows: At, *Arabidopsis thaliana*, white on black; Os, *O. sativa*, black on dark gray; Ppa, *Physcomitrella patens*, black on light gray; Smo, *S. moellendorffii*, black on medium gray. The alignment used for the analysis is 439 amino acids long, and is available from Elias (2008) at http://mbe.oxfordjournals.org/cgi/content/full/25/8/1526. 100 runs with distinct starting trees were used in the default RAxML mode, using the JTT amino acid substitution matrix, using empirical frequencies, and the gamma model of rate heterogeneity, with four discrete rate categories. 100 replicates of nonparametric bootstrapping were used to measure confidence. Filled circles indicate well-supported nodes (bootstrap values 70% or greater), and open circles represent nodes with some support (bootstrap values 65–69%). A complete tree file in Newick format is available for download at the website: oregonstate.edu/~fowlerjo/RopEvolution/

support for the basal nodes of this phylogram, it is not possible to make conclusions regarding the relationships of the angiosperm and basal plant sequences. (Despite strong support, the placement of the AtRopGEF14 group on a branch with an *S. moellendorffii* sequence should probably be viewed with caution, as it is a possible example of the "long branch attraction" artifact – Bergsten 2005.) The addition of gymnosperm and basal angiosperm sequences should help improve resolution at the basal nodes of the phylogram, and also may allow inferences regarding the origin of the RopGEF subgroups relative to the appearance of particular groups. For example, five of the six genes in the large subgroup containing AtRopGEF8 through AtRopGEF13 are associated with pollen-specific or pollen-enriched expression (Zhang and McCormick 2007). Thus, it would be interesting to determine whether gymnosperms express genes in this clade, given that a number of key evolutionary innovations associated with the microgametophyte (e.g., the ability of pollen tubes to deliver the sperm nuclei) first appeared in gymnosperms (Rudall and Bateman 2007).

4.2 Negative Regulators: RhoGDIs and RopGAPs

Plant versions of RhoGDI proteins contain ancient, conserved domains. From an evolutionary standpoint, the RhoGDI family appears simple: one gene in both *P. patens* and *S. moellendorffii*, and three genes in both rice and *Arabidopsis*. Among the proteins in these four species, amino acid identity ranges from 48 to 77%, with the regions of identity distributed across all but a short variable region at the *N*-terminus of each protein. It is not clear whether this level of divergence could generate distinct functions among the RhoGDI isoforms, as homology modeling indicates that plant RhoGDIs probably bind GTPase through a structure similar to those seen for metazoan RhoGDIs (Carol et al. 2005).

The RopGAP family in plants is much more complex, with four recognized types, based on the domain architecture of each family member (Table 4). Each

Table 4 Number of RhoGAP-homologous genes in plant genomes

Plant group	Species	Number of RopGAPs			
		CRIB-RhoGAP[a]	PH-RhoGAP (e.g., REN1)[b]	*C*-terminal RhoGAP[c]	RhoGAP-FH2 (Class III Formin)[d]
Moss	*Physcomitrella patens*	6	2	0	1
Lycophyte	*S. moellendorffii*	2	1	0	1
Dicot – Rosid	*A. thaliana*	5	3	1	0
Monocot – Grass	*O. sativa*	7	4	1	0

[a]Wu et al. 2000
[b]Hwang et al. 2008
[c]Berken 2008
[d]Grunt et al. 2008

type contains a recognizable RhoGAP domain, based on sequence homology, combined with at least one other protein domain. The first type to be identified appears to be the most numerous in plant genomes, present as five paralogs in *Arabidopsis*, and combines a confirmed catalytic RhoGAP domain with the CRIB (Cdc42/Rac interactive binding) domain (Wu et al. 2000). The ancient CRIB (Cdc42/Rac interactive binding) domain was initially documented in yeast and animal models as a specific motif for binding to the active forms of the Cdc42- and Rac-type members of the Rho family (Burbelo et al. 1995), but in these organisms, the CRIB is only present in downstream effector molecules. A second type is represented by the REN1 RopGAP and two other paralogs in *Arabidopsis*, and combines a confirmed catalytic RhoGAP domain with a pleckstrin homology (PH) domain, which may bind phosphoinositides (Hwang et al. 2008). A third type (represented in *Arabidopsis* by a single gene, At5g61530) encodes the RhoGAP domain as its only recognizable motif, near the protein *C*-terminus (Berken and Wittinghofer 2008). No additional data are available on this protein type, although homologs are present in other angiosperm genomes (e.g., rice encodes one – Os07g0486500), but not in moss or *S. moellendorffii*, suggesting a more recent origin. The final type has been documented in basal plants (the two sequenced *Ostreococcus* phytoplankton species, *P. patens* and *S. moellendorffii*) and combines a RhoGAP domain with the FH2 formin homology domain (Grunt et al. 2008; formins are actin-organizing proteins, reviewed in Blanchoin and Staiger 2008). However, this so-called Class III formin was apparently lost in the lineage leading from the lycophyte ancestor to the angiosperms (Grunt et al. 2008). Because the RhoGAP domain in this protein type deviates significantly from the RhoGAP consensus sequence, including the conserved amino acids required for GAP activity, it has been hypothesized that this domain does not inactivate a bound Rho GTPase through enhancing GTP hydrolysis, but rather may merely bind an active Rho and thus directly influence formin activity (Grunt et al. 2008). Although the evolutionary history that gave rise to these varied structures is likely an interesting one, no phylogenetic analysis relating these plant RhoGAPs to each other has been published.

5 Evolution of Output from the ROP Module

Novel functionality has also emerged in the output of the ROP module, as evolutionary innovations have been documented in downstream effector proteins. The ancient CRIB domain is also present in plant effectors, in a family of relatively small proteins termed RICs (Rop-Interactive CRIB motif-containing proteins), which in *A. thaliana* has 11 members (Wu et al. 2001). RICs all contain the conserved CRIB sequence, but also have a diverse set of sequences outside the CRIB domain, unique to the plant kingdom. Thus, there are no clear RIC homologs, or even proposed analogous molecules, in nonplant species. The distinctive non-CRIB sequences, which also allow categorization of RICs into five proposed groups

(Wu et al. 2001), presumably help specific RIC family members to mediate distinct downstream responses. For example, in the pollen tube apex, RIC3 is linked to regulation of cytosolic Ca^{2+} levels and disassembly of F-actin, whereas RIC4 promotes assembly of F-actin (Gu et al. 2005). No cross-species comparison of the RIC family has yet been published to allow assessment of whether the five groups in *A. thaliana* are represented throughout the plant kingdom. However, a barley RIC (HvRIC171) with highest sequence similarity to the *A. thaliana* RIC6–RIC7–RIC8 group (both within and outside of the CRIB domain) has recently been identified (Schultheiss et al. 2008). HvRIC171 interacts with a barley ROP (HvRACB) and appears to be a component of ROP-mediated response to powdery mildew infection, confirming that the RIC family will likely be important for ROP signaling across the flowering plants (Schultheiss et al. 2008).

Another bona fide ROP effector is cinnamoyl-CoA reductase (CCR) (Kawasaki et al. 2005), one of the initial enzymes in the most ancient lignin biosynthetic pathway, thought to be one of the most important evolutionary innovations in land plants (Peter and Neale 2004). Increased lignin biosynthesis helps plant cells respond to many pathogens. Thus, the increase in rice CCR1 activity (encoded by TIGR gene Os02g56460) induced by the binding of CCR to an active ROP (OsRac1) is hypothesized to be an important output from the ROP module upon infection (Kawasaki et al. 2005). Not only is the enzyme specific to plants, but the rice CCR1 ROP-binding site may also be a plant-specific innovation, as no similarity to previously characterized Rho-type GTPase-binding sites has been found (Kawasaki et al. 2005). It will be interesting to determine whether CCR enzymes in lycophytes and gymnosperms (which are vascular plants) also interact with, and are stimulated by, ROP GTPases. This could address whether ROP activity was associated with the origin of the lignin biosynthetic pathway, or is a later regulatory innovation.

Finally, the ICR1/RIP1 protein appears to be part of a novel, plant-specific mechanism to accomplish what may be an ancient Rho GTPase function: to regulate the exocyst complex (Lavy et al. 2007). The exocyst is an eight-protein complex that is crucial for certain forms of exocytosis in plants, fungi, and metazoans (Hála et al. 2008; and reviewed in Zársky et al. 2009). In fungi and metazoans, members of the Rho GTPase family (as well as members of the Rab and Ral GTPases) interact directly with subunits of the exocyst to regulate its localization and/or activity (reviewed in Wu et al. 2008). Although direct interaction of ROPs with a subunit of the plant exocyst has not been demonstrated, the ICR1 protein has been characterized as a scaffold that interacts with both active ROP and the SEC3 subunit of the plant exocyst, and this interaction could help recruit-SEC3 (and the exocyst) to the plasma membrane (Lavy et al. 2007). ICR1 is composed of coiled-coil domains and does not show strong similarity to proteins in nonplant eukaryotes (Lavy et al. 2007). In fact, although ICR1 homologs are in all sequenced angiosperms (Li et al. 2008), and likely ICR1 homologs are present in gymnosperms, BLAST searches do not identify any in *P. patens* or *S. moellendorffii* (J. Fowler, unpublished observations). Thus, a ROP/ICR1/SEC3 mechanism for exocyst regulation could be an evolutionary innovation that is specific for seed plants.

6 Conclusions

Taken together, the evidence discussed in this overview of ROP signaling module evolution supports the general notion suggested in the Introduction: increased complexity associated with the genomes of later-evolving groups correlates with increased complexity in the ROP signaling module. This is apparent in both the greater number of genes per gene family in angiosperms relative to moss and *S. moellendorffii*, and in the increase in genetic diversity in those families. The trend seen in the ROPs themselves (although based on a limited number of genomes and transcriptomes with adequate sequence) appears to be a moderate increase in complexity associated with the gymnosperms, followed by a further increase in the later-evolved angiosperms. However, it is intriguing that the increase in complexity is concentrated in some components of the module (e.g., the PRONE GEFs, the RopGAPs, ROPs themselves to some extent), whereas other components (the SPIKE1-type GEFs, the RhoGDIs) appear more evolutionarily static. This could be due to constraints that limit the ability of particular signaling components to undergo duplication and divergence, and/or the propensity of certain components to be maintained as duplicates, allowing later divergence (Freeling 2009).

The new ROP family phylogeny builds upon earlier analyses (Winge et al. 2000; Christensen et al. 2003; Brembu et al. 2006), providing greater resolution and confidence to the evolutionary framework for understanding the ROP family. For example, the phylogeny could help plant cell and the developmental biologists analyze the functional relevance of ROP-related experimental results by facilitating cross-species comparisons. In addition, the phylogeny, coupled with the protein alignment, may be able to guide domain- or motif-swapping experiments that seek to distinguish between functions for different ROPs encoded by the same genome. Clearly, a more complete picture of gene family evolution is offered by a phylogenetic analysis that includes sequences from across a broader range of species, i.e., through use of gymnosperm and basal angiosperm sequences. Our understanding of other components of the ROP signaling module (e.g., PRONE GEFs, RopGAPs, RICs) would be markedly improved if similar phylogenetic analyses were used to address their evolutionary history. Such analyses might provide a more holistic view of the evolution of the entire module, provoking new questions and experiments. For example, is the origin of type-II ROPs concomitant with the origin of a new clade of PRONE GEFs? If so, do those GEFs show particular affinity for the type-II ROPs? Currently, sequence availability from certain species of particular interest is limited – most notably in gymnosperms, basal angiosperms, nonrosid dicots, and nongrass monocots, especially when confronting the large size of genes encoding some components (e.g., the PRONE GEFs). However, when the ongoing genome sequencing projects for a variety of plant species (e.g., tomato) are completed, and as next-generation sequencing technology is applied to more plant transcriptomes, those limitations will likely disappear or be minimized in the near future.

Acknowledgments I acknowledge the many plant genome and EST sequencing projects, and the comparative genomics portals for providing the data used in this analysis. In addition, I thank Margit Foss for useful critiques and comments on this manuscript, and Drs. Joseph Spatafora and Gi-Ho Sung for assistance with the RAxML software and current methods of phylogeny analysis. Research in my lab is supported by the US National Science Foundation (awards # DBI-0701731 and IOS-0920747).

References

Abbal P, Pradal M, Sauvage FX, Chatelet P, Paillard S, Canaguier A, Adam-Blondon AF, Tesniere C (2007) Molecular characterization and expression analysis of the Rop GTPase family in *Vitis vinifera*. J Exp Bot 58:2641–2652

Adl SM, Simpson AG, Farmer MA, Andersen RA, Anderson OR, Barta JR, Bowser SS, Brugerolle G, Fensome RA, Fredericq S, James TY, Karpov S, Kugrens P, Krug J, Lane CE, Lewis LA, Lodge J, Lynn DH, Mann DG, McCourt RM, Mendoza L, Moestrup O, Mozley-Standridge SE, Nerad TA, Shearer CA, Smirnov AV, Spiegel FW, Taylor MF (2005) The new higher level classification of eukaryotes with emphasis on the taxonomy of protists. J Eukaryot Microbiol 52:399–451

Albert VA, Soltis DE, Carlson JE, Farmerie WG, Wall PK, Ilut DC, Solow TM, Mueller LA, Landherr LL, Hu Y, Buzgo M, Kim S, Yoo MJ, Frohlich MW, Perl-Treves R, Schlarbaum SE, Bliss BJ, Zhang X, Tanksley SD, Oppenheimer DG, Soltis PS, Ma H, dePamphilis CW, Leebens-Mack JH (2005) Floral gene resources from basal angiosperms for comparative genomics research. BMC Plant Biol 5:5

Arthur KM, Vejlupkova Z, Meeley RB, Fowler JE (2003) Maize ROP2 GTPase provides a competitive advantage to the male gametophyte. Genetics 165:2137–2151

Basu D, Le J, Zakharova T, Mallery EL, Szymanski DB (2008) A SPIKE1 signaling complex controls actin-dependent cell morphogenesis through the heteromeric WAVE and ARP2/3 complexes. Proc Natl Acad Sci USA 105:4044–4049

Bergsten J (2005) A review of long-branch attraction. Cladistics 21:163–193

Berken A, Thomas C, Wittinghofer A (2005) A new family of RhoGEFs activates the Rop molecular switch in plants. Nature 436:1176–1180

Berken A (2006) ROPs in the spotlight of plant signal transduction. Cell Mol Life Sci 63:2446–2459

Berken A, Wittinghofer A (2008) Structure and function of Rho-type molecular switches in plants. Plant Physiol Biochem 46:380–393

Bischoff F, Vahlkamp L, Molendijk A, Palme K (2000) Localization of AtROP4 and AtROP6 and interaction with the guanine nucleotide dissociation inhibitor AtRhoGDI1 from *Arabidopsis*. Plant Mol Biol 42:515–530

Blanchoin L, Staiger CJ (2008) Plant formins: diverse isoforms and unique molecular mechanism. Biochim Biophys Acta, in press

Boureux A, Vignal E, Faure S, Fort P (2007) Evolution of the Rho family of ras-like GTPases in eukaryotes. Mol Biol Evol 24:203–216

Bourne HR, Sanders DA, McCormick F (1991) The GTPase superfamily: conserved structure and molecular mechanism. Nature 349:117–127

Brembu T, Winge P, Bones AM (2005) The small GTPase AtRAC2/ROP7 is specifically expressed during late stages of xylem differentiation in *Arabidopsis*. J Exp Bot 56:2465–2476

Brembu T, Winge P, Bones AM, Yang Z (2006) A RHOse by any other name: a comparative analysis of animal and plant Rho GTPases. Cell Res 16:435–445

Burbelo PD, Drechsel D, Hall A (1995) A conserved binding motif defines numerous candidate target proteins for both Cdc42 and Rac GTPases. J Biol Chem 270:29071–29074

Cairney J, Zheng L, Cowels A, Hsiao J, Zismann V, Liu J, Ouyang S, Thibaud-Nissen F, Hamilton J, Childs K, Pullman GS, Zhang Y, Oh T, Buell CR (2006) Expressed sequence tags from loblolly pine embryos reveal similarities with angiosperm embryogenesis. Plant Mol Biol 62:485–501

Carol RJ, Takeda S, Linstead P, Durrant MC, Kakesova H, Derbyshire P, Drea S, Zarsky V, Dolan L (2005) A RhoGDP dissociation inhibitor spatially regulates growth in root hair cells. Nature 438:1013–1016

Christensen TM, Vejlupkova Z, Sharma YK, Arthur KM, Spatafora JW, Albright CA, Meeley RB, Duvick JP, Quatrano RS, Fowler JE (2003) Conserved subgroups and developmental regulation in the monocot rop gene family. Plant Physiol 133:1791–1808

Dovas A, Couchman JR (2005) RhoGDI: multiple functions in the regulation of Rho family GTPase activities. Biochem J 390:1–9

Duvick J, Fu A, Muppirala U, Sabharwal M, Wilkerson MD, Lawrence CJ, Lushbough C, Brendel V (2008) PlantGDB: a resource for comparative plant genomics. Nucl Acids Res 36:D959–D965

Elias M (2008) The guanine nucleotide exchange factors Sec2 and PRONE: candidate synapomorphies for the Opisthokonta and the Archaeplastida. Mol Biol Evol 25:1526–1529

Freeling M (2009) Bias in plant gene content following different sorts of duplication: tandem, whole-genome segmental, or by transposition. Annu Rev Plant Biol 60:433–453

Fu Y, Gu Y, Zheng Z, Wasteneys G, Yang Z (2005) *Arabidopsis* interdigitating cell growth requires two antagonistic pathways with opposing action on cell morphogenesis. Cell 120:687–700

Grunt M, Zárský V, Cvrcková F (2008) Roots of angiosperm formins: the evolutionary history of plant FH2 domain-containing proteins. BMC Evol Biol 8:115

Gu Y, Wang Z, Yang Z (2004) ROP/RAC GTPase: an old new master regulator for plant signaling. Curr Opin Plant Biol 7:527–536

Gu Y, Fu Y, Dowd P, Li S, Vernoud V, Gilroy S, Yang Z (2005) A Rho family GTPase controls actin dynamics and tip growth via two counteracting downstream pathways in pollen tubes. J Cell Biol 169:127–138

Gu Y, Li S, Lord EM, Yang Z (2006) Members of a novel class of *Arabidopsis* Rho guanine nucleotide exchange factors control Rho GTPase-dependent polar growth. Plant Cell 18:366–381

Hála M, Cole R, Synek L, Drdova E, Pecenkova T, Nordheim A, Lamkemeyer T, Madlung J, Hochholdinger F, Fowler JE, Zárský V (2008) An exocyst complex functions in plant cell growth in *Arabidopsis* and tobacco. Plant Cell 20:1330–1345

Hwang JU, Vernoud V, Szumlanski A, Nielsen E, Yang Z (2008) A tip-localized RhoGAP controls cell polarity by globally inhibiting Rho GTPase at the cell apex. Curr Biol 18: 1907–1916

Ivanchenko M, Vejlupkova Z, Quatrano RS, Fowler JE (2000) Maize ROP7 GTPase contains a unique, CaaX box-independent plasma membrane targeting signal. Plant J 24:79–90

Jiang S, Ramachandran S (2006) Comparative and evolutionary analysis of genes encoding small GTPases and their activating proteins in eukaryotic genomes. Physiol Genom 24:235–251

Kaothien P, Ok SH, Shuai B, Wengier D, Cotter R, Kelley D, Kiriakopolos S, Muschietti J, McCormick S (2005) Kinase partner protein interacts with the LePRK1 and LePRK2 receptor kinases and plays a role in polarized pollen tube growth. Plant J 42:492–503

Kawasaki T, Henmi K, Ono E, Hatakeyama S, Iwano M, Satoh H, Shimamoto K (1999) The small GTP-binding protein Rac is a regulator of cell death in plants. Proc Natl Acad Sci USA 96:10922–10926

Kawasaki T, Koita H, Nakatsubo T, Hasegawa K, Wakabayashi K, Takahashi H, Umemura K, Umezawa T, Shimamoto K (2005) Cinnamoyl-CoA reductase, a key enzyme in lignin biosynthesis, is an effector of small GTPase Rac in defense signaling in rice. Proc Natl Acad Sci USA 103:230–235

Keeling PJ (2007) *Ostreococcus tauri*: seeing through the genes to the genome. Trends Genet 23:151–154

Klahre U, Becker C, Schmitt AC, Kost B (2006) Nt-RhoGDI2 regulates Rac/Rop signaling and polar cell growth in tobacco pollen tubes. Plant J 46:1018–1031

Kost B (2008) Spatial control of Rho (Rac-Rop) signaling in tip-growing plant cells. Trends Cell Biol 18:119–127

Lavy M, Bracha-Drori K, Sternberg H, Yalovsky S (2002) A cell-specific, prenylation-independent mechanism regulates targeting of type II RACs. Plant Cell 14:2431–2450

Lavy M, Yalovsky S (2006) Association of *Arabidopsis* type-II ROPs with the plasma membrane requires a conserved *C*-terminal sequence motif and a proximal polybasic domain. Plant J 46:934–947

Lavy M, Bloch D, Hazak O, Gutman I, Poraty L, Sorek N, Sternberg H, Yalovsky S (2007) A novel ROP/RAC effector links cell polarity, root-meristem maintenance, and vesicle trafficking. Curr Biol 17:947–952

Lemichez E, Wu Y, Sanchez JP, Mettouchi A, Mathur J, Chua NH (2001) Inactivation of AtRac1 by abscisic acid is essential for stomatal closure. Genes Dev 15:1808–1816

Li H, Wu G, Ware D, Davis KR, Yang Z (1998) *Arabidopsis* Rho-related GTPases: differential gene expression in pollen and polar localization in fission yeast. Plant Physiol 118:407–417

Li S, Gu Y, Yan A, Lord E, Yang Z (2008) RIP1 (ROP Interactive Partner 1)/ICR1 marks pollen germination sites and may act in the ROP1 pathway in the control of polarized pollen growth. Mol Plant 1:1021–1035

Lyons E, Freeling M (2008) How to usefully compare homologous plant genes and chromosomes as DNA sequences. Plant J 53:661–673

Lyons E, Pedersen B, Kane J, Alam M, Ming R, Tang H, Wang X, Bowers J, Paterson A, Lisch D, Freeling M (2008) Finding and comparing syntenic regions among *Arabidopsis* and the outgroups papaya, poplar, and grape: CoGe with rosids. Plant Physiol 148:1772–1781

Meller N, Merlot S, Guda C (2005) CZH proteins: a new family of Rho-GEFs. J Cell Sci 118:4937–4946

Nibau C, Wu HM, Cheung AY (2006) RAC/ROP GTPases: 'hubs' for signal integration and diversification in plants. Trends Plant Sci 11:309–315

Pereira-Leal JB, Seabra MC (2001) Evolution of the Rab family of small GTP-binding proteins. J Mol Biol 313:889–901

Peter G, Neale D (2004) Molecular basis for the evolution of xylem lignification. Curr Opin Plant Biol 7:737–742

Qiu JL, Jilk R, Marks MD, Szymanski DB (2002) The *Arabidopsis* SPIKE1 gene is required for normal cell shape control and tissue development. Plant Cell 14:101–118

Qiu YL, Li L, Wang B, Chen Z, Knoop V, Groth-Malonek M, Dombrovska O, Lee J, Kent L, Rest J, Estabrook GF, Hendry TA, Taylor DW, Testa CM, Ambros M, Crandall-Stotler B, Duff RJ, Stech M, Frey W, Quandt D, Davis CC (2006) The deepest divergences in land plants inferred from phylogenomic evidence. Proc Natl Acad Sci USA 103:15511–15516

Ralph SG, Chun HJ, Kolosova N, Cooper D, Oddy C, Ritland CE, Kirkpatrick R, Moore R, Barber S, Holt RA, Jones SJ, Marra MA, Douglas CJ, Ritland K, Bohlmann J (2008) A conifer genomics resource of 200,000 spruce (Picea spp.) ESTs and 6,464 high-quality, sequence-finished full-length cDNAs for Sitka spruce (*Picea sitchensis*). BMC Genom 9:484

Rensing SA, Lang D, Zimmer AD, Terry A, Salamov A, Shapiro H, Nishiyama T, Perroud PF, Lindquist EA, Kamisugi Y, Tanahashi T, Sakakibara K, Fujita T, Oishi K, Shin-I T, Kuroki Y, Toyoda A, Suzuki Y, Hashimoto S, Yamaguchi K, Sugano S, Kohara Y, Fujiyama A, Anterola A, Aoki S, Ashton N, Barbazuk WB, Barker E, Bennetzen JL, Blankenship R, Cho SH,

Dutcher SK, Estelle M, Fawcett JA, Gundlach H, Hanada K, Heyl A, Hicks KA, Hughes J, Lohr M, Mayer K, Melkozernov A, Murata T, Nelson DR, Pils B, Prigge M, Reiss B, Renner T, Rombauts S, Rushton PJ, Sanderfoot A, Schween G, Shiu SH, Stueber K, Theodoulou FL, Tu H, Van de Peer Y, Verrier PJ, Waters E, Wood A, Yang L, Cove D, Cuming AC, Hasebe M, Lucas S, Mishler BD, Reski R, Grigoriev IV, Quatrano RS, Boore JL (2008) The *Physcomitrella* genome reveals evolutionary insights into the conquest of land by plants. Science 319:64–69

Rudall PJ, Bateman RM (2007) Developmental bases for key innovations in the seed-plant microgametophyte. Trends Plant Sci 12:317–326

Schultheiss H, Preuss J, Pircher T, Eichmann R, Hückelhoven R (2008) Barley RIC171 interacts with RACB *in planta* and supports entry of the powdery mildew fungus. Cell Microbiol 10:1815–1826

Sørmo CG, Leiros I, Brembu T, Winge P, Os V, Bones AM (2006) The crystal structure of *Arabidopsis thaliana* RAC7/ROP9: the first RAS superfamily GTPase from the plant kingdom. Phytochemistry 67:2332–2340

Stamatakis A (2006) RAxML-VI-HPC: maximum likelihood-based phylogenetic analyses with thousands of taxa and mixed models. Bioinformatics 22:2688–2690

Vernoud V, Horton AC, Yang Z, Nielsen E (2003) Analysis of the small GTPase gene superfamily of *Arabidopsis*. Plant Physiol 131:1191–1208

Wang W, Tanurdzic M, Luo M, Sisneros N, Kim HR, Weng JK, Kudrna D, Mueller C, Arumuganathan K, Carlson J, Chapple C, de Pamphilis C, Mandoli D, Tomkins J, Wing RA, Banks JA (2005) Construction of a bacterial artificial chromosome library from the spikemoss *Selaginella moellendorffii*: a new resource for plant comparative genomics. BMC Plant Biol 5:10

Winge P, Brembu T, Kristensen R, Bones AM (2000) Genetic structure and evolution of RAC-GTPases in *Arabidopsis thaliana*. Genetics 156:1959–1971

Wu G, Li H, Yang Z (2000) *Arabidopsis* RopGAPs are a novel family of rho GTPase-activating proteins that require the Cdc42/Rac-interactive binding motif for rop- specific GTPase stimulation. Plant Physiol 124:1625–1636

Wu G, Gu Y, Li S, Yang Z (2001) A genome-wide analysis of *Arabidopsis* Rop-interactive CRIB motif-containing proteins that act as Rop GTPase targets. Plant Cell 13:2841–2856

Wu H, Rossi G, Brennwald P (2008) The ghost in the machine: small GTPases as spatial regulators of exocytosis. Trends Cell Biol 18:397–404

Yang Z, Watson JC (1993) Molecular cloning and characterization of rho, a ras-related small GTP-binding protein from the garden pea. Proc Natl Acad Sci USA 90:8732–8736

Yang Z (2002) Small GTPases: versatile signaling switches in plants. Plant Cell 14:S375–S388

Yuksel B, Memon AR (2008) Comparative phylogenetic analysis of small GTP-binding genes of model legume plants and assessment of their roles in root nodules. J Exp Bot 59:3831–3844

Zársky V, Cvrcková F, Potocky M, Hála M (2009) Exocytosis and cell polarity in plants – exocyst and recycling domains. New Phytol 183(2):255–272

Zhang Y, McCormick S (2007) A distinct mechanism regulating a pollen-specific guanine nucleotide exchange factor for the small GTPase Rop in *Arabidopsis thaliana*. Proc Natl Acad Sci USA 104:18830–18835

Zheng ZL, Yang Z (2000) The Rop GTPase: an emerging signaling switch in plants. Plant Mol Biol 44:1–9

Index